Agricultural Impacts of Climate Change

Agricultural Impacts of Climate Change

Edited by
Rohitashw Kumar, Vijay P Singh,
Deepak Jhajharia, and Rasoul Mirabbasi

CRC Press
Taylor & Francis Group
Boca Raton London New York

CRC Press is an imprint of the
Taylor & Francis Group, an **informa** business

CRC Press
Taylor & Francis Group
6000 Broken Sound Parkway NW, Suite 300
Boca Raton, FL 33487-2742

International Standard Book Number-13: 978-0-367-34523-5 (Hardback)

Library of Congress Cataloging-in-Publication Data

Names: Rohitashw, Rohitashw, editor.
Title: Agricultural impacts of climate change / edited by Rohitashw
  Rohitashw, Vijay P Singh, Deepak Jhajharia, Rasoul Mirabbasi.
Description: Boca Raton : CRC Press, 2019- | Includes index. | Contents:
  v. 1. Agricultural impacts of climate change — v. 2. Applied agricultural
  practices for mitig
Identifiers: LCCN 2019028533 (print) | LCCN 2019028534 (ebook) | ISBN
  9780367345235 (v. 1 ; hardback) | ISBN 9780367345297 (v. 2 ; hardback) |
  ISBN 9780429326349 (v. 1 ; ebook) | ISBN 9780429326400 (v. 2 ; ebook)
Subjects: LCSH: Crops and climate. | Agriculture—Economic aspects\. |
  Sustainable agriculture. | Climate change mitigation. | Crop yields.
Classification: LCC S600.7.C54 A375 2019 (print) | LCC S600.7.C54 (ebook)
  | DDC 630.2/515—dc23
LC record available at https://lccn.loc.gov/2019028533
LC ebook record available at https://lccn.loc.gov/2019028534

Visit the Taylor & Francis Web site at
http://www.taylorandfrancis.com

and the CRC Press Web site at
http://www.crcpress.com

Contents

Preface

Agriculture is the backbone of the economy in most countries and has a special role in the food chain. In many parts of the world, agriculture is climate restricted and many factors affect crop production, such as water availability and quality, climate, soil productivity and management, degree of farm mechanization, crop variety, and fertilizers. In recent years, climate change has been causing spatial and temporal changes in temperature and precipitation and hence affecting agricultural production. Another important issue is the effect of cultivation of soil, which may reduce the productive capacity of farmlands. These issues are not confined to any one region or country, but are global in nature, requiring multidisciplinary, multi-organizational, and multinational approaches and educational efforts. Conservation agriculture is a concept designed for optimizing crop yields, and reaping economic and environmental benefits. The key elements of conservation agriculture are minimum disturbance of soil, rational organic soil cover using crop residues or cover crops, and the adoption of innovative and economically viable cropping systems and measures to reduce soil compaction through controlled traffic. Conservation agriculture offers a promise in using crop residues for improving soil health, increasing productivity, reducing pollution, and enhancing sustainability and resilience of agriculture.

This handbook, comprising two volumes, is designed to provide a discussion of each of the important aspects of effective factors on crop production, such as climate change, soil management, farm machinery, and different methods for sustainable agriculture. The first volume of the book deals with the effect of climate change on agriculture and mitigation strategies, whereas the second volume focuses on conservation agriculture as an effective strategy for sustainable agriculture and food security. Selected case studies are also provided. The book provides current information which can be utilized when dealing with these issues.

The subject matter of this volume (*Agricultural Impacts of Climate Change*) spans 16 chapters. Introducing the theme of the book in different chapters, the succeeding chapters constitute the section focusing on conservation agriculture to mitigate climate change. Chapter 1 presents the spatial distribution of a daily, monthly, and annual precipitation concentration index. This chapter describes the inherent climatic properties of Iran. The effect of precipitation on water resources including underground water, surface water resources, and snow reserves necessitates the use of indices to demonstrate related changes.

Diffuse Reflectance Fourier Transform Infrared Spectroscopy (DRIFT) is considered to be one of the most sensitive infrared techniques for analyzing the structural composition of soil organic matter, and is discussed in Chapter 2. The objective of this study is to determine the effectiveness of using mid-infrared (MIR) spectroscopic methods to assess the changes in soil quality based on the changes in soil organic C composition in soils with a history of different agricultural management practices from several countries. Chapter 3 describes the effect of climate change on the production of horticultural crops. It discusses that the succulent horticultural crops are highly sensitive to heat, radiation, drought, salinity, and flooding. Elevated CO_2 has a positive effect ranging from 24% to 51% on the productivity of crops like mango, citrus, grapes, guava, fig, annona, tomato, capsicum, onion, cucumber, and melons. However, the rise in temperature affects the crop duration, flowering, fruiting, fruit size, quality, and fruit ripening with reduced productivity and economic yield. Therefore, the overall impact of climate change depends on the interaction effect of elevated CO_2 and rising temperature.

Changing disease scenario due to climate change has been highlighted in Chapter 4, along with the need for better agricultural practices and use of eco-friendly methods in disease management for sustainable crop production. Owing to changing climate and shift in the seasons, the choice of crop management practices based on the prevailing situation becomes important. In such scenarios, weather-based disease monitoring, inoculums monitoring, especially for soil-borne diseases and rapid diagnostics, would play a significant role. The chapter also highlights the need to adopt novel approaches to counter the resurgence of diseases under changed climatic scenario. In addition, monitoring and early warning systems for forecasting disease epidemics should be developed for important host–pathogens which have a direct bearing on the earnings of the farmers and food security at large. Such a diversified crop protection strategy has been highlighted in a comprehensive study on an integrated approach to control all foliar diseases in barley. Chapter 5 describes the solar energy-based greenhouse modeling and its use for the performance evaluation of various applications such as: greenhouse heating, greenhouse cooling, greenhouse drying, and aquaculture greenhouse. Computer simulation and mathematical modeling are important tools for determining the energy-efficient design as well as for predicting the overall system performance. The greenhouse climate may be used for crop drying, distillation, biogas plant heating, space conditioning, and aquaculture. Some other applications, such as covered crop drying and aquaculture, are also discussed. Chapter 6 discusses the development of agriculture under climate and environmental changes in the Brazilian semiarid region. The chapter highlights the agriculture in the Brazilian northeast in the face of climate change and soil usage.

The role of PGPR in sustainable agriculture under the changing scenario of climate change is discussed in Chapter 7. It describes the success of microbial inoculants depending on the availability of microbes as a product or formulation, which facilitates the technology to transfer from laboratory to land. There are several PGPR inoculants currently commercialized that seem to promote growth through at least one mechanism: suppression of plant disease (bioprotectants), improved nutrient acquisition (biofertilizers), or phytohormone production (biostimulants).

Chapter 8 describes the status and prospect of precision farming in India and discusses in detail the precision farming technology. Precision farming is a package of technologies to enhance the input-use efficiency with an aim to sustainably increase productivity without causing any negative impact to the environment. It describes GPS, remote sensing, and GIS technologies which, along with long-term yield monitoring of farms both at a macro and micro level, have been successfully adopted to assist in the decision-making at farm level when embarking on precision farming. Chapter 9 discusses the low-cost on-farm indigenous and innovative technologies of rainwater harvesting. Different indigenous technologies are, in a way, totally self-sustained: They do not require skill, technology, knowledge, management, spare parts, and fuel that need to be brought in from outside. They replenish themselves each year with every drop of monsoon rain and serve the people throughout the year. In this paper, an effort has been made to describe some low-cost on-farm indigenous and innovative technologies of rainwater harvesting which have the potential to increase the productivity of arid and semi-arid areas where water shortage is common because of scanty rainfall and its uneven distribution.

Chapter 10 describes the impact of climate change on food safety. The chapter highlights climate change and variability in environment which have an impact on the occurrence of food safety hazards at different stages of the food chain. There are multiple pathways through which climate-related factors may impact food safety, including: change in temperature and precipitation patterns, increased frequency and intensity of extreme weather events, ocean warming and acidification, changes in contaminants, and greenhouse effects. Climate change may also affect socio-economic aspects related to food systems, such as agriculture, animal production, global trade, and post-harvest quality and human behavior, which ultimately influence food safety. Temperature increase and the effects of greenhouse gases are among the most important issues associated with climate change. Microbial-assisted soil reclamation for sustainable agriculture in climate change is discussed in Chapter 11. The details of soil respiration, indicators of soil health and climate change, exploiting microbial EPS for estimating their role in plant growth management and combating plant pathogens, are also given.

In Chapter 12, the production of temperate fruits in Jammu & Kashmir state under a climate change scenario is described. It also discusses the strategies to overcome the impact of climate changes, evaluation of low-chilling cultivars of different fruit crops, use of rootstocks with specific characters, breeding for development of climate-resilient varieties, rainwater harvesting and moisture conservation strategies, adoption of insect and disease forecasting system, and technology transfer. Chapter 13 describes about impact of climate change on quality of seed production of important temperate vegetable crops. The impact of climate change, and mitigation strategies with reference to vegetable crops, is described in Chapter 14.

Chapter 15 describes the remote sensing and GIS tool of precision agriculture. Different techniques have been discussed for land and water management through remote sensing and GIS. It also highlights precision farming using remote sensing and GIS. Chapter 16 describes the different farm machinery used for conservation agriculture. The chapter recommends mechanization for conservation agriculture to increase the crop productivity. This will help in increased agriculture production, timely sowing, and harvesting of crops which avoids loss of grains.

As editors, we realize that we have just begun to scratch the surface with some of the recent advances in conservation agriculture for climate change mitigation. We would like to take this opportunity to thank the chapter authors for their contributions. All of us (both the editors and authors) are thankful for the valuable and constructive comments that were received on each chapter. This book discusses real-world examples and is based on experiences at different agro climatic regions throughout the world. It is hoped that this book will be useful for conservation agriculture and precision farming for climate change mitigation.

Rohitashw Kumar
Vijay P Singh
Deepak Jhajharia
Rasoul Mirabbasi

Editors

Dr. Rohitashw Kumar (B.E., M.E., Ph. D.) is a Professor in the College of Agricultural Engineering and Technology, She-e-Kashmir University of Agricultural Sciences and Technology of Kashmir, Srinagar, India. He is also Professor Water Chair (Sheikkul Alam Shiekh Nuruddin Water Chair), Ministry of Water Resources, Govt. of India, at the National Institute of Technology, Srinagar (J&K). He obtained his Ph.D. degree in Water Resources Engineering from NIT, Hamirpur, and Master of Engineering Degree in Irrigation Water Management Engineering from MPUAT, Udaipur. He received a Special Research Award in 2017 and Student Incentive Award-2015 (Ph.D. Research) from the Soil Conservation Society of India, New Delhi. He also got the first prize in India for best M. Tech thesis in Agricultural Engineering in year 2001. He graduated from Maharana Pratap University of Agricultural and Technology, Udaipur, India, in Agricultural Engineering. He has published over 80 papers in peer-reviewed journals, one book, four practical manuals, and 20 chapters in books. He has guided ten post-graduate students in Soil and Water Engineering. He has handled more than ten research projects as a principal or co-principal investigator. Since 2011, he has been Principal Investigator of the All India Coordinated Research Project on Plasticulture Engineering and Technology.

Prof. Vijay P Singh is a Distinguished Professor, a Regents Professor, and the inaugural holder of the Caroline and William N. Lehrer Distinguished Chair in Water Engineering at Texas A&M University. His research interests include Surface-Water Hydrology, Groundwater Hydrology, Hydraulics, Irrigation Engineering, Environmental Quality, Water Resources, entropy theory, copula theory, and mathematical modeling. He graduated with a B.Sc. in Engineering and Technology with an emphasis on Soil and Water Conservation Engineering in 1967 from U.P. Agricultural University, India. He earned an MS in Engineering with specialization in Hydrology in 1970 from the University Of Guelph, Canada, a Ph.D. in Civil Engineering with specialization in Hydrology and Water Resources in 1974 from the Colorado State University, Fort Collins, USA, and a D.Sc. in Environmental and Water Resources Engineering in 1998 from the University of the Witwatersrand, Johannesburg, South Africa. He has published extensively on a wide range of topics. His publications include more than 1,200 journal articles, 30 books, 70 edited books, 305 book chapters, and 315 conference proceedings papers. For seminar contributions, he has received more than 90 national and international awards, including three honorary doctorates. Currently, he serves as President-Elect of the American Academy of Water Resources Engineers, the American Society of Civil Engineers, and previously he served as President of the American Institute of Hydrology. He is Editor-in-Chief of two book series and one journal and serves on the editorial boards of more than 20 journals. He has served as Editor-in-Chief of three other journals. He is a Distinguished Member of the American Society of Civil Engineers, an Honorary Member of the American Water Resources Association, and a fellow of five professional societies. He is also a fellow or member of 11 national or international engineering or science academies.

Dr. Deepak Jhajharia (B. Tech., M. Tech. and Ph. D.) is currently working as a Professor in the Department of Soil & Water Conservation Engineering, College of Agricultural Engineering & Post Harvest Technology, Ranipool, Gangtok, Sikkim, India. He is also acting as Principal Investigator of the All India Coordinated Research Project on Plasticulture Engineering Technology CAEPHT (Gangtok) center, which is funded entirely by ICAR Central Institute of Post Harvest Engineering and Technology, Ludhiana, India, since 2016. He graduated from the College of Technology and Engineering (MPUAT), Udaipur, Rajasthan, India, in Agricultural Engineering in 1998, and did post-graduation from the Indian Institute of Technology Delhi, India, in Water Resources Engineering, the Department of Civil Engineering. He obtained his Ph.D. degree from the Department of Hydrology, Indian Institute of Technology Roorkee, Uttarakhand, India. He was awarded Young Talent Attraction – BJT of Science without Borders Program for International Scientists, National Council for Scientific and Technological Development (CNPq), Brazil, as Research Collaborator at Universidade Federal Rural de Pernambuco (UFRPE), Recife, PE, Brazil, in 2013. He has published over 50 papers in peer-reviewed journals, three extension bulletins, and seven chapters in books. He has guided eight M. Tech. theses in the field of Soil and Water Conservation Engineering along with many undergraduate theses in the field of Agricultural Engineering. He has handled nine research projects sponsored by governmental agencies as principal or co-principal investigator. He also conducted one 21-day summer school for scientists from ICAR and faculty members from different universities and one 90-day skill development training program on greenhouse technology for school drop-outs and unemployed rural youth from six states of northeast India. He was awarded the CSIRO Land and Water Publication Award 2013, CSIRO Australia for a global review paper published in the *Journal of Hydrology*. He is a Fellow of the Indian Association of Hydrologists, Roorkee (in 2015) and Society of Extension Education, Agra (in 2018). He is a recipient of the Distinguished Alumni Award (in 2016) from the College of Technology and Engineering Alumni Society, CTAE (MPUAT), Udaipur. He has also adjudged the Best Extension Scientist (2017–18) of the AICRP on PET in recognition of outstanding contribution to the extension and popularization of Plasticulture Technologies in Sikkim. He is also a life member of 14 different professional societies from India and abroad.

Dr. Rasoul Mirabbasi is an Associate Professor of Hydrology and Water Resources Engineering at Shahrekord University, Iran. His research focuses mainly on Statistical and Environmental Hydrology and Climate Change. In particular, he is working on Modeling Natural Hazards, including floods, droughts, winds, and pollution, toward a sustainable environment. Formerly, he was a Visiting Researcher at the University of Connecticut, United States. He has contributed to more than 150 publications in journals, books, or technical reports. He is the reviewer of about 20 Web of Science (ISI) journals. He is currently the Head of the Water Resources Center of Shahrekord University.

Contributors

Farshad Ahmadi
Department of Hydrology and Water
 Resources Engineering
Shahid Chamran University
Ahwaz, Iran

Renato Dantas Alencar
Federal Institute of Rio Grande do
 Norte
Mossoró, Brazil

F. A. Banday
Division of Fruit Science
Sher-e-Kashmir University of
 Agricultural Sciences and
 Technology of Kashmir
 (SKUAST-K)
Srinagar, India

Desh Raj Choudhary
Department of Vegetable Science
Chaudhary Charan Singh Haryana
 Agricultural University (CCS HAU)
Hisar, India

Gaussuddin
Department of Industrial Microbiology,
 Jacob Institute of Biotechnology and
 Bio-engineering
Sam Higginbottom University of
 Agriculture, Technology and
 Sciences
Allahabad, India

Harender Raj Gautam
Department of Plant Pathology
Dr. Y.S. Parmar University of
 Horticulture and Forestry
Solan, India

Jagan Singh Gora
Fruit Science
ICAR-Central Institute for Arid
 Horticulture
Bikaner, India

Magda Maria Guilhermino
Federal University of Rio Grande do
 Norte
Mossoró, Brazil

Shweta Gupta
Department of Basic Sciences, College
 of Forestry
Dr. Y.S. Parmar University of
 Horticulture and Forestry
Solan, India

Nayeema Jabeen
Division of Vegetable Science
Sher-e-Kashmir University of
 Agricultural Sciences and
 Technology of Kashmir
 (SKUAST-K)
Srinagar, India

Dilip Jain
Division of Agricultural Engineering
 and Renewable Energy
ICAR-Central Arid Zone Research
 Institute
Jodhpur, India

Deepak Jhajharia
Department of Soil and Water
 Conservation Engineering
College of Agricultural Engineering &
 Post Harvest Technology
Gangtok, India

Bunjirtluk Jintaridth
Department of Soil, Environmental and
  Atmospheric Sciences
University of Missouri
Columbia, Missouri

Arjun Karmakar
Department of Industrial Microbiology,
  Jacob Institute of Biotechnology and
  Bio-engineering
Sam Higginbottom University of
  Agriculture, Technology and
  Sciences
Allahabad, India

Rajesh Kaushal
Department of Soil Science and Water
  Management, College of Forestry
Dr. Y.S. Parmar University of
  Horticulture and Forestry
Solan, India

Keivan Khalili
Department of Water Engineering
Urmia University
Urmia, Iran

Rohitashw Kumar
Department of Soil and Water
  Engineering, College of Agricultural
  Engineering and Technology
Sher-e-Kashmir University of
  Agricultural Sciences and
  Technology of Kashmir
  (SKUAST-K)
Srinagar, India

Amit Kumar
Division of Fruit Science
Sher-e-Kashmir University of
  Agricultural Sciences and
  Technology of Kashmir
  (SKUAST-K)
Srinagar, India

Juliana Espada Lichston
Department of Botany and Zoology
Research Laboratory of Energy Crops
Federal University of Rio Grande do
  Norte
Mossoró, Brazil

Emile Rocha de Lima
Federal University of Rio Grande do
  Norte
Mossoró, Brazil

Shiv Kumar Lohan
Department of Farm Machinery and
  Power Engineering
Punjab Agricultural University
Ludhiana, India

Ajaz A. Malik
Division of Vegetable Science
Sher-e-Kashmir University of
  Agricultural Sciences and
  Technology of Kashmir
  (SKUAST-K)
Srinagar, India

Rasoul Mirabbasi
Department of Water Engineering
Shahrekord University
Shahrekord, Iran

Peter Motavalli
Department of Soil, Environmental and
  Atmospheric Sciences
University of Missouri
Columbia, Missouri

H. R. Naik
Food Science and Technology
Sher-e-Kashmir University of
  Agricultural Sciences and
  Technology of Kashmir
  (SKUAST-K)
Srinagar, India

Mahesh Kumar Narang
Department of Farm Machinery and
   Power Engineering
Punjab Agricultural University
Ludhiana, India

Uday Shankar Pandey
Department of Industrial
   Microbiology, Jacob Institute
   of Biotechnology and
   Bio-engineering
Sam Higginbottom University of
   Agriculture, Technology and
   Sciences
Allahabad, India

Neelam Patel
Water Technology Centre
Indian Agriculture Research Institute
New Delhi, India

Anjulata Suman Patre
Department of Industrial
   Microbiology, Jacob Institute
   of Biotechnology and
   Bio-engineering
Sam Higginbottom University of
   Agriculture, Technology and
   Sciences
Allahabad, India

Jyotsna Kiran Peter
Department of Industrial
   Microbiology, Jacob Institute
   of Biotechnology and
   Bio-engineering
Sam Higginbottom University of
   Agriculture, Technology and
   Sciences
Allahabad, India

I. M. Sharma
Department of Plant Pathology
Dr. Y.S. Parmar University of
   Horticulture and Forestry
Solan, India

M. K. Sharma
Division of Fruit Science
Sher-e-Kashmir University of
   Agricultural Sciences and
   Technology of Kashmir
   (SKUAST-K)
Srinagar, India

Jitendra Singh
Departmet of Horticulture, College of
   Horticulture and Forestry
Agricultural University
Kota, Rajasthan

P. K. Singh
Department of Soil and Water
   Engineering, College of Technology
   and Engineering
Maharana Pratap University of
   Agriculture and Technology
Udaipur, India

Pradeep Kumar Singh
Division of Vegetable Science
Sher-e-Kashmir University of
   Agricultural Sciences and
   Technology of Kashmir
   (SKUAST-K)
Srinagar, India

Gaurav Sood
Department of Basic Sciences, College
   of Forestry
Dr. Y.S. Parmar University of
   Horticulture and Forestry
Solan, India

Mohammad Nazeri
Department of Water Engineering
University of Birjand
Birjand, Iran

Ajay Kumar Verma
ICAR-Central Institute for Arid
   Horticulture
Bikaner, Rajasthan

Dinesh Kumar Vishwakarma
Soil and Water Engineering, College
  of Agricultural Engineering and
  Technology
Sher-e-Kashmir University of
  Agricultural Sciences and
  Technology of Kashmir
  (SKUAST-K)
Srinagar, India

Mushtaq A. Wani
Division of Soil Science
Sher-e-Kashmir University of
  Agricultural Sciences and
  Technology of Kashmir
  (SKUAST-K)
Srinagar, India

1 Spatial Distribution of a Daily, Monthly and Annual Precipitation Concentration Index

Farshad Ahmadi
Shahid Chamran University of Ahvaz

Mohammad Nazeri
University of Birjand

Rasoul Mirabbasi
Shahrekord University

Keivan Khalili
Urmia University

Deepak Jhajharia
College of Agricultural Engineering &
Post Harvest Technology, CAU

CONTENTS

1.1   INTRODUCTION

Climate changes are one of the most important environmental challenges at present. Our understanding of the human impacts on the environment, particularly those associated with the warming due to the increasing greenhouse gases, shows that several parameters are most likely changing. According to scientific reports, the average surface temperature of Earth has increased about 6°C during the 20th century. It is expected that the amount of evaporation will have an ascending trend. Therefore the atmosphere will be able to move greater amounts of water vapor and hence the amount of precipitation will be affected (Tabari 2011). Insufficient precipitation and its severe fluctuations on daily, seasonal, and annual scale are amongst the inherent climatic characteristics of Iran. The precipitation patterns have changed under the influence of global warming and led to the occurrence of extreme weather events such as floods, drought, rain, storm, etc. For instance a significant reduction in the number of rainy days has been confirmed in many parts of the world, including China (Zhang et al. 2008, Ren et al. 2000, Gong and Ho 2002, Zhai et al. 2005). One of the most important aspects of climate change that requires closer examination is to review the temporal distribution of precipitation and its historical changes. With regard to the effect of precipitation on water resources like groundwater, surface water, and snow reserves, it seems necessary to use indices for the expression of changes. Some of these indicators include Standardized Precipitation Index (SPI), Precipitation Concentration Index (PCI), and Density Index (DI).

Concentration Index (CI) is actually an index to examine the statistical characteristics of daily precipitation. PCI index is a part of the famous Fournier index used for the analysis of natural resources such as soil erosion (Luis et al. 2011). The results of PCI calculations could be hydraulic management programs, water and environmental resources as a warning instrument to be prepared against flood or erosion (Adegun et al. 2012). This concept could also be used in irrigation planning and designing new systems. An unbalanced distribution of precipitation could reduce the agricultural crops yield through the reduction of moisture in soil and increase the irrigation periods. Besides, an unbalanced distribution of precipitation means drought. Precipitation less than the regional average may cause drought of varying intensity since soil moisture will reduce, and vegetation will be destroyed.

This will eventually lead to increased protection precautions in the basin in order to maintain water structures. Martin-Vide (2004) calculated the CI for daily precipitation in Spain and divided the results into two regions with highest dispersion and regular dispersion of precipitation. Zhang et al. (2009) calculated the CI for serial precipitation for Pearl River basin. Alijani et al. (2008) examined Iran's rainfall intensity in 90 synoptic stations and indicated that the rainfall dispersion has been erratic in Iran. The stations located in the Caspian Sea, the Zagros Mountains, and the North West of Iran had the most significant amount of rainfall. Li et al. (2011) calculated the CI values for Kaidu River basin. Luis et al. (2011) studied the annual and seasonal average amounts and wet and dry periods of PCI in the vicinity of Spain for two (1976–2005 and 1964–1975) time spans. The analysis of the two sub periods revealed that significant rainfall changes occurred in Spain from 1946 to 2005. Adegun et al. (2012) evaluated the PCI index on two annual and seasonal scales during 1974–2011 in two areas in Nigeria. The results of PCI analysis in this area showed that 87% and 71% of studied years were located in the first and second area, respectively, within an average concentration domain. Valli et al. (2013) used the PCI index to show rainfall pattern in Andhra Pradesh State during 1981–2010 on two annual and seasonal scales. The results indicate that there is an irregular distribution of rainfall (with values ranging from 16 to 35) in this area. Khalili et al. (2016) studied the dispersion of monthly and annually rainfall of synoptic stations in Iran for the last 50 years in a 25-year period. Scientists believe that the changes in concentrations of the greenhouse gases resulting from fossil fuel consumption lead to drastic changes in some components of the hydrological cycle, such as precipitation in different parts of the world. In the following sections of this chapter, the fundamentals of PCI and CI indices theory will be explained. Then the Lake Urmia basin will be introduced and the aforesaid indices will be used to study the distribution and concentration of rainfall in the basin of Lake Urmia. The results are presented and discussed here and the overall results are presented in the final section.

1.2 METHODOLOGY

In this section, two indices of PCI and CI are completely described. Necessary examples are also given for further explanation of the calculation method.

1.2.1 CI

The method of calculating CI is based on the principle that the overall ratio of rainy days to total rainfall could be adjusted by a negative exponential distribution (Brooks and Carruthers 1953, Martin-Vide 2004). According to the geographic features and time span, the chance of low daily precipitation is probably higher than the large amounts of precipitation. So the initial lowest precipitation class will reduce the ultimate absolute frequency (Martin-Vide 2004). To study the effect of different doses of daily precipitation and the ratio of high amounts of precipitation to total precipitation, cumulative precipitation percent (Y) and cumulative daily percent (X) were studied during Y events. According to the studies by Olascoaga (1950), daily precipitation data in the range of 1 mm/day were firstly classified to examine the CI. The number of days with rainfall was specified in each category and their cumulative values were also calculated. Finally, the cumulative percentage of rainy days and the relationship between precipitation and rainy days were established. According to the steps outlined, an exponential curve of the cumulative percentage of rainy days (X) in contrast to the percentage of cumulative rainfall (Y) was obtained. Martin-Vide (2004) recommended $Y = (a \times X)\exp(b \times X)$ for this curve in which a, b are the regression coefficients. Gini CI $2S/10,000$ will be applied as an index to measuring the concentration in which S is equal to the area enclosed by the first quarter bisector and polygon line or Lorenz curve. In fact, the precipitation concentration is derived from the Gini coefficient. The Lorenz curve is demonstrated by $Y = (a \times X)\exp(b \times X)$ in which a, b coefficients are calculated by the least squares method (Martin-Vide 2004). When a, b coefficients were determined, the definite integral of exponential curve between 0 and 100 shows the area under the curve or A'.

$$A' = \left[\frac{a}{b}e^{bX}\left(X - \frac{1}{b}\right)\right]_0^{100} \tag{1.1}$$

Based on the above equation, the area enclosed by the curve and X = 100 distribution line between 5,000 and amount calculated from 1.6 relation are different (Martin-Vide 2004). The concentration values of precipitation which is similar to the Gini coefficient can be calculated using the following equation:

$$CI = S' / 5000 \tag{1.2}$$

We conclude that the CI index values are a fraction of the amount S' and triangle formed by the bottom of the chart. Coefficients a and b can be obtained from the following equation:

$$\ln a = \frac{\sum X_i^2 \sum \ln Y_i + \sum X_i \sum X_i \ln X_i - \sum X_i \sum \ln X_i - \sum X_i \sum X_i \ln Y_i}{n \sum X_i^2 - \left(\sum X_i\right)^2} \tag{1.3}$$

$$b = \frac{n \sum X_i \ln Y_i + \sum X_i \sum \ln X_i - n \sum X_i \ln X_i - \sum X_i \sum \ln Y_i}{n \sum X_i^2 - \left(\sum X_i\right)^2} \tag{1.4}$$

In relations 1.3 and 1.4, X and Y values are demonstrated in Table 1.3.

1.2.2   PCI

PCI has been proposed as an index both for concentration and dispersion of precipitation. Seasonal and annual scales of these indices are respectively calculated on the basis of 1.5 and 1.6 relations (Oliver 1980):

$$PCI_{Seasonal} = \frac{\sum_{i=1}^{3} p_i^2}{\left(\sum_{i=1}^{3} p_i\right)^2} * 25 \tag{1.5}$$

$$PCI_{Annual} = \frac{\sum_{i=1}^{12} p_i^2}{\left(\sum_{i=1}^{12} p_i\right)^2} * 100 \tag{1.6}$$

where p_i is the monthly precipitation in ith month.

Based on the formula and the preliminary results by Oliver (1980), the minimum amount of PCI theory was 3/8 which showed the complete uniformity in the distribution of precipitation (in other words the same amount of rainfall has occurred in every month). The 16.7 value for PCI shows that total precipitation has occurred in half of the time span, and a value of 25 for this index indicated that total precipitation has occurred in one third of the time span (in other words total precipitation has occurred during 4 months). According to the preliminary results of the investigation, Oliver (1980) suggested that the PCI values lower than 10 show a uniform distribution of precipitation (low concentration precipitation). The PCI values between 11 and 15 show an average concentration of precipitation and values between 16 and 20 also show an irregular distribution of precipitation. According to the Oliver classification, values higher than 20 for PCI index represent a high level of irregular distribution of precipitation (high concentration of precipitation) (Luis et al. 2011).

Now, to review the aforesaid approaches and their practical uses, PCI and CI indices were calculated and presented using daily, monthly, and annual data of stream-gaging stations in the Lake Urmia basin in the North West of Iran.

Lake Urmia as the focus of surface currents surplus of all the rivers of enclosed basin of Lake Urmia, with an approximate area of 5,750 km and an average elevation of 1,276 m above sea level, is located in the middle of Northern District of the basin. There are 16 wetlands surrounding Lake Urmia with an area of 5–120 ha (some have dried) and mostly having sweet or salt-and-sweet water which bear high ecosystem value. The drainage basin of Lake Urmia is situated in EL: 44-14 to 47-53 and NL: 35-40 and 38-30. The precipitation changes in the drainage basin of Lake Urmia are 220–900 mm and the average precipitation is 263 mm. The precipitation level increases from the central areas of the basin toward the surrounding highlands. The position of Lake Urmia and the stream-gaging stations under study in the basin of Lake Urmia are demonstrated in Figure 1.1, and their statistical characteristics are also defined in Table 1.1.

1.3 RESULTS AND DISCUSSION

1.3.1 Results of Lake Urmia Basin PCI

According to the monthly and annual precipitation data of selected stream-gaging stations in the basin of Lake Urmia in statistical 1984–2013 period and 1.5 and 1.6 relations, the PCI index in seasonal and annual scales was calculated and the average results are presented in Table 1.2. The box plot of PCI changes in basin area are presented in Figures 1.2–1.6.

The results of precipitation concentration of the Lake Basin showed that there was no regular concentration (PCI < 10) in annual scale in the basin area of stream-gaging stations. In annual scale, stations no. 4, 3, 2, 18, 11, 22, and 24 showed average precipitation distribution. More than 50% of the years studied had irregular concentration of monthly precipitation which means that there is irregular distribution of precipitation in 12 months of the year. There was also a severely irregular distribution in this scale observed in the Tipak, Moosh Abad, Gerd Yaghoob, and Dashband stations.

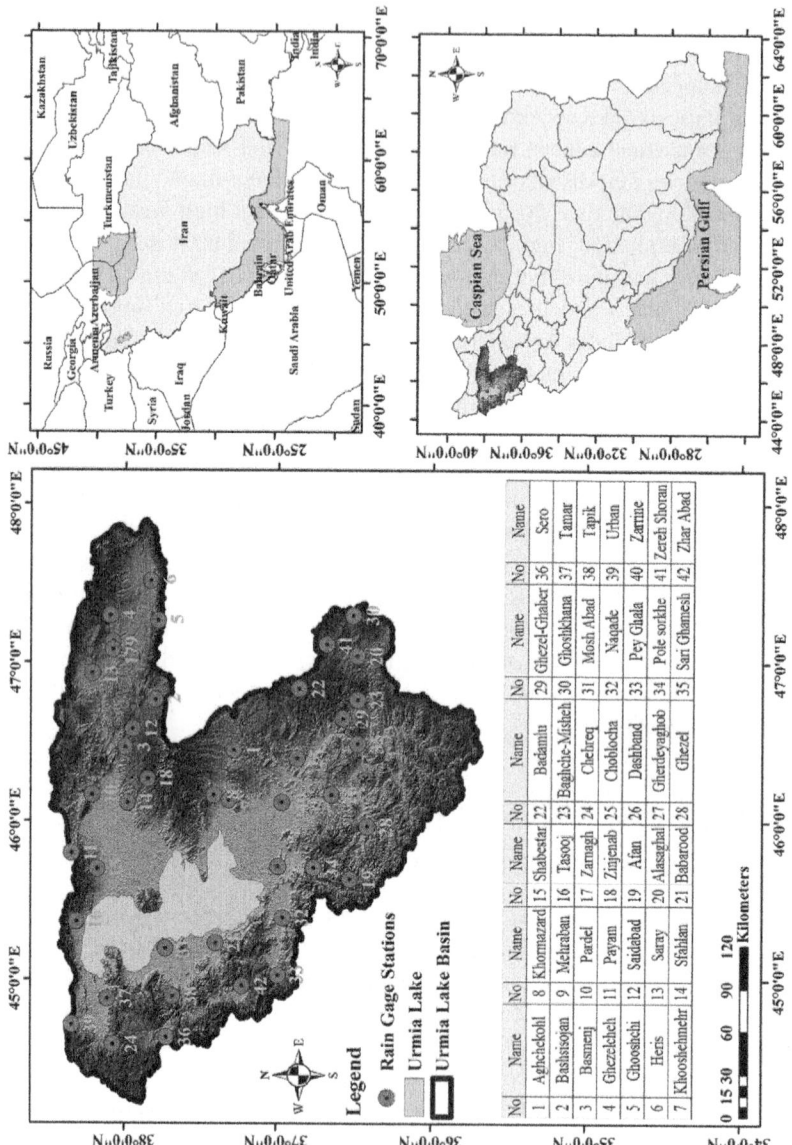

FIGURE 1.1   Urmia Lake basin highland and studied rain gauge stations.

TABLE 1.1

Statistical Properties of Rain Gauge Stations Located in Urmia Lake Basin in the Period 1984–2013

Station Number	Station Name	Elevation (m)	Latitude (m)	Longitude (m)	Annual Pre (mm/y)	Station Number	Station Name	Elevation (m)	Latitude (m)	Longitude (m)	Annual Pre (mm/y)
1	Agchekol	1,710	4,127,289	628,538	424.0	22	Badamloo	2,119	408,136	663,820	638.4
2	Bashsizojan	1,850	4,185,094	655,525	352.6	23	Bagch	1,898	403,901	658,254	348.3
3	Basmanj	1,700	4,206,829	628,772	172.0	24	Chehreq	1,611	421,451	464,363	366.3
4	Ghezelche	1,844	4,219,408	701,672	339.5	25	Chobchole	1,361	408,209	624,549	306.3
5	Ghoshchi	1,980	4,184,195	699,604	768.3	26	Dashband	1,318	405,698	604,229	397.3
6	Haris	1,690	4,190,307	721,473	463.7	27	Gerdeyaghob	1,280	409,459	563,788	268.7
7	Khoshehmehr	1,320	4,130,604	600,422	309.0	28	Ghezel Gonbad	1,374	4,303,110	586,755	402.5
8	Khormazard	1,556	4,141,734	603,238	398.3	29	Ghezel-Ghaber	1,657	4,304,935	647,580	322.6
9	Mehraban	1,608	4,217,210	687,099	327.2	30	Ghoshkhana	2,260	404,350	705,960	379.8
10	Pardel	1,415	4,230,498	602,129	238.1	31	Moshabad	1,281	417,599	517,436	248.4
11	Payam	1,790	4,244,952	569,903	457.7	32	Naqade	1,306	409,099	534,795	323.2
12	Saeedabad	1,950	4,201,449	639,111	381.3	33	Pey Ghala	1,306	409,428	502,663	486.1
13	Saray	1,545	4,231,622	669,248	287.6	34	Pole sorkhe	1,350	406,948	563,126	35.0
14	Isfahlan	1,400	4,204,554	598,064	272.6	35	Sari Ghamesh	1,391	4,303,843	633,518	322.5
15	Shabestar	1,400	4,226,388	561,305	297.4	36	Sero	1,628	417,517	468,231	360.8
16	Tasoj	1,390	4,241,014	532,053	372.5	37	Tamar	1,387	421,856	489,613	212.6
17	Zarnaq	1,600	4,217,110	682,713	297.7	38	Tapik	1,398	417,071	490,988	368.6
18	Zenjanab	2,110	4,189,928	611,483	332.9	39	Urban	1,840	424,417	475,445	310.0
19	Afan	1,620	404,314	557,052	578.8	40	Zarineh	1,390	4,089,831	593,482	422.5
20	Alasagel	1,700	403,978	683,041	392.7	41	Zereshoran	2,100	406,223	689,112	427.6
21	Babarood	1,282	413,958	520,490	342.8	42	Zharabad	1,569	411,990	496,768	542.3

TABLE 1.2

Results of Calculation of the PCI Index in Rain Gauge Stations Located in Urmia Lake Basin

Station Number	Station	X_UTM	Y_UTM	Spring	Summer	Autumn	Winter	Annual
1	Aghchekohlzaman	628,539	4,127,290	11.86	20.02	12.13	9.88	15.29
2	Bashsisojan	655,535	4,185,095	11.67	17.06	12.71	9.79	14.98
3	Basmenj	628,773	4,206,830	11.05	16.84	12.43	10.17	14.95
4	Ghezelchehsadat	701,672	4,219,409	11.51	17.81	11.13	10.83	15.88
5	Ghooshchisarab	699,604	4,184,196	11.39	19.7	11.77	9.32	15.29
6	Herissarab	721,474	4,190,308	11.6	16.64	13.17	10.22	16.91
7	Koshehmehr	600,422	4,130,604	12.79	23.21	12.96	10.57	17.41
8	Khormazard	603,238	4,141,735	12.72	21.53	13.18	10.31	17.11
9	Mehraban	687,099	4,217,210	11.87	16.75	11.9	10.7	16.9
10	Pardel	602,130	4,230,498	11.19	18.76	12.15	11.1	15.6
11	Payam	569,903	4,244,952	10.92	18.67	11.59	10.21	15.44
12	Saidabad	639,111	4,201,450	11.96	17.29	13.95	10.29	16.97
13	Saray	669,248	4,231,622	11.41	17.49	12.46	10.06	14.87
14	Sfahlan	598,064	4,204,554	11.75	20.09	12.61	10.67	17.07
15	Shabestar	561,305	4,226,388	12.25	20.3	12.38	10.4	17.36
17	Zarnaghheris	682,714	4,217,111	11.24	15.13	12.37	10.52	15.5
17	Tasooj	532,053	4,241,014	12.08	17.72	13.3	11.08	17.6
18	Zinjenab	611,438	4,189,928	10.68	19.16	13.04	10.28	15.99
19	Afan	557,052	4,043,148	13.61	22.05	13.03	9.6	17.08
20	Alasaghal	683,041	4,039,877	12.92	21.65	14.31	9.46	16.06
21	Babarood	520,490	4,139,581	13.36	21.49	13.82	10.84	17.38
22	Badamlu.xls	663,820	4,081,365	12.3	21.99	11.49	9.3	15.57
23	Baghche-Misheh	658,254	4,039,018	12.64	23.71	13.79	10.17	18.13
24	Chehreq	464,363	4,214,549	10.38	14.85	12.34	10.45	14.23
25	Choblocha	624,549	4,082,095	14.24	25	12.36	10.61	18.54
26	Dashband	604,229	4,056,986	13.51	23.25	13.52	10.2	18.12
27	Gherdeyaghob	563,788	4,094,597	13.3	22.06	13.61	10.76	19.69
28	Ghezel	586,755	4,031,106	14.82	23.79	14.51	10.28	18.61
29	Ghezel-Ghaber	647,580	4,049,351	13.58	23.03	13.97	9.66	16.97
30	Ghoshkhana	705,960	4,043,501	13.26	22.95	12.85	9.48	16.86
31	Mosh Abad	517,436	4,175,992	12.53	21.18	13.79	11.58	18.54
32	Naqade	534,795	4,090,995	12.96	21.15	12.66	10.55	16.99
33	Pey Ghala	502,663	4,094,287	13.62	21	12	9.63	15.89
34	Pole sorkhe	563,126	4,069,480	12.82	21.53	13.01	10.47	17.81
35	Sari Ghamesh	633,518	4,038,435	12.96	21.15	12.66	10.55	16.99
36	Sero	468,231	4,175,171	11.46	17.3	11.95	10.74	15.95
37	Tamar	489,613	4,218,563	12.18	21.06	15.12	12.77	19.08
38	Tapik	490,988	4,170,717	12.97	20.21	14.81	11.4	18.33
39	Urban	475,445	4,244,172	11.4	18.05	12.33	13.27	17.52
40	Zarrine	599,592	4,092,413	13.12	20.95	13.1	10.15	17.44
41	Zereh Shoran	689,112	4,062,234	12.21	21.75	14.88	9.35	16.93
42	Zhar Abad	496,768	4,119,904	13.65	20.65	13.34	10.56	16.75

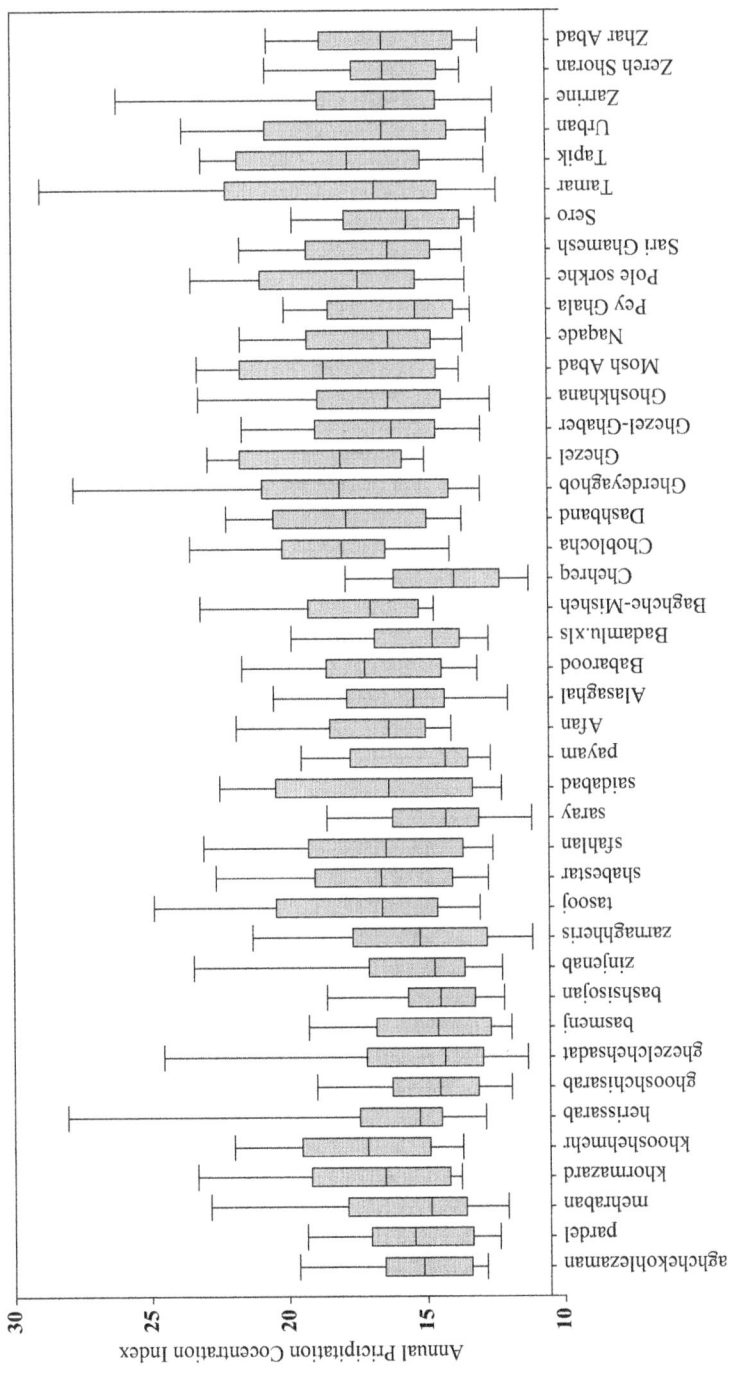

FIGURE 1.2 Results of calculation of the annual PCI in the period 1984–2013.

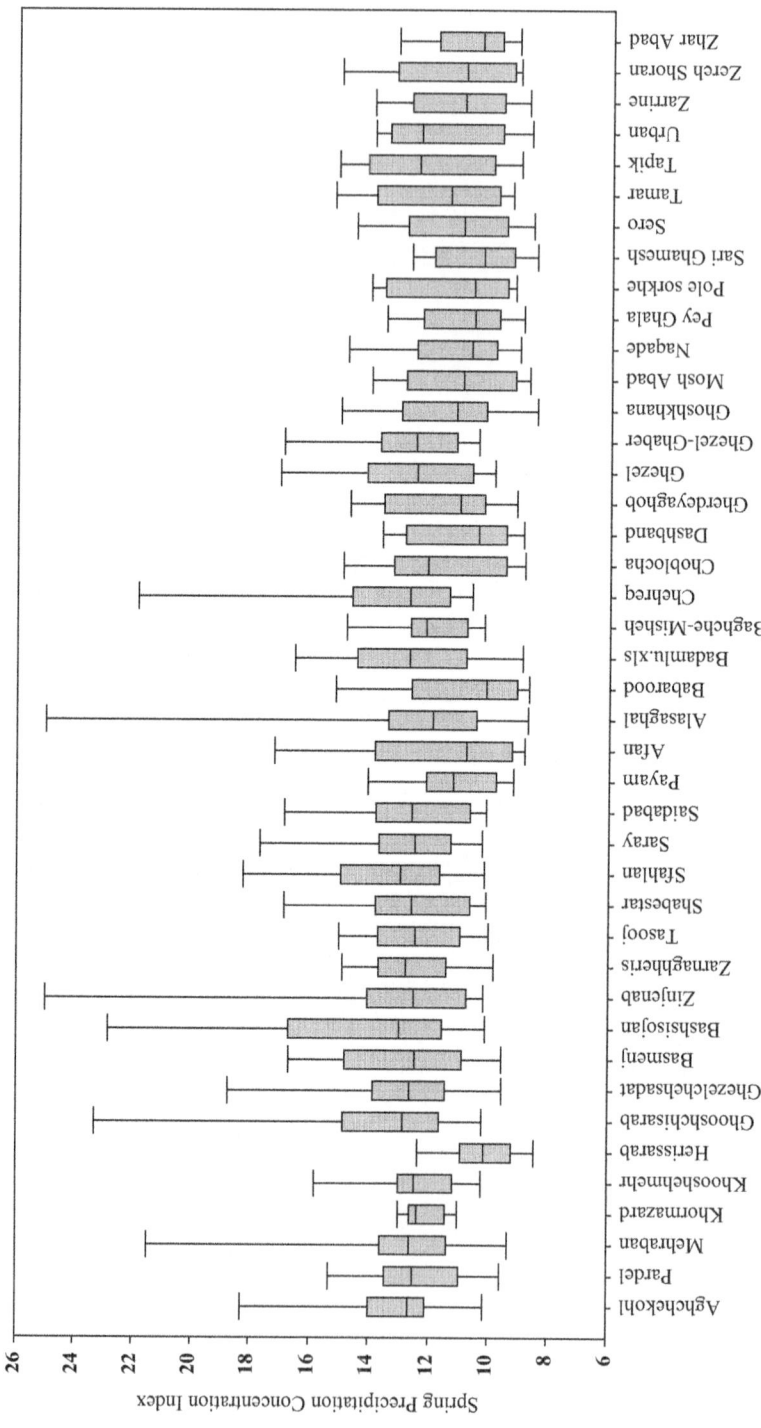

FIGURE 1.3   Results of calculation of the spring PCI in the period 1984–2013.

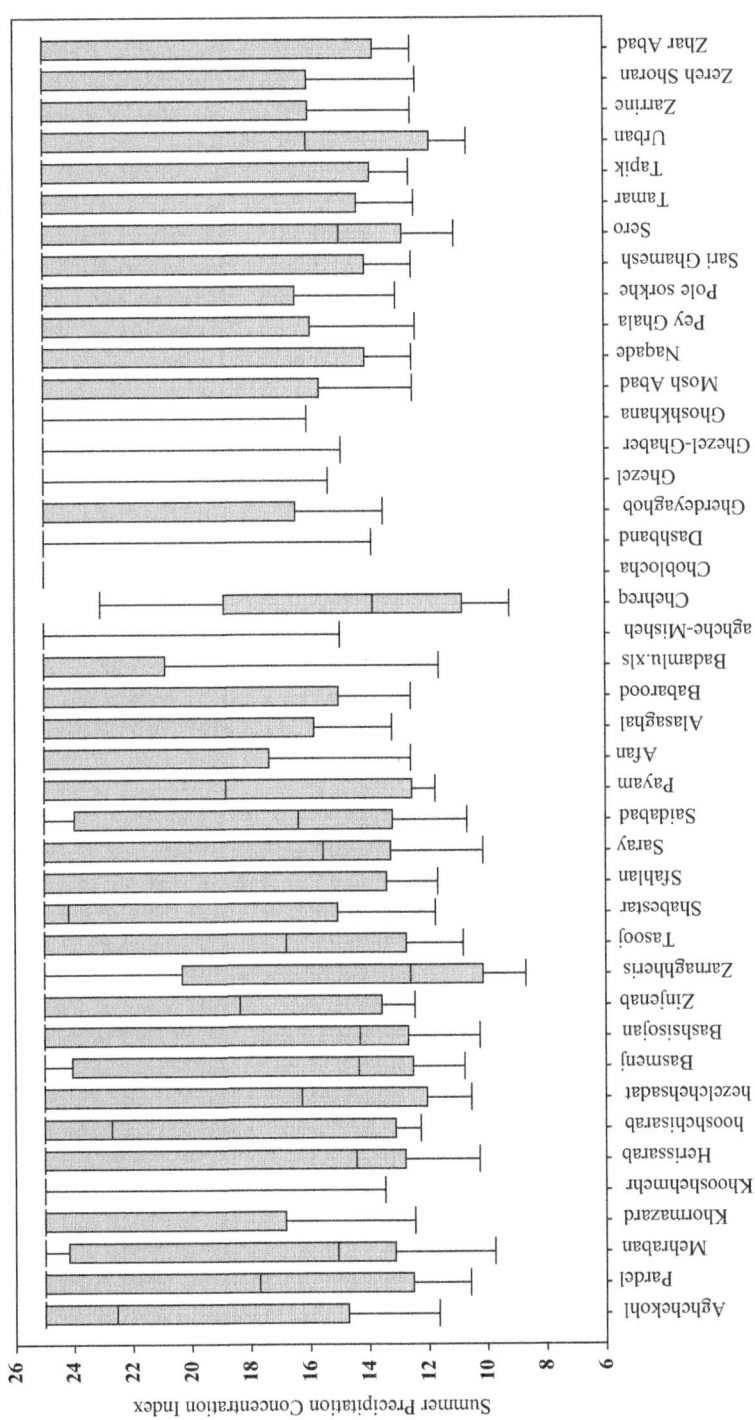

FIGURE 1.4 Results of calculation of the summer PCI in the period 1984–2013.

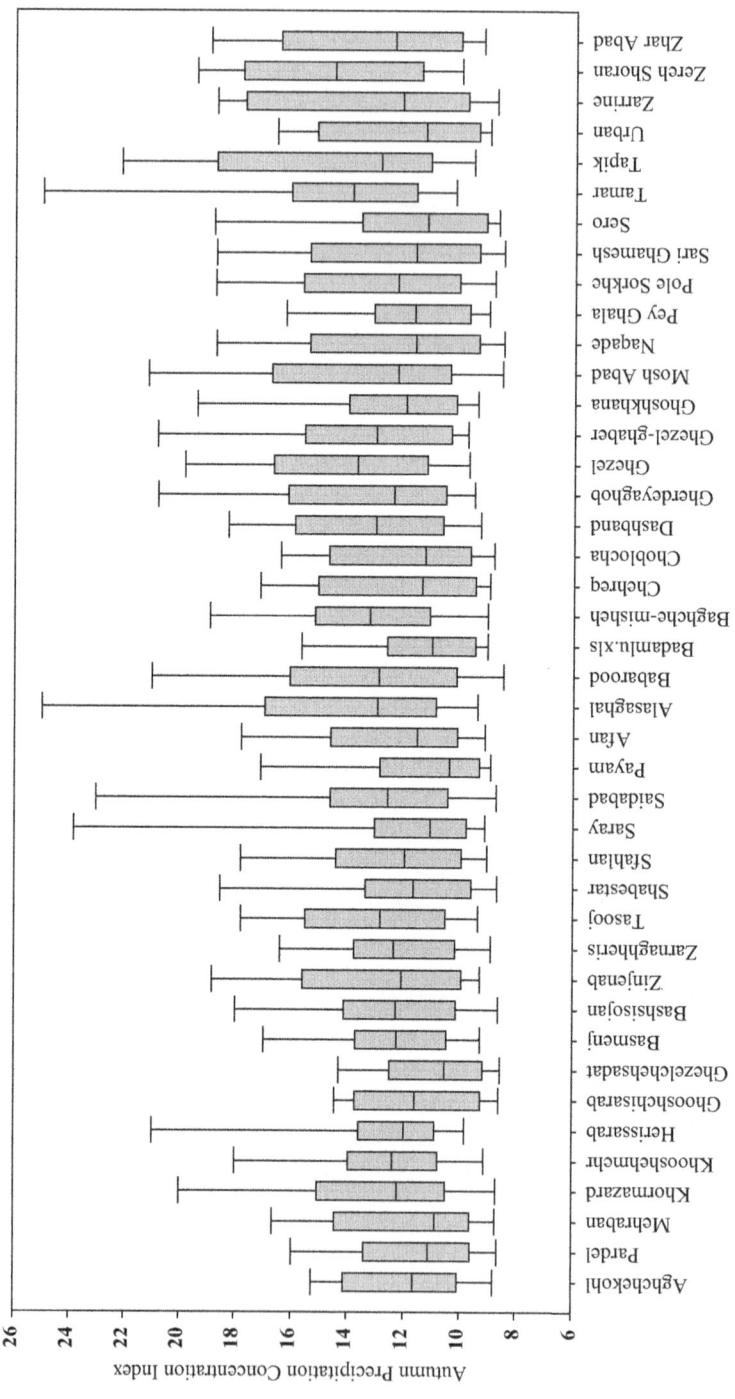

FIGURE 1.5 Results of calculation of the autumn PCI in the period 1984–2013.

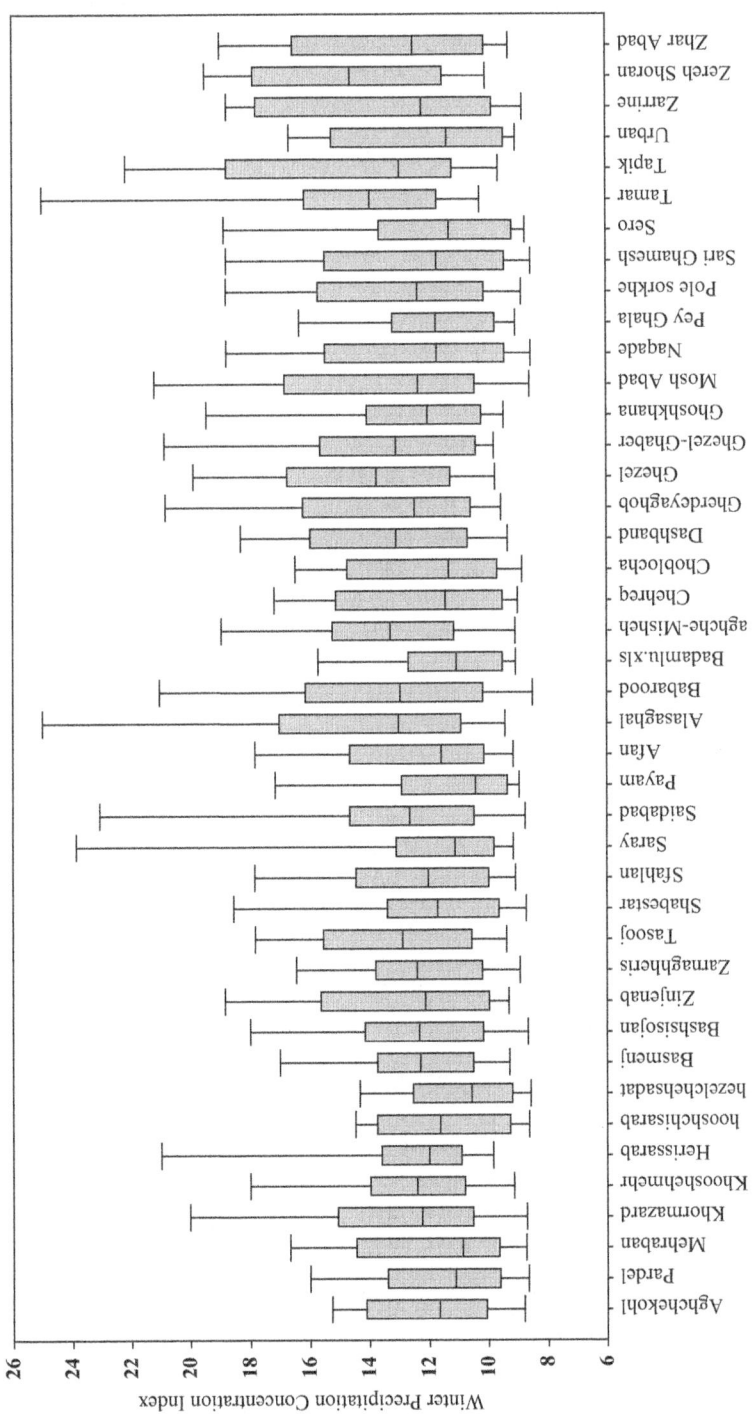

FIGURE 1.6 Results of calculation of the winter PCI in the period 1984–2013.

At seasonal scale in spring, there was at least one case of regular distribution of precipitation during the years studied at basin level (PCI < 10). Almost 50% of the years under study showed average concentration and dispersion in all the stations in spring. Stations no. 22 and 41 experienced a uniform dispersion of precipitation in spring. Stations no. 37 and 39 also experienced an irregular concentration of precipitation in most of the years. It means that the distribution of precipitation was irregular in 3 months during the studied years. There was also severely irregular distribution in this scale in stations no. 2, 12, and 7 in the Eastern area of Lake Urmia. The stream-gaging stations no. 42, 41, 40, 39, 38, 35, 33, 32, 31, 29, 28, 27, 26, 25, 23, 21, 20, and 19 in the Western area of Lake Urmia were observed during 1 to 3 years under study. The results showed that there was an irregular increase of precipitation in the Western stations of Lake Urmia compared to the Eastern stations of Lake Urmia during the last 30 years.

There was no regular concentration of precipitation observed in seasonal scale in summer in the basin area. In summer, the prevailing concentration of precipitation in the stations under study was placed in the severely irregular category. The stations in the West Lake Urmia had higher percentage of precipitation and exhibited more irregular concentration compared to the other stations in East Lake Urmia. The stations in the East Lake Urmia in summer had a lower level of average and irregular concentration of precipitation. The results showed that concentration of precipitation in the different months of summer showed severely irregular distribution.

Regular concentration was seen in a few years from any station during the studied years. In autumn, concentration of precipitation of most stations under study was categorized into two categories of average and irregular concentration. Similar to summer, the stations in West Lake Urmia had higher percentage of precipitation and more irregular distribution compared to the stations located in East Lake Urmia. The severely irregular concentration of precipitation was observed in 1–3 years in most of the stations.

In seasonal scale in winter in Lake Urmia basin during the years under study, no severely irregular concentration was observed. The concentration of precipitation from most stations under study in winter was divided into two categories of average and irregular concentration. The concentration of precipitation in January, February, and March was more regular than that in the other months. To study the concentration distribution index in Lake Urmia basin, the results of the index related to basin are presented and zoned in two seasonal and annual scales and are demonstrated in Figure 1.7.

As shown in Figure 1.7, small areas in the North West and North East of Lake Urmia basin showed regular concentration and distribution in spring. A small area in the South West of Lake Urmia basin was located in the category of irregular concentration of precipitation. It is eventually seen that most of the areas in Urmia Lake basin in spring showed average concentration and distribution of precipitation. Rainfall around the Lake Urmia during spring and in the months of April, May, and June showed average distribution of precipitation. The dominant concentration of stream-gaging stations in Lake Urmia basin in summer was severely irregular. As shown in Figure 1.7, the areas surrounding Lake Urmia, Southern areas, and Lake Urmia Lake basin showed severely irregular distribution of precipitation.

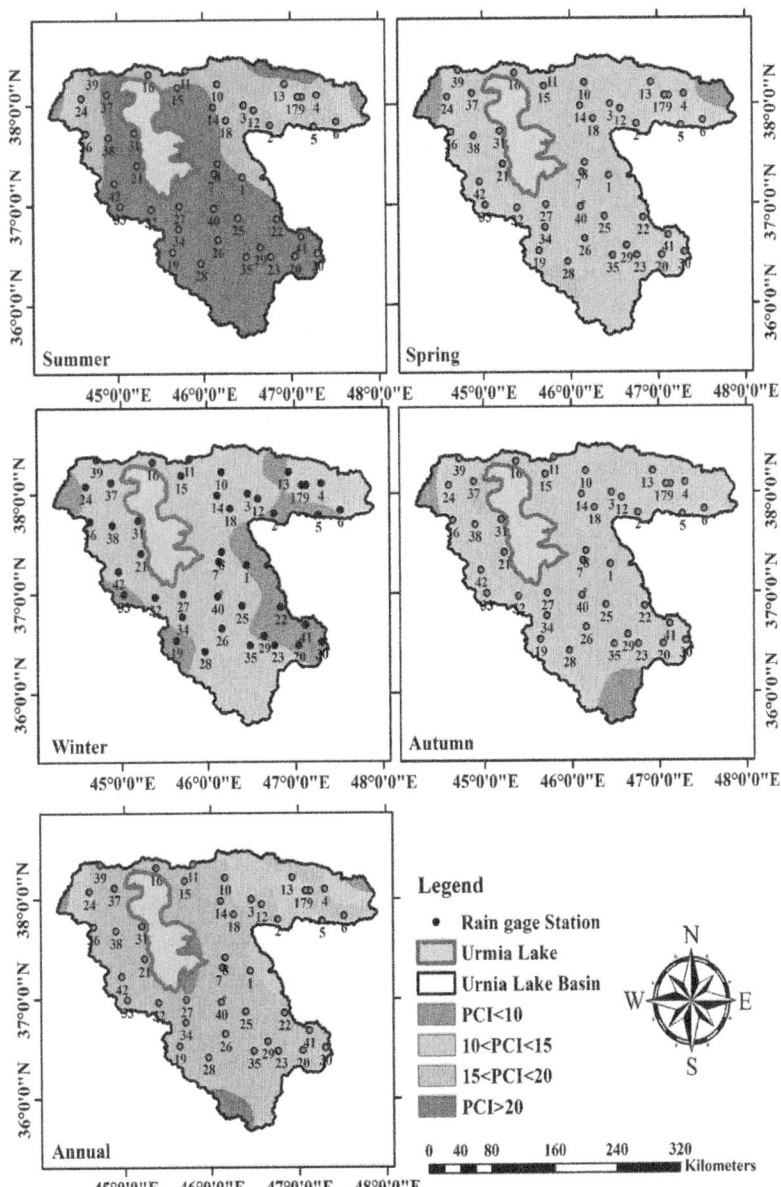

FIGURE 1.7   Delineation of calculated PCI in annual and seasonal scales in the period 1984–2013.

The results showed severely irregular and reduced precipitation in the summer months (July, August, September). The North West, North East, and North West of the Lake Urmia basin in summer had irregular concentration of precipitation. In the more distant areas of Lake Urmia, we observed more irregular precipitation. The results of concentration precipitation zonation in autumn also showed that nearly all regions of Lake Urmia basin had average concentration and distribution precipitation in this season. A small area in the North West Lake Urmia in this season (autumn) had irregular distribution in the monthly precipitations. In winter, as compared to other seasons, the precipitations had more irregular distribution. Most of the stream-gaging stations in Lake Urmia basin showed average precipitation in winter. The areas around Lake Urmia had average precipitation in this season. The border areas in Lake Urmia in winter had regular distribution of precipitation. In this season (winter), irregular and severely irregular distribution was not observed. The precipitation distribution in annual scale of most of the stations in Lake Urmia was irregular. In other words, distribution of precipitation was irregular within a year. The South East of Lake Urmia in annual scale had severely irregular concentration of precipitation. The areas in the North West and North East of Lake Urmia basin had more irregular concentration. The results of the present study were consistent with studies by Khalili et al. (2014, on the concentration and precipitation trend of synoptic stations in Iran.

1.3.2 A STUDY ON CI OF PRECIPITATION DISTRIBUTION IN LAKE URMIA

As mentioned in the materials and methods section, the precipitations in the basin area were classified in 1 mm categories in order to measure the concentration in the basin area. This classification is presented in the first column of Table 1.3.

TABLE 1.3

Step-by-Step Results of the Calculation of the CI of Afan Station

1	2	3	4	5	6	7	8
Class	0.50	27.00	27.00	13.50	13.50	1.31	0.06
0.1–0.9	1.50	151.00	178.00	226.50	240.00	8.62	1.08
1–1.9	2.50	196.00	374.00	490.00	730.00	18.12	3.28
2–2.9	3.50	165.00	539.00	577.50	1,307.50	26.11	5.87
3–3.9	4.50	158.00	697.00	711.00	2,018.50	33.77	9.07
4–4.9	5.50	132.00	829.00	726.00	2,744.50	40.16	12.33
5–5.9	6.50	126.00	955.00	819.00	3,563.50	46.27	16.01
6–6.9	7.50	104.00	1,059.00	780.00	4,343.50	51.31	19.51
7–7.9	8.50	90.00	1,149.00	765.00	5,108.50	55.67	22.95
8–8.9	9.50	70.00	1,219.00	665.00	5,773.50	59.06	25.94
9–9.9	10.50	117.00	1,336.00	1,228.50	7,002.00	64.73	31.46
10.0–10.9	11.50	60.00	1,396.00	690.00	7,692.00	67.64	34.56
11.0–11.9	12.50	88.00	1,484.00	1,100.00	8,792.00	71.90	39.50
12.0–12.9	13.50	72.00	1,556.00	972.00	9,764.00	75.39	43.87
13.0–13.9	14.50	48.00	1,604.00	696.00	10,460.00	77.71	46.99
14.0–14.9	15.50	58.00	1,662.00	899.00	11,359.00	80.52	51.03
15.0–15.9	16.50	40.00	1,702.00	660.00	12,019.00	82.46	54.00
16.0–16.9	17.50	31.00	1,733.00	542.50	12,561.50	83.96	56.43
17.0–17.9	18.50	30.00	1,763.00	555.00	13,116.50	85.42	58.93
18.0–18.9	19.50	16.00	1,779.00	312.00	13,428.50	86.19	60.33

(Continued)

TABLE 1.3 (*Continued*)

Step-by-Step Results of the Calculation of the CI of Afan Station

1	2	3	4	5	6	7	8
19.0–19.9	20.50	27.00	1,806.00	553.50	13,982.00	87.50	62.82
20.0–20.9	21.50	11.00	1,817.00	236.50	14,218.50	88.03	63.88
21.0–21.9	22.50	13.00	1,830.00	292.50	14,511.00	88.66	65.19
22.0–22.9	23.50	11.00	1,841.00	258.50	14,769.50	89.20	66.35
23.0–23.9	24.50	22.00	1,863.00	539.00	15,308.50	90.26	68.77
24.0–24.9	25.50	32.00	1,895.00	816.00	16,124.50	91.81	72.44
25.0–25.9	26.50	15.00	1,910.00	397.50	16,522.00	92.54	74.23
26.0–26.9	27.50	14.00	1,924.00	385.00	16,907.00	93.22	75.96
27.0–27.9	28.50	10.00	1,934.00	285.00	17,192.00	93.70	77.24
28.0–28.9	29.50	10.00	1,944.00	295.00	17,487.00	94.19	78.56
29.0–29.9	30.50	14.00	1,958.00	427.00	17,914.00	94.86	80.48
30.0–30.9	31.50	4.00	1,962.00	126.00	18,040.00	95.06	81.05
31.0–31.9	32.50	6.00	1,968.00	195.00	18,235.00	95.35	81.92
32.0–32.9	33.50	10.00	1,978.00	335.00	18,570.00	95.83	83.43
33.0–33.9	34.50	4.00	1,982.00	138.00	18,708.00	96.03	84.05
34.0–34.9	35.50	12.00	1,994.00	426.00	19,134.00	96.61	85.96
35.0–35.9	36.50	2.00	1,996.00	73.00	19,207.00	96.71	86.29
36.0–36.9	38.50	17.00	2,013.00	654.50	19,861.50	97.53	89.23
38.0–38.9	39.50	1.00	2,014.00	39.50	19,901.00	97.58	89.41
39.0–39.9	41.50	22.00	2,036.00	913.00	20,814.00	98.64	93.51
41.0–41.9	42.50	2.00	2,038.00	85.00	20,899.00	98.74	93.89
42.0–42.9	43.50	2.00	2,040.00	87.00	20,986.00	98.84	94.28
43.0–43.9	46.50	5.00	2,045.00	232.50	21,218.50	99.08	95.33
46.0–46.9	49.50	4.00	2,049.00	198.00	21,416.50	99.27	96.22
49.0–49.9	50.50	5.00	2,054.00	252.50	21,669.00	99.52	97.35
50.0–50.9	54.50	0.00	2,054.00	0.00	21,669.00	99.52	97.35
51.0–54.9	53.00	10.00	2,064.00	590.00	22,259.00	100.00	100.00
55.0–64	59.k.50	27.00	27.00	13.50	13.50	1.31	0.06

The second column of the table represents the mean classifications. The number of rainy days for each class was identified and registered in front of the related class. The fourth column in Table 1.3 represents the sum of the cumulative precipitations for each class or, in other words, the total cumulative of the third column. The multiplication of the values given in the second and third columns is shown in the fifth column obtained from the product of multiplying mean categories by the number of rainfall days. The total cumulative value of the fifth column is given in column 6. Finally, the cumulative percentage of column 4 is equal to X mentioned in column 7. The cumulative percentage of the fourth column is equal to Y mentioned in column 8. Now, the regression of a, b is obtained by X and Y values, as well as relations 1.3 and 1.4. The S value (sub-area of Lorenz curve) is calculated on the basis of a, b coefficients. Then the CI value is calculated by relation 1.2 for every station. The results presented in Table 1.3 are related to annual CI of Afan station in statistical period under study. The coefficients a, b for this station in annual scale were 0.102 and 0.022, respectively. According to a, b coefficients, the CI value for this basin is equal to 0.473. The CI was similarly calculated for the other stations and the results are presented in Tables 1.4 and 1.5. The results of studying daily concentration indices in seasonal scale are presented in Table 1.6.

TABLE 1.4
Results of Calculation of the Annual CI of Rain Gauge Stations in the Period 1984–2013

No.	En-Name	X_UTM	Y_UTM	a	b	S	Annual CI
1	Aghchekohlezaman	628,538.699	4,127,289.58	0.17	0.02	2,969.68	0.41
2	Bashsisojan	655,535.104	4,185,094.93	0.1	0.02	2,638.43	0.47
3	Basmenj	628,772.852	4,206,829.86	0.17	0.02	3,053.47	0.39
4	Ghezelchehsadat	701,672.386	4,219,408.73	0.17	0.02	3,053.47	0.39
5	Ghooshchisarab	699,604.463	4,184,195.57	0.16	0.02	3,028.96	0.39
6	Herissarab	721,473.539	4,190,307.78	0.09	0.02	2,526.12	0.49
7	Khooshehmehr	600,421.574	4,130,604.31	0.17	0.02	2,969.68	0.41
8	Khormazard	603,238.13	4,141,734.88	0.14	0.02	2,811.98	0.44
9	Mehraban	687,099.259	4,217,210.11	0.12	0.02	2,697.35	0.46
10	Pardel	602,129.79	4,230,498.39	0.12	0.02	2,654.22	0.47
11	Payam	569,903.257	4,244,952.21	0.12	0.02	2,654.22	0.47
12	Saidabad	639,111.125	4,201,449.67	0.1	0.02	2,638.43	0.47
13	Saray	669,248.49	4,231,621.96	0.06	0.03	2,276.26	0.54
14	Sfahlan	598,064.055	4,204,553.98	0.1	0.02	2,638.43	0.47
15	Shabestar	561,305.177	4,226,388.06	0.07	0.03	2,391.7	0.52
17	Zarnaghheris	682,713.648	4,217,110.53	0.12	0.02	2,697.35	0.46
17	Tasooj	532,053.483	4,241,014.38	0.12	0.02	2,654.22	0.47
18	Zinjenab	611,438.353	4,189,928.33	0.14	0.02	2,811.98	0.44
19	Afan	557,052	4,043,148	0.1	0.02	2,606.73	0.48
20	Alasaghal	683,041	4,039,877	0.11	0.02	2,628	0.47
21	Babarood	520,490	4,139,581	0.1	0.02	2,606.55	0.48
22	Badamlu.xls	663,820	4,081,365	0.14	0.02	2,896.44	0.42
23	Baghche-Misheh	658,254	4,039,018	0.13	0.02	2,780.97	0.44
24	Chehreq	464,363	4,214,549	0.1	0.02	2,585.03	0.48
25	Choblocha	624,549	4,082,095	0.15	0.02	2,905.52	0.42
26	Dashband	604,229	4,056,986	0.11	0.02	2,655.65	0.47
27	Gherdeyaghob	563,788	4,094,597	0.1	0.02	2,538.01	0.49
28	Ghezel	586,755	4,031,106	0.13	0.02	2,824	0.44
29	Ghezel-Ghaber	647,580	4,049,351	0.14	0.02	2,896.44	0.42
30	Ghoshkhana	705,960	4,043,501	0.11	0.02	2,628	0.47
31	Mosh Abad	517,436	4,175,992	0.11	0.02	2,645.33	0.47
32	Naqade	534,795	4,090,995	0.12	0.02	2,665.45	0.47
33	Pey Ghala	502,663	4,094,287	0.16	0.02	2,912.26	0.42
34	Pole sorkhe	563,126	4,069,480	0.05	0.03	2,163.22	0.57
35	Sari Ghamesh	633,518	4,038,435	0.14	0.02	2,791.31	0.44

(Continued)

TABLE 1.4 (*Continued*)

Results of Calculation of the Annual CI of Rain Gauge Stations in the Period 1984–2013

No.	En-Name	X_UTM	Y_UTM	a	b	S	Annual CI
36	Sero	468,231	4,175,171	0.12	0.02	2,675.28	0.46
37	Tamar	489,613	4,218,563	0.17	0.02	3,053.47	0.39
38	Tapik	490,988	4,170,717	0.11	0.02	2,704.35	0.46
39	Urban	475,445	4,244,172	0.16	0.02	2,958.55	0.41
40	Zarrine	599,592	4,092,413	0.11	0.02	2,660.5	0.47
41	Zereh Shoran	689,112	4,062,234	0.12	0.02	2,743.48	0.45
42	Zhar Abad	496,768	4,119,904	0.13	0.02	2,787.26	0.44

TABLE 1.5

Results of Calculation of the Seasonal CI of Rain Gauge Stations in the Period 1984–2013

Station No.	Station	Autumn CI	Winter CI	Spring CI	Summer CI
1	Aghchekohlezaman	0.39	0.42	0.41	0.4
2	Bashsisojan	0.48	0.47	0.45	0.48
3	Basmenj	0.4	0.41	0.38	0.4
4	Ghezelchehsadat	0.4	0.41	0.38	0.4
5	Ghooshchisarab	0.42	0.4	0.39	0.42
6	Herissarab	0.47	0.45	0.52	0.55
7	Khooshehmehr	0.39	0.42	0.41	0.4
8	Khormazard	0.43	0.4	0.44	0.43
9	Mehraban	0.46	0.46	0.46	0.46
10	Pardel	0.45	0.43	0.49	0.48
11	Payam	0.45	0.43	0.49	0.48
12	Saidabad	0.48	0.47	0.45	0.48
13	Saray	0.55	0.52	0.53	0.53
14	Sfahlan	0.48	0.47	0.45	0.48
15	Shabestar	0.5	0.51	0.52	0.56
16	Zarnaghheris	0.46	0.46	0.46	0.46
17	Tasooj	0.45	0.43	0.49	0.48
18	Zinjenab	0.43	0.4	0.44	0.43
19	Afan	0.46	0.48	0.48	0.56
20	Alasaghal	0.48	0.47	0.48	0.44
21	Babarood	0.47	0.47	0.48	0.54
22	Badamlu.xls	0.4	0.42	0.44	0.42
23	Baghche-Misheh	0.43	0.46	0.41	0.52
24	Chehreq	0.5	0.46	0.49	0.45
25	Choblocha	0.42	0.41	0.41	0.38
26	Dashband	0.46	0.47	0.46	0.45

(Continued)

TABLE 1.5 (*Continued*)

Results of Calculation of the Seasonal CI of Rain Gauge Stations in the Period 1984–2013

Station No.	Station	Autumn CI	Winter CI	Spring CI	Summer CI
27	Gherdeyaghob	0.51	0.47	0.5	0.48
28	Ghezel	0.44	0.45	0.41	0.52
29	Ghezel-Ghaber	0.4	0.42	0.44	0.42
30	Ghoshkhana	0.48	0.47	0.48	0.44
31	Mosh Abad	0.48	0.45	0.48	0.41
32	Naqade	0.47	0.46	0.42	0.4
33	Pey Ghala	0.44	0.4	0.42	0.4
34	Pole sorkhe	0.56	0.57	0.56	0.49
35	Sari Ghamesh	0.45	0.44	0.42	0.45
36	Sero	0.47	0.47	0.46	0.44
37	Tamar	0.4	0.41	0.38	0.4
38	Tapik	0.47	0.46	0.47	0.44
39	Urban	0.42	0.4	0.42	0.48
40	Zarrine	0.55	0.47	0.46	0.5
41	Zereh Shoran	0.53	0.46	0.44	0.34
42	Zhar Abad	0.44	0.46	0.42	0.42

TABLE 1.6

Summary Results of CI Index in the Period 1984–2013

CI	Annual	Autumn	Winter	Spring	Summer
Max	0.57	0.56	0.57	0.56	0.56
Min	0.39	0.39	0.40	0.38	0.34
Ave	0.45	0.46	0.45	0.45	0.45

To evaluate the accuracy of a, b coefficients, $Y = (aX) \exp (bX)$ and Lorenz curve could be applied. The Lorenz chart related to stream-gaging stations no. 19, 20, 21, and 23 are presented in Figures 1.8–1.11.

The results of CI showed that the mean index in annual and seasonal scales in all the stations surveyed was 0.45. The lowest and highest CI values were respectively 0.39 and 0.57 in annual scale. A brief review of the stations' results in terms of the daily concentration precipitation is presented in Table 1.6. According to the results it can be concluded that the maximum CI values for all scales surveyed were 0.57 and 0.56.

According to Martin-Vide (2004), the high CI values (CI > 0.61) signify high daily concentration. For the stations with CI > 0.61, almost 70% of the precipitation occurred in 25% of the rainfall days. This indicates the occurrence of aggressive and heavy rainfalls in stations with CI > 0.61. The CI < 0.61 values showed more regularity in the distribution of daily precipitation and lower values of this criterion means that the stations under study possessed more regularity in daily precipitations and the number of rainy days. In the present study, no station showed the CI > 0.61

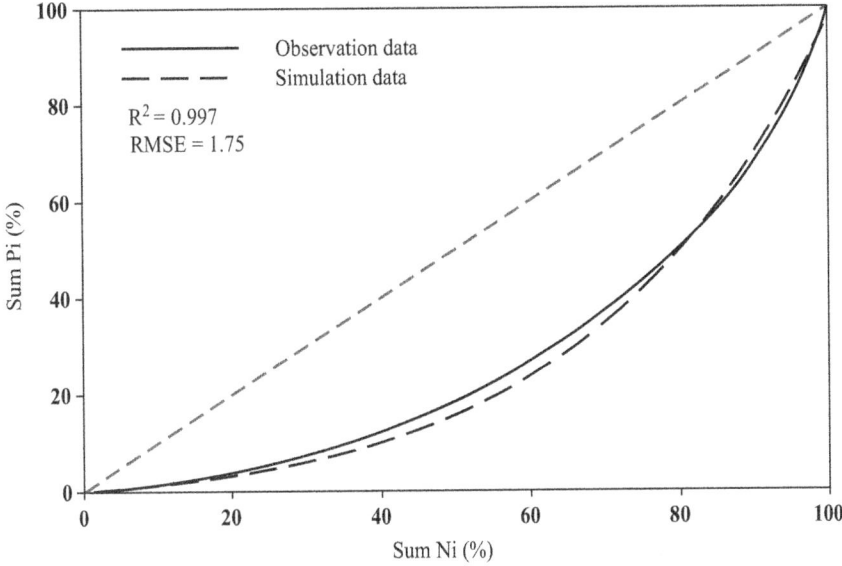

FIGURE 1.8 Concentration, or Lorenz curves for two observations and estimation of CI values of Afan station (19).

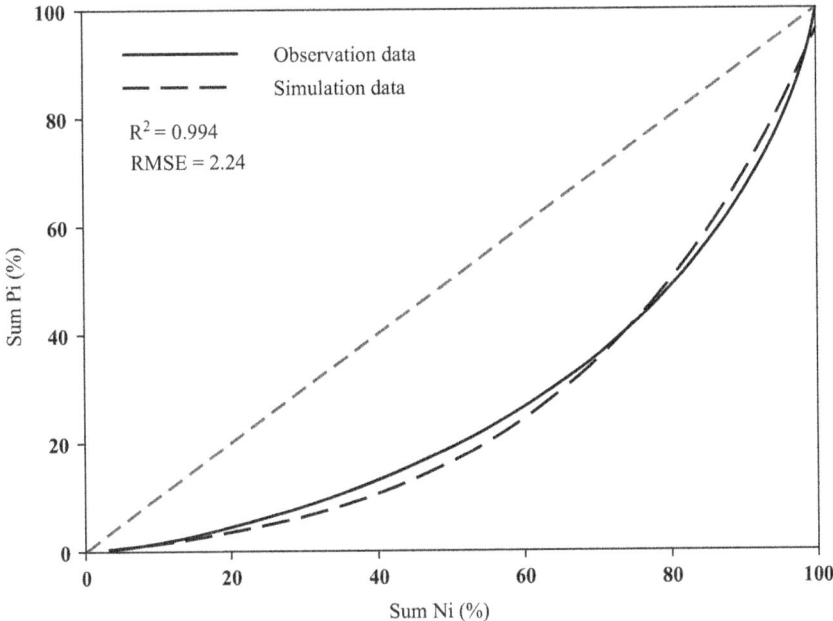

FIGURE 1.9 Concentration, or Lorenz curves for two observations and estimation of CI values of Alasagel station (20).

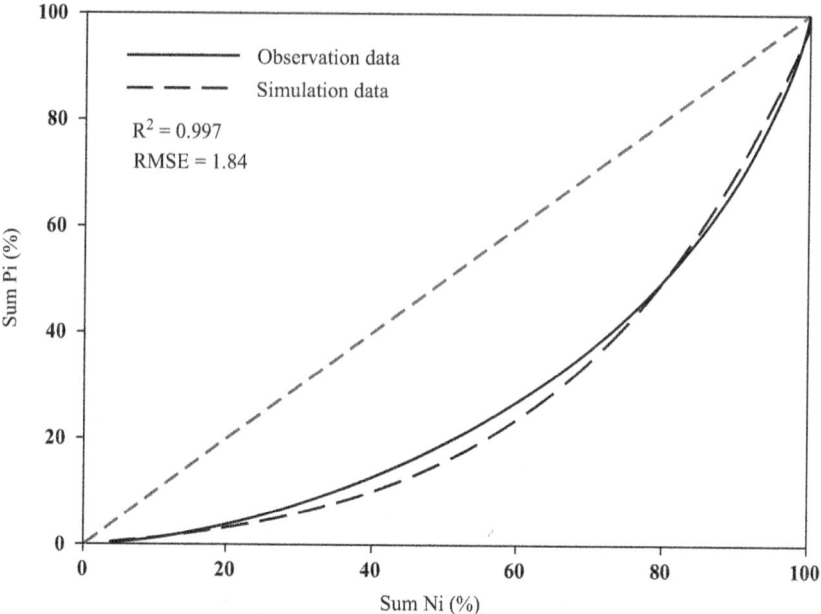

FIGURE 1.10 Concentration, or Lorenz curves for two observations and estimation of CI values of Babarood station (21).

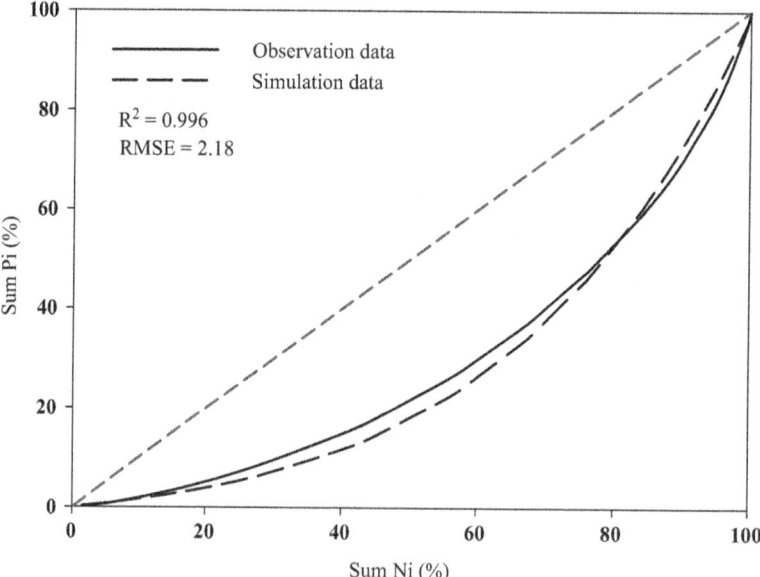

FIGURE 1.11 Concentration, or Lorenz, curves for two observation and estimation CI values of Baghcheh-Misheh station (23).

value. According to the annual scale, stations no. 4, 3, 5, and 37 had the lowest and
Pol Sorkh station had the highest CI value. To evaluate regional CI, the CI values in
the basin areas were zoned with different scales and are presented in Figure 1.12.

The results of CI index in Lake Urmia basin during the statistical period in spring
showed that CI > 0.6 is not observed in any of the areas under study. The majority
of Lake Urmia areas in spring had CI values between 0.4 and 0.5 (yellow colored

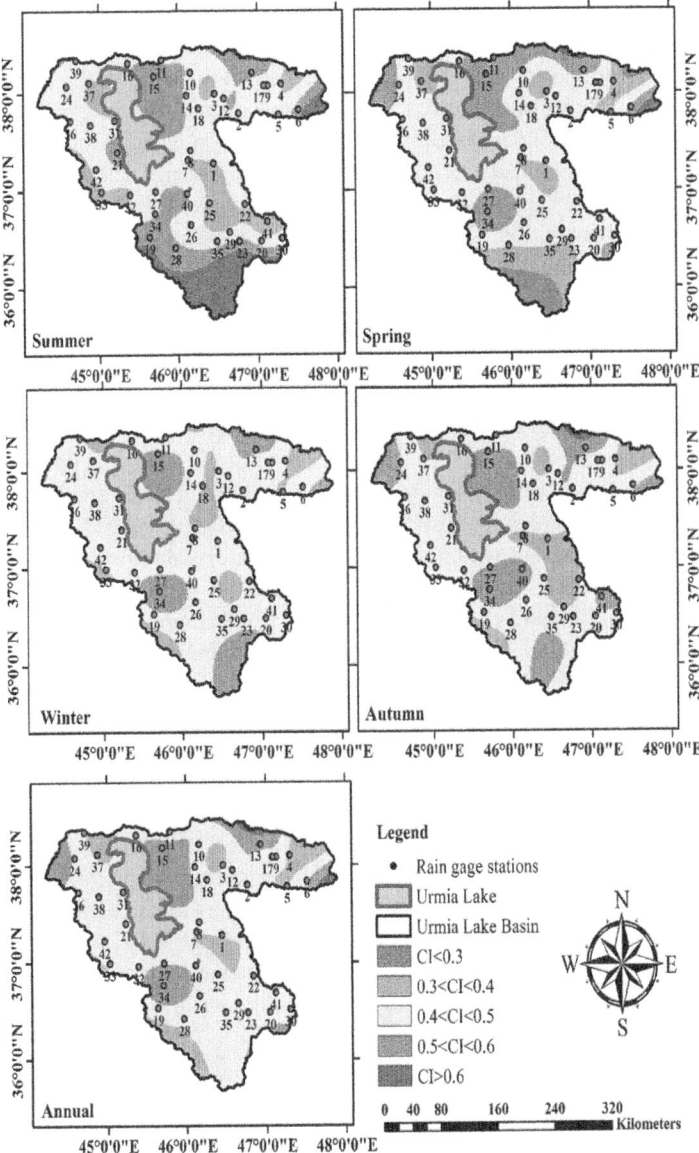

FIGURE 1.12 Delineation of the calculated CI values in annual and seasonal scales in the
period 1984–2013.

areas). In this season (spring), aggressive and terrifying precipitations were observed in none of the basin areas. In North East Lake Urmia and one part of Lake Urmia, the CI value was between 0.5 and 0.6 which indicates the irregularity in the daily precipitation in these areas. The southern part of the Lake Urmia had better regulation in the distribution of daily precipitation and the precipitation data showed regular exponential distribution in rainfall days. It was observed that the southern areas of Lake Urmia were more regular in the distribution of daily precipitation in spring.

As with spring in the areas under study, in autumn also CI > 0.6 was not observed in any of the stations studied. The areas of East and South Lake Urmia and Northeast and Southeast regions of Lake Urmia basin had more irregularity compared to the other areas. The results showed that aggressive rainfalls and rainstorms were observed during rainfall days both in spring and autumn in these areas and the distribution of precipitation had less regularity than other areas. The East and Northeast areas of the study area (green areas) had relatively fair regularity in daily precipitation and related distribution. Other areas (yellow areas) showed average distribution of average daily precipitation. Regular concentration was not observed in this season.

In winter, as well as spring and autumn, a part of the South and the Northeast of Lake Urmia and also an area in the Northeast of the basin had average CI between 0.5 and 0.6 in the statistical period and shows the high concentration of daily precipitation and irregularity in distribution of daily precipitation in these areas. It can be concluded that the amounts of daily rainfall in the listed areas were not distributed evenly among the rainfall days. The CI values in most of the areas under study were between 0.4 and 0.5. In this season, regular distribution (CI < 0.3) and irregular daily distribution (CI > 0.6) were not observed. Several small areas in the East Lake Urmia had CI between 0.3 and 0.4 which showed the regular distribution of daily precipitation during winter days. In other words, the amount of precipitation in this season amongst the rainfall days was distributed in a relatively regular way.

The southern areas of Lake Urmia basin had severe irregular precipitation in summer and it could be concluded that 70% of the rainfall in this season occurred only in 25% of the rainfall days. In this season the Northeast and South regions of Lake Urmia, an area in Northeast of Lake Urmia, as well as a station in the Southwest of Lake Urmia, had irregular distribution of precipitation in summer. In this season (summer), irregular distribution of daily precipitation in the basin under study was not observed. Several stations in the Northeast, Southeast, and Southwest of Lake Urmia had relatively regular distribution of daily precipitation.

In annual scale, most of the Lake Urmia areas had an average concentration (CI between 0.4 and 0.5). In annual scale, similar to seasonal scale, the Northeast and South of Lake Urmia had more irregular concentration compared to other areas. In annual scale, none of the stations under study in the Lake basin had regular daily distribution of precipitation.

1.4   CONCLUSION

The statistical analysis of data on daily, monthly, or annual precipitation in enclosed basins such as the Lake Urmia basin in terms of precipitation distribution bears great significance from a climatic aspect. Therefore, examination of the indices revealing the distribution is a noteworthy issue. PCI is an index to determine the precipitation changes in a specified area and the results of its analysis could demonstrate the accessibility to water in an area. CI is also an index used to examine the daily distribution and statistical structure of precipitation. The statistical structure of precipitation could be demonstrated using precipitation concentration curves related to the percent of accumulative percent ratio to accumulative percent of rainfall days. CI is a curve which defines the contrast or concentration of different values of daily precipitation.

In this study, two indices of PCI and CI were used to analyze the distribution and precipitation structure in two annual and seasonal scales for PCI index and daily index for CI in Lake Urmia (statistical period: 1984–2013). The results of annual precipitation concentration in Lake Urmia basin showed that the mean precipitation distribution was irregular in the basin area. The values of precipitation distribution index were 17.55 for East Lake Urmia stations and 16.37 for West Lake Urmia stations which signify the irregular precipitation in the eastern areas of Lake Urmia as compared to the western areas. The results also showed that during the last 10 years, as compared to the first 10 years of the statistical period, PCI in stations in the eastern and western parts of the Lake respectively increased by 0.38 and 4.06. Almost 2.28% increase in annual precipitation concentration is observed in the basin area. The precipitation distribution in spring in the basin was average. In this season (spring), PCI during the last 10 years, as compared to the first 10 years of the statistical period, had a 3.90% and 5.50% increase in the western and eastern parts of the Lake, respectively. The results showed that the distribution precipitation in the Lake Urmia basin was more regular during the statistical period in spring. The results of precipitation distribution in summer during the statistical period showed that PCI in summer in the surveyed basin during the last 10 years, as compared to the first 10 years of the statistical period, had a 14.20% decrease, while the share of the stations in West Lake Urmia was almost 13% and the share of the eastern stations was 15.50%. Reduced PCI in Lake Urmia basin in this season (summer) showed a better regularity in precipitation during the summer months. The results of precipitation distribution in autumn during the statistical period showed that PCI in the surveyed basin during the last 10 years, as compared to the first 10 years of the statistical period, had a 20.54% decrease, while the share of the stations in West Lake Urmia was almost 19.91% and the share of the eastern stations was 21.22%. Increased PCI in Lake Urmia basin in autumn showed that precipitation distribution moved from average toward irregular. The results of precipitation distribution in winter during the statistical period showed that PCI in surveyed basin during the last 10 years, as compared to the first 10 years of the statistical period, had a 2% increase and 6.45% decrease in the eastern areas of Lake Urmia. The results showed that the western areas of Lake Urmia during the statistical period had increased irregularity of precipitation, while the eastern areas of Lake Urmia experienced a better situation. With regard to the climatic changes and reduced precipitation that occurred around Lake Urmia during the last few years, precipitation distribution in spring, summer, and winter was more desirable. However, in annual scale during autumn, precipitation distribution was more irregular in the basin area.

The results of CI in the basin area showed that the daily precipitation in Lake Urmia basin was not in regular or severely irregular situation in any of the stations under study. The entire stations surveyed were in relatively regular, average concentration and relatively irregular situation in terms of the daily precipitation distribution. Most of the studied stations were in average concentration category in terms of daily precipitation concentration. In autumn, winter, spring, and summer in annual scale, almost 74%, 90%, 81%, 74%, and 84% of the stream-gaging stations showed average precipitation concentration. The CI generally confirms the fact that the real share of rainfall ratio to the total amount of rainfall was properly described by an exponent distribution or not. This method included collecting daily rainfall with 1 mm increase and classification and definition of relative influence of various classes by analyzing the relative contribution of in precipitation (1%). The CI values are basically between 0 and 1, which geometrically show the percent of triangle area between X = Y and the exponent curve. When the share of each rainfall category is identical to the total precipitation, CI is closer to zero and when the precipitation share is attributed to one category of precipitation and the exponent curve changed to Y = 0, the index value will be equal to one. In the analysis of PCI and CI, we could examine the irregular and severely irregular events as well as the inequality of daily, monthly, and annual precipitations. Therefore, more research studies are necessary to examine the relation between PCI and the occurrence of disasters such as floods, erosion, and drought. The indices under study as statistical indices could determine different precipitation weights including distribution, share of severe precipitations, and spatial distribution ratio to total rainfall. One of the advantages of examining drought indices such as PCI is that we could observe the precipitation distribution during the months of a season or even the months of a year and gather information on their distributions. The results could also be used to confront the droughts or severe precipitations.

REFERENCES

Adegun, O., Balogun, I. and Adeaga, O. 2012. Precipitation concentration changes in Owerri and Enugu. *Special Publication of the Nigerian Association of Hydrological Sciences*, pp. 383–391.

Alijani, B., O'brien, J. and Yarnal, B. 2008. Spatial analysis of precipitation intensity and concentration in Iran. *Theoretical and Applied Climatology*, 94(1–2), pp. 107–124.

Brooks, C.E.P. and Carruthers, N. 1953. *Handbook of Statistical Methods in Meteorology*. London: Government publication.

Gong, D.Y. and Ho, C.H. 2002. Shift in the summer rainfall over the Yangtze River valley in the late 1970s. *Geophysical Research Letters*, 29(10), pp. 78-1–78-4.

Khalili, K., Nazeri Tahrudi, M. and Khanmohammadi, N. 2014. Trend analysis of precipitation in recent two decade over Iran. *Journal of Applied Environmental and Biological Sciences*, 4(1s), pp. 5–10.

Khalili, K., Tahoudi, M.N., Mirabbasi, R. and Ahmadi, F. 2016. Investigation of spatial and temporal variability of precipitation in Iran over the last half century. *Stochastic Environmental Research and Risk Assessment*, 30(4), pp. 1205–1221.

Li, X., Jiang, F., Li, L. and Wang, G. 2011. Spatial and temporal variability of precipitation concentration index, concentration degree and concentration period in Xinjiang, China. *International Journal of Climatology*, 31(11), pp. 1679–1693.

Luis, M.D., Gonzalez-Hidalgo, J.C., Brunetti, M. and Longares, L.A. 2011. Precipitation concentration changes in Spain 1946–2005. *Natural Hazards and Earth System Sciences*, 11(5), pp. 1259–1265.

Martin-Vide, J. 2004. Spatial distribution of a daily precipitation concentration index in peninsular Spain. *International Journal of Climatology: A Journal of the Royal Meteorological Society*, 24(8), pp. 959–971.

Olascoaga, M.J. 1950. Some aspects of Argentine rainfall. *Tellus*, 2(4), pp. 312–318.

Oliver, J.E. 1980. Monthly precipitation distribution: a comparative index. *The Professional Geographer*, 32(3), pp. 300–309.

Ren, G.Y., Wu, H. and Chen, Z.H. 2000. Spatial patterns of change trend in rainfall of China. *Quarterly Journal of Applied Meteorology*, 11(3), pp. 322–330.

Riehl, H. 1949. Some aspects of Hawaiian rainfall. *Bulletin of the American Meteorological Society*, 30, pp. 176–187.

Tabari, H., Marofi, S., Aeini, A., Talaee, P.H. and Mohammadi, K. 2011. Trend analysis of reference evapotranspiration in the western half of Iran. *Agricultural and Forest Meteorology*, 151(2), pp. 128–136.

Valli, M., Shanti Sree, K. and Murali Krishna, I.V. 2013. Analysis of precipitation concentration index and rainfall prediction in various agro-climatic zones of Andhra Pradesh, India. *International Research Journal of Environmental Sciences*, 2(5), pp. 53–61.

Zhai, P., Zhang, X., Wan, H. and Pan, X. 2005. Trends in total precipitation and frequency of daily precipitation extremes over China. *Journal of Climate*, 18(7), pp. 1096–1108.

Zhang, Q., Xu, C.Y., Zhang, Z., Chen, Y.D., Liu, C.L. and Lin, H. 2008. Spatial and temporal variability of precipitation maxima during 1960–2005 in the Yangtze River basin and possible association with large-scale circulation. *Journal of Hydrology*, 353(3–4), pp. 215–227.

Zhang, Q., Xu, C.Y., Gemmer, M., Chen, Y.D. and Liu, C. 2009. Changing properties of precipitation concentration in the Pearl River basin, China. *Stochastic Environmental Research and Risk Assessment*, 23(3), pp. 377–385.

2 Use of Mid-Infrared Diffuse Reflectance to Assess the Effects of Soil Management on Soil Quality in Tropical Hill Slope Agro Eco-Systems

Bunjirtluk Jintaridth and Peter Motavalli
University of Missouri

CONTENTS

2.1 INTRODUCTION

Soil organic matter (SOM) is one of the most important components in soil because it affects many soil functions such as nutrient cycles, water movement, thermal properties, and the fate of chemical use for soil management (Carter, 2002). The assessment of changes in SOM is a key issue for soil quality. Soil organic carbon consists of humic and non-humic substances, which are formed by the decomposition of natural organic matter. Humic substances are powerful complexing and chelating entities whose sorption characteristics depend on their chemical composition. They contribute significantly to the chemical, physical, and biological properties of soil (Garcia et al., 1992; Chefetz et al., 1998; Varanini and Pinton, 1995). Due to the relatively slow biodegradation rate of some components of SOM, their influence on soil properties and performance lasts over a long time.

Infrared (IR) spectroscopy is the study of the molecular vibrations of bonded atoms. The frequency of the absorption is characteristic of the atom in the bond and the type of motion associated with the vibration. The observed frequency can be used to distinguish the component atoms as well as the bonding characteristics of those atoms (Niemeyer et al., 1992). According to Painter et al. (2012) and Niemeyer et al. (1992), Diffuse Reflectance Fourier Transform Infrared Spectroscopy (DRIFT) technique offers several advantages over transmission IR spectroscopy: (i) a simple sample preparation procedure; (ii) high resolution of the spectra because of reduction in the sensitivity towards light scattering; and (iii) a more reliable method for quantitative estimations of functional groups. Wander and Traina (1996) used DRIFT spectroscopy to examine the functional groups of SOM fractions. They also reported that the characteristics and distribution of individual SOM fractions may provide a means for assessing management impacts on SOM quality that can influence soil and plant productivity.

Reflectance characteristic of soils are related to chemical functional groups known as "chromophores". Soil mineral and organic matter will show the results of absorption in specific IR spectral signature due to the vibrations of functional groups within the mineral and organic substances. In the near infrared (NIR) region (700–2,500 nm), the spectra have sensitive vibrational absorbance due to –OH, –CH, and –NH organic functional groups in SOM.

Wave numbers and assignments for peaks in the DRIFT spectrum are the same as in the IR spectrum (Baes and Bloom, 1989; Niemeyer et al., 1992). The characteristic peaks for humic acid (HA) and fulvic acid (FA) appear distinctly and sharply at the wave numbers.

Wander and Traina (1996) characterized the physically and chemically isolated SOM fractions collected from organic matter obtained from organically and conventionally managed soils. They used ratios of reactive (O-containing) and recalcitrant (C, H, and/or N) functional group heights to characterize SOM fraction composition. The range of soils and cropping practices that tested this technique were limited. The diffuse reflectance signal can be described by the Kubelka-Munk function (KM) where KM is the ratio of the diffuse reflectance from the sample and the diffuse reflectance from a non-absorbing power. This ratio is related to sample absorbance (a), concentration (c), and a scattering coefficient (s) in the following manner: $KM = (2.303 \, ac/s)$. This relationship is useful for quantification of the functional group composition in spectra of heterogeneous organic materials.

The objective of this study was to determine the effectiveness of using mid-infrared (MIR) spectroscopic methods in assessing soil quality based on the status of soil organic C using O/R ratios across a range of soils from the agricultural management practices from different countries. It was hypothesized that the O/R ratios from MIR would provide a sensitivity indicator of soil quality changes due to the agricultural management and landscape positions.

2.2   MATERIALS AND METHODS

2.2.1   STUDY SITE

Soils with diverse properties and management were collected from various locations, including (i) Umala municipality in Bolivia; (ii) Cochabamba in Bolivia; (iii) Kecamatan Nanggung, West Java, Indonesia, and (iv) Lantapan region of Mindanao, The Philippines Location 1 was located in the Umala municipality of the Central Altiplano region of Bolivia. Four locations in Umala, Bolivia, namely: Circa, San José, Kellhuiri, and Vinto Coopani, were selected for this study. Soil samples were taken from cultivated lands with different fallow length ranging from 1 year, 6 years, 8 years, 10 years, 20 years, and more than 40 years. They are exposed to frequent adverse climatic conditions, mainly frost and drought events, which limit agricultural production to just a few crops, such as potato (*Solanum tuberosum* L.), quinoa (*Chenopodium quinoa* Wild.), and barley (*Hordeum vulgare* L.). The communities of Kellhuiri and Vinto Coopani are situated at relatively higher elevations (approximately 4,070 and 4,013 m above sea level), and San Juan Circa and San José de Llanga are located at relatively lower elevations (approximately 3,771 and 3,806 m above sea level). At the higher elevations, the land has more hills and steeper slopes and soil tillage practices are generally dependent on animal traction and their soils mostly consist of rocks. Farmers do not use adequate amounts of manure and manufactured fertilizers for optimizing crop growth. In the lower elevations communities are situated in relatively flat areas. People in this area frequently use mechanical traction for tillage. Soils in the lower elevation are generally sandier.

Location 2 was in Cochabamba, Bolivia in the localities of Toralapa Baja, Waylla Pujru, and Sancayani Alto, in the province of Tiraque, Department of Cochabamba, Bolivia. To get this information, 60 soil samples (30 degraded soils and 30 non-degraded soils) were collected in the farmers' fields. This area has never had soil classification before, so separating for degraded and non-degraded land depended on the professional criteria and agricultural soil problems in the communities where they work. The main problems in this area were soil erosion, low soil fertility, monoculture, pests, contamination, and poor soil conservation practices.

Location 3 in Indonesia was situated in Kecamatan Nanggung, West Java. The study site was originally covered with a typical humid tropical forest. Soil samples were collected using alley cropping agroforestry and non-agroforestry areas. These areas received different types and rates of amendments which included manure, compost, and chemical fertilizer. The climate of the area was tropical with monsoons occurring in the rainy season. Soils were poorly-to-moderately drained and occurred on undulating-to-flat uplands. The elevation was less than 200–1,800 m above sea level.

Location 4 was in the Lantapan region of the Philippines. Soil samples were taken from the Lantapan region of Mindanao. The soil is mostly a reddish soil which has mostly fine texture, poor drainage, is low in SOM, and has a heavy weight. It was expected that the landscape position and cultivation would influence SOM. In this study, soil samples were taken along atransect from the top to the bottom of a cultivated alley and a non-cultivated area. Each transect was divided into summit, shoulder, backslope, footslope, and toeslope. Fifteen soil samples from the non-cultivated area and 15 soil samples from the cultivated area were collected from the farmers' fields.

2.3   EQUIPMENTS AND PROCEDURES

2.3.1   HA EXTRACTION AND FRACTIONATION PROCEDURE

The soil was processed by crushing and air-drying it before passing it through a 2 mm mesh sieve. Sieve soils (30 g) were weighed into a 250-mL plastic polyethylene centrifuge bottle. The sample was equilibrated to a pH value between 1 and 2 with 1 M HCl at room temperature and the solution volume was adjusted with 0.1 M HCl to provide a final concentration with a ratio of 10 mL liquid/1 g dry sample. The suspension was shaken with a shaker for 1 h, and the residue was separated from the supernatant and saved by decantation after allowing the solution to settle for 15 min. The soil residue was then neutralized with 1 M NaOH to pH 7.0 and 0.1 M NaOH was added to obtain a ratio of 10:1. The air in the bottle with the solution was displaced by N gas (N_2) to minimize the chemical changes due to auto-oxidation, and the plastic bottle with the soil extraction was shaken for 4 h at room temperature. The alkaline suspension was then allowed to settle overnight and the supernatant was collected by centrifugation. The supernatant was separated from the residue by centrifugation at 4,000 rpm for 15 min. The supernatant was then collected and acidified with 6 M HCl by constant stirring to pH 1.0 and the suspension was allowed to stand for 12–16 h. The suspension was then centrifuged at 4000 rpm to separate the HA (precipitate) fractions and the HA fraction was redissolved by adding a minimum volume of 0.1 M KOH under N_2 gas. Solid KCl was added to attain 0.3 M (K^+) and it was then centrifuged at a high speed (8,000 rpm) to remove suspended solids. The dark colored suspended HAs were precipitated in 0.1 M HCl and 6% HF in plastic test tubes with a 10 mL solution/1 g of ground organic horizon to remove possible silicates. The HA were shaken for 24 h at room temperature with an HCl/HF treatment. The insoluble residues (HA) were separated from the supernatant by centrifuging at 10,000 rpm for 20 min. The HA precipitates were transferred to a Visking dialysis tube [Spectrum*Spectra/Por* biotech cellulose ester (CE) dialysis membrane] with 10,000 Dalton molecular weight cut off (MWCO). Before transferring the HA, the precipitates were slurried with water and dialyzed against distilled water for 4–5 days until the dialysis water gave a negative Cl-test with 1 M of silver nitrate (no white color appeared). The purified HA was then freeze-dried, ground to powder, and stored at room temperature prior to chemical analysis (Stevenson, 1994).

2.3.2 SAMPLE PREPARATION AND ANALYSIS FOR DRIFT

DRIFT was performed in an IR spectrophotometer with a DRIFT Smart Collector accessory (Thermo Electron Corp.). All HA fractions were powdered with a sapphire mortar and pestle, and stored in a drying box. The HA fractions were mixed with ground KBr for diluting which samples were prepared of 8% in pre-ground KBr. Then the samples were ground with an agate mortar and pestle, and spectra of the samples were collected using a Nicolet TM 4700 IR spectrophotometer (Thermo-Fisher, Madison, WI).

2.3.3 SPECTRA DATA ACQUISITION AND ANALYSIS

To obtain DRIFT data, samples were scanned from 4,000 to 400 cm^{-1} and 400 scans were collected at a resolution of 4 cm^{-1} and spectra. The percent reflectance spectrum was converted to KM units. Absorption spectra were converted to a KM function and also the baseline was corrected using Gram/32 software package (Galactic Corporation). Transformation to KM units typically results in better spectra, especially in the region of strong absorption below 1,500 cm^{-1} (Childers et al., 1986). Before spectra data acquisition, the spectra were baseline normalized. During spectral data acquisition, the silica band was removed at around 1,800; 1,900 and 2,000 cm^{-1}.

The peak assignments and intensity were calculated for determining the O/R ratio, in which the reactive groups are O-containing, and recalcitrant groups are R-containing. Heights of peaks of O/R ratio were characterized for SOM fraction composition. Peak assignments and intensity (by height) ratios were calculated and based on the methods of Baes and Bloom (1989), Niemeyer et al. (1992), Wander and Traina (1996), and Ding et al. (2002) (Table 2.1). Based on these approximate peak assignments, the actual peaks for each set of soil samples from the different locations were identified and then used for O/R ratio calculations.

2.4 RESULTS AND DISCUSSION

2.4.1 DRIFT SPECTROSCOPY AND COMPARISON OF PEAK RATIO (O/R RATIO) OF HAs ACROSS LOCATIONS

2.4.1.1 Location 1

All spectra of HA of soils of different fallow length looked similar in all samples with minor differences (Figure 2.1).

The comparison of O/R ratio in fallow length shows that the O/R ratio increased with increasing fallow length. In samples from all four locations in Umala, a ratio of the fallow length in each community was compared to the ratio of selected peak intensities. The O/R ratio in soil samples from San José increased with increasing fallow length as did samples from San Juan Circa, where the O/R ratio increased from 1.04 to 1.26, 1.28, and 1.71 from 1 year to 10, 20, and >40 years, respectively.

Greater fallow length only had a small impact on the increase in O/R ratio in samples from Kellhuiri and Vinto Coopani. However, the fallow length samples from these communities were only from 1 to 8 years in both locations. During the fallow period,

TABLE 2.1

Approximate DRIFT Infrared Spectra Peak Assignments and Occurrences in Physical and Chemical Organic Matter Fractions

Assignments	Wave Number (cm^{-1})
Mineral OH	3,690
Mineral OH	3,621
Phenol OH, amide N–H	3,279–3,340
CH$_2$ symmetric stretch	2,962–2,950
CH$_2$ asymmetric stretch	2,924–2,930
CH$_2$ symmetric stretch	2,850
CO–OH H bonded	2,500
C=O stretch	1,850
C=O ketonic, COOH	1,735–1,713
C=O, C=O–H bonded, amide H	1,650
C=C aromatic	1,630–1,608
Aromatic ring, amide	1,509
CH, asymmetric stretch, CH bend	1,457
Aromatic ring stretching	1,420
COO salt, COOH	1,400
COO–, CH, symmetric stretch	1,379–1,327
CO, COOH, COC, phenol OH	1,260–1,240
Aliphatic, alcoholic OH	1,190–1,127
CO aliphatic alcohol	1,080–1,050
Aliphatic COC, aromatic ether, Si-O	1,030
OH, COOH, Al–OH	918–912
CH aromatic bend, Al–O–Si	850–830
Fe–O–Si	797
CH aromatic bend	779
Unknown mineral peak	750
Unknown mineral peak	794
COO salt, Mg/Si–O aliphatic	560
Si–O	530–520
Aromatic ring	480

Peak assignments based on van der Marel and Beutlespacher (1976), Griffiths (1983), Stevenson et al. (1989), Baes and Bloom (1989), and Inbar et al. (1989).

residues produced from native vegetation decompose and degrade, possibly allowing for an increase in SOM, the restoration of nutrients, and an enhancement of the population and activity of indigenous microbe consortia (Aguilera, 2010). Fallowing may increase the bioavailability of SOM leading to the change of HA composition as well as the O/R ratio. Singh et al. (2005) showed that fallow and cropping land had an impact on soil organic carbon concentrations, and also reported that Fourier transform infrared (FTIR) spectra were different among spectra from different land management

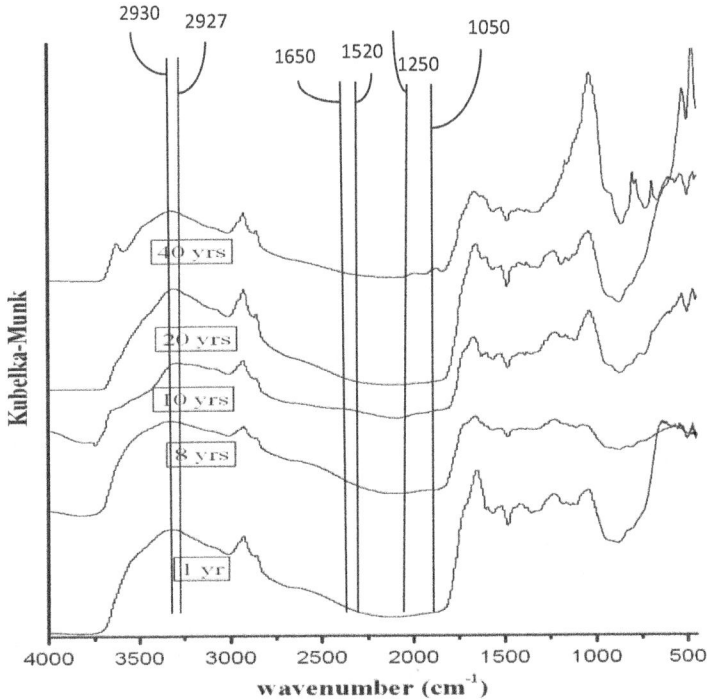

FIGURE 2.1   Diffuse Reflectance Fourier Transform Infrared spectra of humic acids due to different fallow length (0, 8, 10, 20, and 40 years) in soils collected from Location #1 (Umala, Bolivia).

practices. Our results suggest that an increase in HA bioavailability could be achieved by a greater fallow length in the Bolivian Altiplano environment.

2.4.1.2   Location 2

In the degraded soils and non-degraded soils collected in Cochabamba, Bolivia, DRIFT spectra showed similar peaks for all three communities (Figures 2.2 and 2.3).

Among degraded soils, the O/R ratios could be ranked as Sanyacani > Waylla Pujru > Toralapa Baja which are 1.06, 1.08, and 1.09, respectively. The O/R ratios ranked the same results (Sanyacani > Waylla Puiru > Toralapa Baja) at 1.10, 1.13, and 1.14 in non-degraded soils, respectively.

Comparing the ratio of peak heights intensity (O/R) showed that no statistical difference ($P < 0.05$) was observed for the ratio of peak heights intensity (O/R) between degraded soils and non-degraded soils. However, the non-degraded soils had O/R ratios higher than degraded soil for all communities. This trend was also the same for each location, which was Sanyacani > Waylla Puiru > Toralapa Baja (Figure 2.4). It is probable that in the area of degraded soil there are problems such as low soil fertility, poor soil management practices, monoculture, pests, contamination, and lack of soil conservation practices which could lower the active component

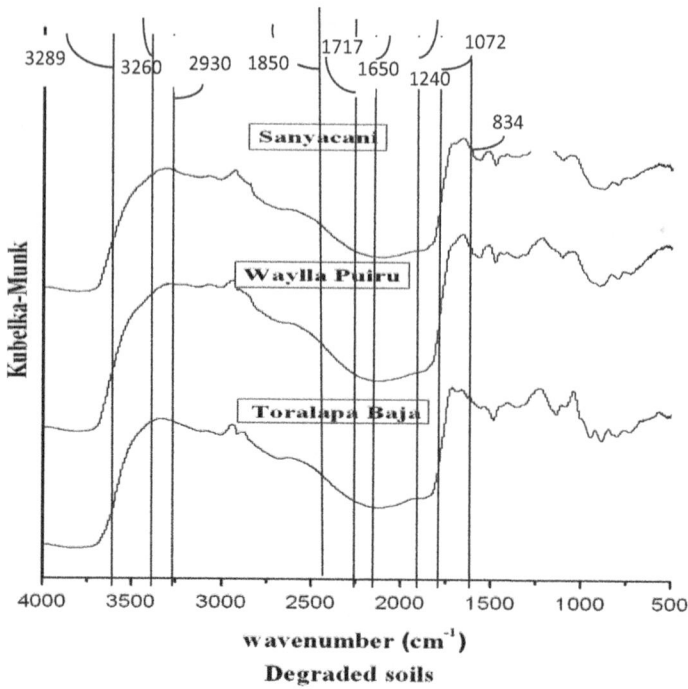

FIGURE 2.2 Diffuse Reflectance Fourier Transform Infrared spectra of humic acids in degraded land in three communities at different elevations [Toralapa Baja (low elevation), Waylla Puiru (medium elevation), and Sanyacani (high elevation)] located in Location #2 (Cochabamba, Bolivia).

of HA. Soils collected from lower elevations had higher O/R ratio than those of higher elevations. It is probable that agricultural practices, such as tillage methods, affected the soil quality. In non-degraded lands, soils were more fertile compared to degraded lands. Campbell et al. (2000) found that the rate of residue decomposition affected soil organic carbon in soils, not only by the levels of organic residue input, but also due to the biological activity, and overall nutrient status in soils (Kirchmann and Gerzabek, 1999). Management practices such as crop rotation, soil tillage, and fertilization affect the soil C inputs and turnover in arable cropping systems (Six et al., 2000).

2.4.1.3   Location 3

In Indonesia, soil samples were collected from Kecamatan Nanggung. The study site is located around Kecamatan Nanggung, West java.

Under agroforestry systems in Indonesia, the spectra of low and medium inputs of manure were similar compared to the spectrum of the control (no manure) area indicating that the application of manure might have an impact on the composition of HA (Figure 2.5).

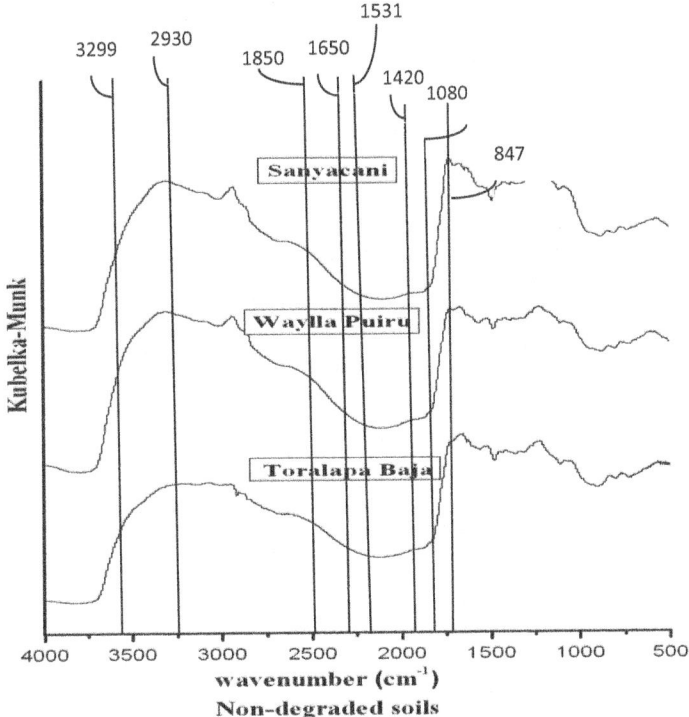

FIGURE 2.3 Diffuse Reflectance Fourier Transform Infrared spectra of humic acids in non-degraded land in three communities at different elevations [Toralapa Baja (low elevation), Waylla Puiru (medium elevation), and Sanyacani (high elevation)] located in Location #2 (Cochabamba, Bolivia).

In the non-agroforestry area, the results were similar to those of the agroforestry area (Figure 2.6).

Results of comparison of selected peak intensities in agroforestry and non-agroforestry areas are shown in Figure 2.7. The O/R ratio was highest in medium manure (3.30 ± 0.04). The lowest was in the control (no manure) and the average O/R ratio was 1.29 ± 0.39. The average HA fraction in low manure was 2.05 ± 1.46. Based on the peak ratio comparison of HA in the non-agroforestry area, the O/R ratio was similar to that in the agroforestry area. The lowest to highest average HA fraction in agroforestry was in the control (no manure), low manure, and medium manure and the average O/R ratio was 1.29 ± 0.39. The HA fraction with low manure inputs was 2.05 ± 1.46. The quantity of plant residues produced by the different vegetation under the two systems may have affected the soil organic carbon. Sanchez et al. (2003) suggested that different management systems affect the soil organic carbon content by changing the annual inputs of organic residues. In comparing agroforestry and non-agroforestry, there were different HA peak height ratios in which the peak height ratio was higher in agroforestry than in non-agroforestry, and the HA peak height ratio decreased with decreasing manure

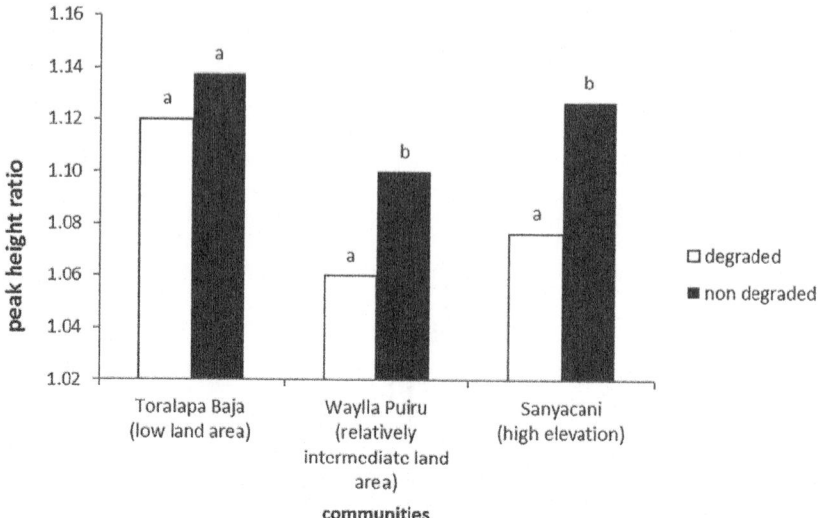

FIGURE 2.4 Comparing the ratio of peak heights intensity (O/R) of humic acids with degraded and non-degraded land in Location #2 (Cochabamba, Bolivia). Within a bar, the numbers followed with the same letter are not significantly different to the LSD0.05 value.

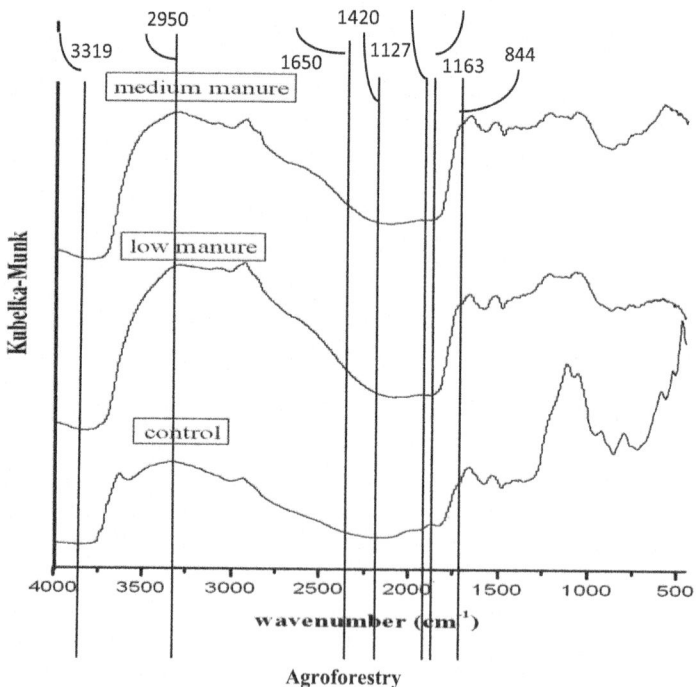

FIGURE 2.5 Diffuse Reflectance Fourier Transform Infrared spectra of humic acids in agroforestry area receiving different amounts of manure in Location #3 (Indonesia).

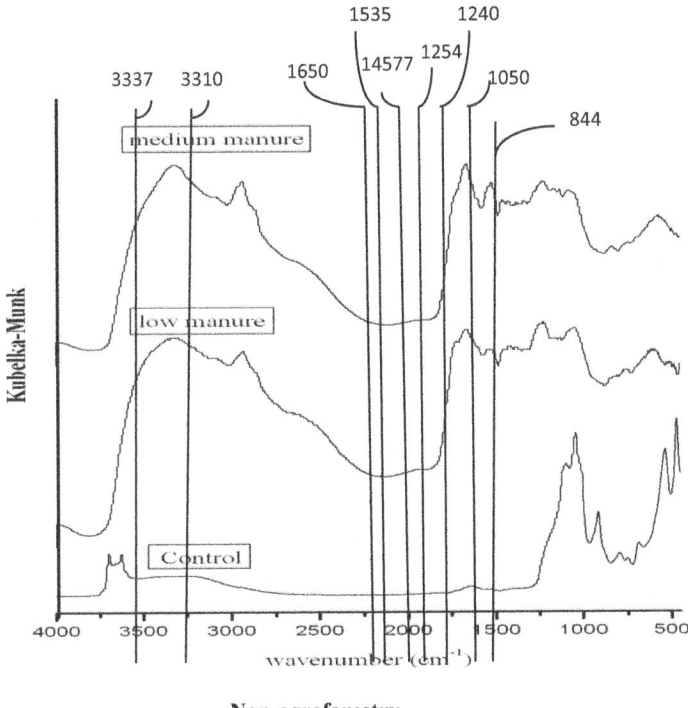

Non-agroforestry

FIGURE 2.6 Diffuse Reflectance Fourier Transform Infrared spectra of humic acids in non-agroforestry area receiving different amounts of manure in Location #3 (Indonesia).

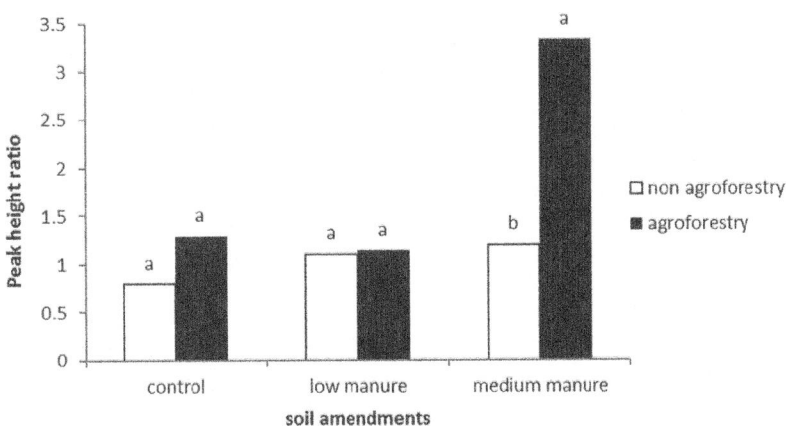

FIGURE 2.7 Comparing ratio of peak heights intensity (O/R) of HA under agroforestry, and non-agroforestry systems and different manure amendments in Location #3 (Indonesia). Comparing within a soil amendment treatment, bars with the same letter are not significantly different based on the LSD0.05 value.

amendments. Therefore, the reductions in HA could be related to differences in both cropping systems and soil amendments.

2.4.1.4   Location 4

In the Philippines, soil samples were taken from the Lantapan region of Mindanao. The soil is mostly a reddish soil which has mostly fine texture, poor drainage, is low in SOM, and has a heavy weight. It was expected that the landscape position and cultivation would influence SOM. Thus, in this study, a set of soil samples would be compared from a transect of samples collected from the top to the bottom of the cultivated and non-cultivated area.

All results indicated that the landscape position and cultivation might have impacted the composition of HA. Comparing different landscape positions under the cultivated area, it was found that the highest to lowest of O/R ratios was in summit (1.21 ± 0.03), toeslope (1.19 ± 0.08), footslope (1.15 ± 0.07), shoulder (1.07 ± 0.05), and backslope (0.74 ± 0.22), respectively. For the peak ratio comparison of HA in non-cultivated area, O/R ratio was in footslope (1.16 ± 0.04), toeslope (1.13 ± 0.01), summit (1.08 ± 0.01), shoulder (1.08 ± 0.02), and backslope (1.05 ± 0.03), respectively. The O/R ratio of HA under non-cultivated area was higher than cultivated area except for the toeslope (Figures 2.8 and 2.9). It was probable that there was an addition of SOC from up-hill to down-hill during erosion, or addition of amendments during cultivation. The O/R ratio in soil samples taken from the cultivated

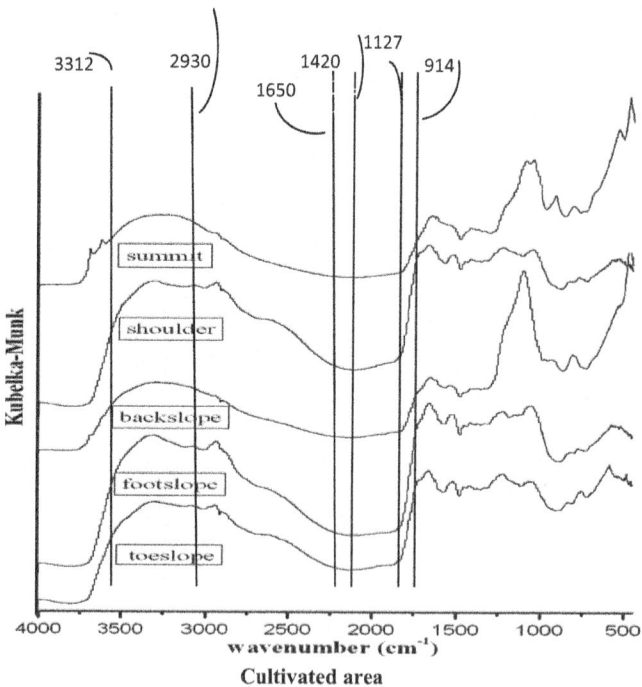

FIGURE 2.8 Diffuse Reflectance Fourier Transform Infrared spectra of humic acids in cultivated area at different landscape positions in Location #4 (Philippines).

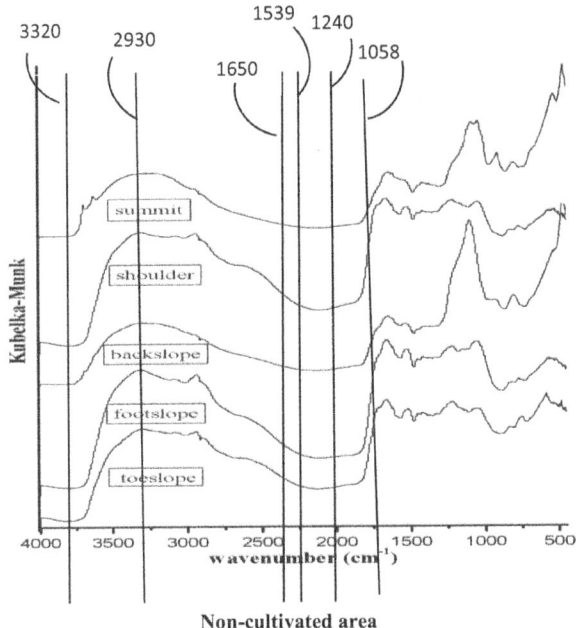

FIGURE 2.9 Diffuse Reflectance Fourier Transform Infrared spectra of humic acids in non-cultivated area at different landscape positions in Location #4 (Philippines).

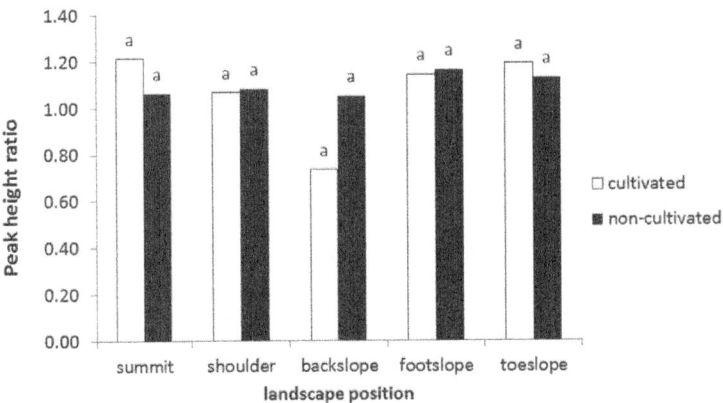

FIGURE 2.10 Comparing ratio of peak heights intensity (O/R) of humic acid with different landscape positions in cultivated, and non-cultivated areas in Location #4 (Philippines). Comparing within a landscape position, bars with the same letter are not significantly different based on the LSD0.05 value.

area was lower than samples from the non-cultivated area for each landscape position except toeslope. Summit, footslope, and toeslope tend to have higher peak height ratio than shoulder and backslope, and particularly backslope showed the lowest peak height ratio in both cultivated and non-cultivated area (Figure 2.10). It is likely due to soil erosion along the landscape. Smith et al. (2001) reported that patterns of soil organic C vary widely across the agricultural landscape where water, tillage, and wind erosion occurred. Upland areas have less SOC than soils in deposition areas. The SOC decreased as the gradient slope increased and soils on concave slopes had higher SOC than soils on convex slopes (Ritchie et al., 2007). Erosion was greatest on the area with the steepest slope which led to lower SOC compared to areas with less slope (Pennock and Frick, 2001).

2.5 CONCLUSIONS

The results suggest that use of the O/R ratio calculated from DRIFT spectra of HA is a sensitive technique to detect the differences caused by land use and landscape position under a wide range of environments and may be possibly related to changes in the soil quality. This method can be used to characterize the functional group of heterogeneous materials which may be a more direct method to determine the changes in SOM and soil quality caused by soil management practices, compared to other chemical and spectral techniques. The high resolution of the spectra, and a more reliable method for quantitative estimations of functional groups, could be used to analyze soil organic carbon composition. Therefore, in future work this technique is interesting as an accurate and simple method for helping to understand the composition of soil organic carbon, and identifying soil C as the indicator of soil quality. Some drawbacks to this technique for use in these environments include the high cost of purchasing and maintaining the MIR instrument, the need for extensive training, the lengthy sample preparation time, and possible interferences caused by minerals and water contained in the soil.

REFERENCES

Aguilera, J. 2010. Adaptive evolution of baker's yeast in a dough-like environment enhances freeze and salinity tolerance. Microbial *Biotechnology*, 3(2), pp. 210–221.

Alcón, A. 2010. Impacts of soil management practices on soil fertility in potato-based cropping systems in the Bolivian Andean highlands (Doctoral dissertation, University of Missouri--Columbia).

Baes, A.U. and Bloom, P.R. 1989. Diffuse reflectance and transmission Fourier transform infrared (DRIFT) spectroscopy of humic and fulvic acids. *Soil Science Society of America Journal*, 53(3), pp. 695–700.

Campbell, C.A., Zentner, R.P., Liang, B.C., Roloff, G., Gregorich, E.C. and Blomert, B. 2000. Organic C accumulation in soil over 30 years in semiarid southwestern Saskatchewan–effect of crop rotations and fertilizers. *Canadian Journal of Soil Science*, 80(1), pp. 179–192.

Carter, M.R. 2002. Soil quality for sustainable land management. *Agronomy Journal*, 94(1), pp. 38–47.

Chefetz, B., Adani, F., Genevini, P., Tambone, F., Hadar, Y. and Chen, Y. 1998. Humic-acid transformation during composting of municipal solid waste. *Journal of Environmental Quality*, 27(4), pp. 794–800.

Childers, J.W., Röhl, R., and Palmer, R.A. 1986. Direct comparison of the capabilities of photoacoustic and diffuse reflectance spectroscopies in the ultraviolet, visible, and near-infrared regions. Analytical Chemistry, 58, pp. 2629–2636.

Ding, G., Novak, J.M., Amarasiriwardena, D., Hunt, P.G. and Xing, B. 2002. Soil organic matter characteristics as affected by tillage management. *Soil Science Society of America Journal*, 66(2), pp. 421–429.

Ding, G., Mao, J.D., and Herbert, S., Amarasiriwardena, D. and Xing, B. 2007. Spectroscopic evaluation of human changes in response to soil management. In E.A. Ghabbour and G. Davis (ed.) *Humic Substances: Structure, Models and Functions*. The Royal Society of Chemistry, Cambridge, London.

Garcia, C., Hernandez, T., Costa, F. and Ayuso, M. 1992. Evaluation of the maturity of municipal waste compost using simple chemical parameters. *Communications in soil science and plant analysis*, 23(13–14), pp. 1501–1512.

Griffiths, P.R. 1983. Fourier transform infrared spectrometry. *Science*, 222, pp. 297–302.

Inbar, Y., Chen, Y. and Hadar, Y. 1989. Solid-state carbon-13 nuclear magnetic resonance and infrared spectroscopy of composted organic matter. *Soil Science Society of America Journal*, 53(6), pp. 1695–1701.

Kirchmann, H. and Gerzabek, M.H. 1999. Relationship between soil organic matter and micropores in a long-term experiment at Ultuna, Sweden. *Journal of Plant Nutrition and Soil Science*, 162(5), pp. 493–498.

Niemeyer, J., Chen, Y. and Bollag, J.M. 1992. Characterization of humic acids, composts, and peat by diffuse reflectance Fourier-transform infrared spectroscopy. *Soil Science Society of America Journal*, 56(1), pp. 135–140.

Painter, P.C., Starsinic, M. and Coleman, M.M. 2012. Determination of functional groups in coal by Fourier transform interferometry. *Fourier Transform Infrared Spectroscopy*, 4, pp. 169–240.

Pennock, D.J. and Frick, A.H. 2001. The role of field studies in landscape scale applications of process models: an example of soil redistribution and soil organic carbon modeling using CENTURY. *Soil and tillage Research*, 58, pp. 183–191.

Ritchie, J.C., McCarty, G.W., Venteris, E.R. and Kaspar, T.C. 2007. Soil and soil organic carbon redistribution on the landscape. *Geomorphology*, 89(1–2), pp. 163–171.

Sanchez, F.G., Carter, E.A. and Klepac, J.F. 2003. Enhancing the soil organic matter pool through biomass incorporation. *Biomass and Bioenergy*, 24(4–5), pp. 337–349.

Singh, A.R., Gerzabek, M.H., Haberhauer, G. and Eder, G. 2005. Long-term effects of cropped vs. fallow and fertilizer amendments on soil organic matter II. Nitrogen. *Journal of Plant Nutrition and Soil Science*, 168(2), pp. 212–218.

Six, J., Paustian, K., Elliott, E.T. and Combrink, C. 2000. Soil structure and organic matter I. Distribution of aggregate-size classes and aggregate-associated carbon. *Soil Science Society of America Journal*, 64(2), pp. 681–689.

Smith, S.V., Renwick, W.H., Buddemeier, R.W. and Crossland, C.J. 2001. Budgets of soil erosion and deposition for sediments and sedimentary organic carbon across the conterminous United States. *Global Biogeochemical Cycles*, 15(3), pp. 697–707.

Stevenson, F.J. 1994. *Humus Chemistry: Genesis, Composition, Reactions*. John Wiley & Sons, New York.

Stevenson, F.J., Elliott, E.T., Cole, C.V., Ingram, J., Oades, J.M., Preston, C. and Sollins, P.J. 1989. Methodologies for assessing the quantity and quality of soil organic matter. In D.C. Coleman, J.M. Oades, and G. Urehara (Eds.), *Dynamics of Soil Organic Matter in Tropical Ecosystems*, NiFTAL, Honolulu, pp. 1–249

Van der Marel, H.W. and Beutelspacher, H. 1976. *Atlas of Infrared Spectroscopy of Clay Minerals and their Admixtures*. Elsevier Publishing Company, Amsterdam.

Varanini, Z., and Pinton, R. 1995. Humic substance and plant nutrition. In H.D. Bahnke, U. Luttge, K. Esser, J.W. Kadereit, and M. Runge (Eds.), *Progress in Botany*, Springer-Verlag, Berlin.

Wander, M.M. and Traina, S.J., 1996. Organic matter fractions from organically and conventionally managed soils: I. Carbon and nitrogen distribution. *Soil Science Society of America Journal*, 60(4), pp. 1081–1087.

3 Climate Change and Production of Horticultural Crops

Jagan Singh Gora and Ajay Kumar Verma
ICAR-Central Institute for Arid Horticulture

Jitendra Singh
Agricultural University, Kota

Desh Raj Choudhary
Chaudhary Charan Singh Haryana
Agricultural University (CCS HAU)

CONTENTS

3.1 INTRODUCTION

Climate can be defined as an average weather condition over a long period (typically 30 years). Change in the climate that persists for decades or longer, arising from either natural causes or human activity is referred to as climate change (IPCC, 2007). It includes increases in temperature, changes in rainfall pattern, sea level rise, salt-water intrusion, generation of floods and droughts etc. (Bates et al., 2008; Shetty et al., 2013; Pathak et al., 2012). Climate change per se is not necessarily harmful. The problems that arise from extreme events are difficult to predict (FAO, 2001). More erratic rainfall patterns and unpredictable high temperature spells will consequently reduce the crop productivity. The resulting anthropogenic activities are responsible for an increase in gases, viz. carbon dioxide (CO_2), methane (CH_4), nitrous oxide (N_2O) and chlorofluorocarbons (CFCs) popularly known as the "greenhouse gases" (GHGs). Specially, CO_2 concentration in the atmosphere has increased drastically from 280 ppm to 370 ppm and is likely to be doubled in the 21st century (IPCC,

2007). The Indian climate has undergone significant changes showing increasing trends in annual temperature with an average of 0.56°C rise over the last 100 years (Rao et al., 2009; IMD, 2010).

Climate change poses serious impacts on agriculture, horticulture, environment, and health all over the world. Environmental stresses severely affect the soil organic matter decomposition, nutrient recycling, nutrient availability, limited soil moisture, low yield, and water availability to the plant. It is predicted that by 2080 the cereal production could be reduced by 2%–4%, meanwhile the price will increase by 13%–45%, and about 36%–50% of the population will be affected by hunger (FAO, 2009). Despite several negative impacts, there are a few beneficial aspects of enhanced GHGs e.g., higher atmospheric concentration of CO_2 may enhance the crop production of onion, leafy vegetables, rice, wheat, and soybean (Spaldon et al., 2015; Shetty et al., 2013). A significant change in climate on a global scale impacts agriculture and consequently affects the world's food supply (Afroza et al., 2010; Pathak et al., 2012). As per an analysis of ongoing temperature conducted by scientists at NASA's Goddard Institute for Space Studies (GISS), the average global temperature on Earth has increased by about 0.8°C (1.4°F) since 1880. The sea level rise has been estimated to be on an average between +2.6 mm and +2.9 mm/year ± 0.4 mm since 1993. Additionally, sea level rise has accelerated in recent years.

Some prominent consequences of climate change are as under:

1. Rise in global mean temperatures by 0.74°C during the last 100 years
2. Global warming.
3. Rise in sea level.
4. Increase in frequency and intensity of wild fires, floods, droughts, and tropical storms.
5. Changes in the amount, timing and distribution of rain, snow, run-off and disturbance of coastal marine and other ecosystems.
6. More acidic ocean-disrupting marine plankton.
7. Vulnerable production of vegetables.
8. Increase in abiotic stresses like extreme temperature (low/high), soil salinity, droughts and floods.
9. Detrimental influence on vegetative growth; flowering and fruiting are significantly influenced by the vagaries of climate.

3.2   CAUSES OF CLIMATE CHANGE

Natural processes: Natural changes in the components of the Earth's climate system and their interactions are the cause of internal climate variability or internal forcing. Generally the five types of the Earth's climate systems, namely atmosphere, hydrosphere, cryosphere, lithosphere, and biosphere (Weis and Berry, 1988; Shetty et al., 2013) form the very basis of climate change.

1. **Ocean variability:** The Ocean is a fundamental part of the climate system. Some changes occur in it at a longer time scale than in the atmosphere, massing hundreds of times more and having very high thermal inertia such

as the ocean depths still lagging today in temperature adjustment from the little Ice Age.

2. **Orbital variation:** Slight variations in the Earth's orbit lead to changes in the seasonal distribution of sunlight reaching the Earth's surface and how it is distributed across the globe. There is very little change to the area in respect to annually averaged sunshine but there can be strong changes in the geographical and seasonal distribution.

3. **Changes in the sun:** The sun is the predominant source of energy input to the Earth. Both long and short-term variations in solar intensity are known to influence the global climate. Three to four billion years ago the sun emitted only 70% as much power as it does today. If the atmospheric composition had been the same as today, liquid water should not have existed on Earth.

4. **Volcanic eruptions:** Volcanic eruptions release gases and particulate matter into the atmosphere. The eruptions are large enough to affect the climate several times per century and cause cooling for a period of a few years.

5. **Tectonic plate movement:** Over the course of millions of years, the motion of tectonic plates reconfigures global land, ocean areas, and generates topography. This can affect both global and local patterns of climate and ocean circulation. The positions of the continents determine the geometry of the oceans and therefore influence patterns of ocean circulation. The locations of the seas are important in controlling the transfer of heat and moisture across the globe and therefore in determining global climate.

6. **Human activities:** The scientific consensus on climate change is that the climate is changing and that these changes in large part are caused by human activities and they are largely irreversible. The biggest concern in these anthropogenic factors is the increase in CO_2 levels due to emissions from fossil fuel combustion, followed by aerosols (particulate matter in the atmosphere), and cement manufacturing. Other factors like land use, ozone depletion, agriculture, and deforestation etc. separately and in conjunction with other factors affect the climate, micro-climate and measures of climate variables. The human activities include power plants (40% of carbon emissions), automobiles (33% of carbon emissions), deforestation (responsible for 20%–25% of carbon emissions), buildings (12% of carbon emissions) and aeroplanes (3.5% of global warming) etc.

Due to rising pressure on land aggravated by various anthropogenic factors, there has been a continuous rise in the level of CO_2 since 2006 till the recorded period of 2014 as furnished below This is the prime cause of global warming and the associated ill effects on the entire Earth. The CO_2 (ppm) level for different years is summarized in Table 3.1.

TABLE 3.1
CO_2 Level in Different Years

Year	2014	2013	2012	2011	2010	2009	2008	2007	2006
CO_2 (ppm)	398.5	396.4	393.8	391.6	389.8	387.3	385.5	383.7	381.9

7. **Greenhouse effect:** Greenhouse effect is one of the main reasons for climate change. The Earth is the only planet in our solar system that supports life, because of its unique environmental conditions with the presence of water, an oxygen-rich atmosphere, and a suitable surface temperature. It has an atmosphere of proper depth and chemical composition. About 30% of the incoming energy from the sun is reflected back to space while the rest which reaches the earth has a role in warming the air, oceans, and land, maintaining an average surface temperature of about 15°C. The concentration of nitrogen and oxygen in the atmosphere is 78% and 21%, respectively, which all animals need to survive. Only a small portion, i.e., 0.036% is made up of carbon dioxide which is required by plants for photosynthesis. The solar energy is absorbed by the land, sea, mountains etc. and simultaneously released in the form of infrared waves. All this released heat is not lost to space but is partly absorbed by some gases present in very small quantities in the atmosphere called GHGs, consisting of carbon dioxide, methane, nitrous-oxide, water vapor, ozone and a few others and leads to greenhouse effect (Kricksen, 2008).

3.2.1 CHALLENGES DUE TO CLIMATE CHANGE IN HORTICULTURE PRODUCTION

The abiotic stress arises due to change in climate. It affects several morphological, anatomical, physiological, and biochemical parameters of the plants. Environmental stress is the primary cause of low production for most of the fruits and vegetables worldwide. Some of the important environmental stresses which affect crop production have been reviewed below:

Temperature: A constantly high temperature causes an array of morpho-anatomical changes in plants which affect the seed germination, plant growth, flower shedding, pollen viability, gametic fertilization, fruit setting, fruit size, fruit weight, and fruit quality etc. Exposure of horticulture crops to increased heat stress is the cause of physiological disorders and their associated problems. Poor pollination, especially under low humidity and high temperatures, will occur in many crops (e.g., sweet corn, lettuce, carrot, cucurbits, tomato, and avocado), together with a reduction in the number of pollinator insect species (Deuter, 2008). Pollen germination in tomato is affected at temperatures above 27°C. It results in the reduced fruit set, smaller size, and lower quality fruits (Stevens, 1978). Floral abortion will occur in capsicum when temperatures exceed 30°C (Erickson and Markhart, 2002). In peas, temperature above 25.6°C during bloom and pod set reduce flower, pod number, and yield. In beans, high temperature delays flowering because they enhance the short day photoperiod (Davis, 1997). In cucumber, sex expression is affected leading to production of more male flowers. In bulb crops, 1°C increase in temperature will decrease bulb yield by 3.5%–15%. Increasing carbon dioxide could increase the incidence of brown fleck disorder and due to high temperature it is likely to reduce tuber initiation in potatoes. Out of season high temperatures cause bolting of lettuce and celery, resulting in poor quality heads and reduced yields.

Color development in apples occurs through the production of anthocyanin. Anthocyanin production is reduced at high temperatures (Figure 3.1). Similarly, in capsicum red color development during ripening is inhibited above 27°C.

Sun scorching in sweet orange	Bark splitting on lime	Fruit drop in lime
Pink berry formation in grape	Fruit cracking in pomegranate	
Declining plants of mandarin	Cracking in Litchi fruits	

FIGURE 3.1 High temperature incited problems in fruit production.

In banana, temperature below 10°C leads to impedance of inflorescence and malformation of bunches. For avocados, increased heat stress will adversely affect the fruit size and the capacity to "store" a mature crop on the tree. Pome and stone fruits require a specific amount of winter chilling to develop fruitful buds and break dormancy satisfactorily in the spring (Table 3.2). Increasing minimum temperatures under climate change may induce insufficient chilling accumulation resulting in

TABLE 3.2

List of Physiological Disorders Caused by High Temperature

S. No.	Crops	Physiological Disorders
1.	Carrot	Low carotene content
2.	Asparagus, Bean	High fiber content in stalks/pods
3.	Cauliflower, Broccoli	Hollow stem, leafiness, no head
4.	Cole crops, Lettuce	Tip burns, bolting, loose puffy heads
5.	Tomato, Pepper	Fruit cracking, sun scald
6.	Tomato, Pepper, Watermelon	Blossom end rot (BER)
7.	Mango	Spongy tissue
8.	Citrus	Granulation
9.	Grape	Pink berry formation
10.	Litchi, pomegranate	Fruit cracking
11.	Apricot	Tip burning
12.	Pineapple	Sun scald
13	Rose	Blackening of petals
14.	Chrysanthemum	Bleached petals, bract buds

Source: Bahadur et al. (2011), Muthukumar and Selvakumar (2013).

uneven or delayed bud break. A number of fruit commodities (e.g., pome fruit, stone fruit, and avocadoes) require cross-pollinating cultivars for effective fruit set. Increasing temperatures may adversely affect the synchronization of flowering of these cultivars, resulting in inefficient pollination and reduced yield and quality (Deuter, 2008).

Drought: Water availability is highly sensitive to climate change and severe water stress conditions will affect crop productivity. In combination with elevated temperatures, decreased precipitation could cause a reduction in the availability of irrigation water and increase in evapotranspiration, leading to severe crop water stress conditions (IPCC, 2001). Drought-stress causes an increase in solute concentration in the environment (soil), leading to an osmotic flow of water out of plant cells. This leads to an increase in the solute concentration in plant cells, thereby lowering the water potential and disrupting membranes and cell processes such as photosynthesis. Water-stress condition affects the plants in terms of narrow leaf orientation, lesser germination, delayed maturity, small and delayed flowering, decline in chlorophyll content, reduced rate of transpiration, less uptake of nutrients, and severe reduction in yield (Bhardwaj, 2012). The soil moisture stress during the vegetative stage of banana causes the plant to extend its life cycle, and leads to poor bunch formation, lesser number of fingers, and small sized fingers. The water stress during flowering causes poor filling of fingers and unmarketable bunches.

Salinity: Salinity is a serious problem that reduces the growth and productivity of horticulture crops in many salt-affected areas. It is estimated that about 20% of cultivated lands and 33% of irrigated agricultural lands worldwide are affected by high salinity (Foolad, 2004). In addition, the salinity affected areas are increasing at a rate of 10% annually. Low precipitation, high surface evaporation, weathering of native rocks, irrigation with saline water, and poor cultural practices are the major hazards of increasing soil salinity. The high solute concentration in the soil imposes an initial water deficit. Altered K^+/Na^+ ratios, and build up due to Na^+ and Cl^- concentrations are detrimental to plants (Yamaguchi and Blumwald, 2005). Soil salinity can affect seed germination through osmotic effects; loss of turgor, growth reduction, wilting, leaf curling, epinasty, decreased photosynthesis, respiratory changes, loss of cellular integrity, tissue necrosis, and death of the plant (Jones, 1984; Cheeseman, 1988). Pea shows poor seed germination under saline condition (Kumar et al., 2012). Onion and Kagzi lime are sensitive to saline soils, while cucumbers, eggplants, peppers, beet root, beet leaf, and tomatoes are moderately sensitive.

Flooding: The excess amount of water than its optimum requirement is known as flooding/water logging. Crop production is often limited during the rainy season due to excessive moisture brought about by heavy rains. In general, the damage to crops by flooding is due to reduction of oxygen in the root zone, which inhibits aerobic processes (Table 3.3). Flooded plants accumulate endogenous ethylene that causes damage to the plants (Drew, 1981). Low oxygen levels stimulate an increased production of an ethylene precursor, 1-amino cyclopropane-1-carboxylic acid (ACC), in the roots (Kawase, 1981). It disturbs the physiological functioning, and the vegetative and reproductive growth of plants. Flooding symptoms such as rapid wilting and death of plants in tomato and increased incidence of pathogens viz. late blight, root pathogens increases with rising temperatures (Kuo et al., 1982). Some insects are killed or removed from crops by heavy rains e.g., onion thrips. Most fruits and vegetables are highly sensitive to water-logging or over-irrigation, particularly tomato, chilli, papaya, and early cauliflower.

Increase of CO_2 concentration: The change in CO_2 concentration in the atmosphere can alter plant tissues in terms of growth and physiological behavior (Pathak et al., 2012). Srinivasa and Bhatt (1992) studied the effect of tomato cv. Arka Ashish at higher CO_2 (550ppm) which influenced growth and development and increased

TABLE 3.3
Abiotic Stress Susceptible Horticultural Crops

S. No.	Abiotic Stress	Crops
1	High temperature	Peas, tomato, potato, beans, capsicum, banana, papaya, litchi, citrus, bael, rose.
2	Low temperature	Tomato, brinjal, onion, drumstick, Indian gooseberry, ber, senna, phalsa, gonad, rose, jasmine, tropical orchids, carnation.
3	Drought	Chilli, turnip, tomato, onion, pomegranate, custard apple, fig, grape, mango, banana, guava, black pepper, cardamom.
4	Salinity	Onion, radish, potato, beans, melons, peas, mango, fig, citrus, grape, guava, custard apple, apple, pear, strawberry.
5	Flooding/excess moisture	Chilli, onion, tomato, papaya, early cauliflower, banana.

Source: Singh (2010), Muthukumar and Selvakumar (2013).

yield by 24.4%. Higher CO_2 influenced overall growth and development of onion cv. Arka Kalyan, producing higher dry matter content in leaves, stems, and bulbs. At the bulb development stage, the photosynthesis rate was higher at elevated CO_2 levels as compared to ambient level. In tuber crops every 100 ppm increase in CO_2 increases tuber initiation, flowering, tuber weight, tuber number, and tuber yield by approximately 10% (Miglietta et al., 1998). The few negative effects of elevated CO_2 concentration include reduction in chlorophyll content in leaves particularly during late growing season after tuber initiation (Bindi et al., 2001). Root crop yield increased by 34% for an increase in CO_2 from 325 to 530 ppm (Kimball, 1983). Nederhoff (1994) found that yield increased by 34% for an increase in CO_2 from 364 to 620 μ/mol in cucumber. In coconut, arecanut, and cocoa, increased CO_2 led to higher biomass production and total dry matter content (Singh et al., 2010).

Outbreak of insect-pest: Changes in temperature and variability in rainfall, unseasonal rains, and heavy dew during the flowering and fruiting period would affect incidence of insect-pests, diseases, and virulence of major crops. Higher temperature generally results in increased insect-pest activity e.g., an extra generation of insect-pests such as halitosis may be possible in most locations. Higher temperatures may result in a longer period of pest activity; especially where production is extended e.g., Diamondback Moth (DBM) is a pest of worldwide significance wherever *Brassica* vegetables are grown. With a warming climate DBM will have an increased impact in all *Brassica* growing regions, particularly sub-tropical regions and increasingly so in temperate regions (Deuter, 2008). Pollinating insect activities were also reduced to the minimum, resulting in poor setting of fruits, vegetables, and nuts during that period. Due to the climate change, the number of crops (host) affected by a particular pest has increased (Singh et al., 2010). See Table 3.4 for such details.

TABLE 3.4
Crops Infested by a Particular Pest

S. No.	Insect-Pest	Major Crop	Crops Infested
1	Serpentine leaf miner (*Liriomyza trifolii*)	Tomato	Brinjal, Cow pea, French bean, Leafy vegetables, Cucurbits
2	Spiraling whiteflies (*A. macfarlanei*)	Guava, Citrus	Okra
3	Mealy bug (*C. insolita, P. solanpisis*)	Cotton, Jute	Brinjal, Tomato, Chilli, Okra
4	Hadda beetle (*H. vigitioctopuntata*)	Brinjal	Bitter gourd
5	Fruit borer (*H. armigera*)	Gram, Cotton, Tomato	Peas, Chilli, Brinjal, Okra
6	Cabbage butterfly (*Pieris brasicae*)	Cabbage, Cauliflower, and Mustard	Knol-khol, Radish
7	Red spider mite (*Tetranychus* sp.)	Okra	Brinjal, Cowpea, Indian bean

Source: Sharma (2012), Singh et al. (2010).

3.3 MITIGATION STRATEGIES FOR ENHANCING HORTICULTURAL PRODUCTION

Various management practices have the potential to increase the yield of fruits and vegetables grown under adverse conditions. The World Vegetable Center, Taiwan, has developed technologies to alleviate production challenges such as limited irrigation water and flooding to mitigate the effects of salinity, and also to ensure appropriate availability of nutrients to the plants. Strategies include modifying fertilizer application to enhance nutrient availability to plants, direct delivery of water to roots (drip irrigation), grafting to increase disease tolerance, and use of soil amendments to improve soil fertility and enhance nutrient uptake by plants (AVRDC, 2009; Schwarz et al., 2010).

Water harvesting: Collection of rainwater is termed as water harvesting. It can be collected both in-situ and ex-situ modes. In the former method, the rain drop is collected wherever it falls. In ex-situ conservation, water which otherwise forms run-off, is collected in a suitable structure. It may be a farm pond, reservoir or other suitable structure known locally by different names. A country like India where the annual rainfall is more than 500 mm, barring arid region, receives a calculated amount of more than 5,000,000 lakh liters water per ha. However, after cessation of rains, no moisture is left for successful cultivation. This calls for collection of the rainwater. A farmer by constructing a farm pond of 1 m^3 size can store 1.0 lakh liter of water. This much water is sufficient for life saving irrigation, especially in a moderately spaced new orchard for almost one month which continues to be very critical during summer month.

Water saving irrigation: The quality and efficiency of water management determine the yield and quality of the products. Too much or too little water causes abnormal plant growth, predisposes plants to infection by pathogens, and causes nutritional disorders. The timely irrigation and conservation of soil moisture reserves are the most important agronomic interventions to maintain yield during drought stress (Phene, 1989). The World Vegetable Center and other institutions promote affordable, small-scale drip irrigation technologies. The water use efficiency by chilli was significantly higher in drip irrigation as compared to furrow irrigation (AVRDC, 2005, 2006). For drought tolerant crops like watermelon, the yield difference was non-significant between furrow and drip irrigation, but there was a reduction in the incidence of *Fusarium* wilt in drip irrigation method. In general, the use of low cost drip irrigation is cost-effective, labor-saving, and allows more plants to be grown per unit of water, thereby saving water and increasing the farmers' incomes at the same time.

1. **Cultural practices to conserve water:** Several crop management practices such as mulching, use of shelters and raised beds help to conserve soil moisture, prevent soil degradation, and protect plants from heavy rains, high temperatures, and flooding. The use of mulches helps reduce evaporation, soil temperature, erosion, and minimizes weed growth. Mulching has been found to improve the growth of brinjal, okra, bottle gourd, round melon, ridge gourd, and sponge gourd as compared to

the non-mulched crop (Pandita and Singh, 1992). Planting of fruit and vegetable seedlings in raised beds can ameliorate the effects of flooding during the rainy season (AVRDC, 1979, 1981). Additional effects on yield were observed when the seedlings were planted in raised beds with rain shelters.

2. **Improved stress tolerance through grafting:** Grafting involves uniting of two living plant parts (rootstock and scion) to produce a single growing plant. It has been used primarily to control soil-borne diseases affecting the production of fruit vegetables such as tomato, brinjal, and cucurbits. However, it can provide tolerance to soil-related environmental stresses such as drought, salinity, low soil temperature, and flooding if appropriate tolerant rootstocks are used (Edelstein, 2004; Singh, 2005; AVRDC, 2009; Schwarz et al., 2010). Romero et al. (1997) reported that melons grafted onto hybrid squash rootstocks were more salt tolerant than the non-grafted melons. However, rootstocks from Cucurbita spp. are more tolerant of salt than rootstocks from *Lagenaria siceraria* (Matsubara, 1989). Okimura et al. (1986) found that *Solanum lycopersicum* x *S. habrochaites* rootstocks provide tolerance of low soil temperatures (100°C–130°C) for their grafted tomato scions, while eggplants grafted onto *S. integrifolium* x *S. melongena* rootstocks grew better at lower temperatures (180°C–210°C) than non-grafted plants (Table 3.5).

3. **Use of resilient source:** An improved and adapted germplasm is the most cost-effective option for farmers to meet the challenges of a changing climate. However, most modern cultivars represent a limited sampling of the available genetic variability, including tolerance to environmental stresses. New breeding varieties particularly for intensive, high input production and tolerance to different biotic and abiotic stresses are worth using (Pena and Hughes, 2007).

TABLE 3.5
Suitable Rootstocks for Different Purposes in Vegetables

S. No.	Vegetables	Rootstock	Uses
1	Tomato	Brinjal and wild spp. of Tomato	Tolerance to corky rot, fungal diseases, cold hardiness, and enhanced yield
2	Brinjal	Brinjal and its wild spp.	Tolerance to bacterial wilt, *Verticilium* wilt, low temperature, nematodes, vigor, and yield
3	Pepper	Pepper and its wild spp.	Tolerance to bacterial wilt, *Verticilium* wilt, nematodes, induced vigor, and yield
4	Cucumber	Squash and fig leaf gourd	Tolerance to *Fusarium* wilt, cold hardiness, and favorable sex ratio
5	Watermelon	Bottle gourd and Squash	Tolerance to *Fusarium* wilt, wilting, and drought

Source: Singh (2005).

i. **Varieties tolerant to high temperatures:** The key to obtaining high yield with heat tolerant cultivars is the broadening of their genetic base through crosses between heat tolerant tropical lines and disease resistant temperate or winter varieties (Opena and Lo, 1981). The heat tolerant tomato lines were developed using heat tolerant breeding lines and landraces from the Philippines (VC11-3-1-8, VC 11-2-5, Divisoria-2) and the United States (Tamu Chico III, PI289309) (Hazra et al., 2007; Opena et al., 1989). However, lower yields in the heat tolerant lines are still a concern.

ii. **Drought tolerance sources:** Some rootstocks like Dogridge (*Vitis champine*) of grape, *Zizyphus rotundifolia* of ber, MM-111 and MM-104 of apple, Mahaleb of cherry, *Rosa canina* and *Rosa indica* var. *odorata* of rose were found promising both for improvement in vigor, yield and quality as well as for tolerance to drought and salinity (Muthukumar and Selvakumar, 2013). Genetic variability for drought tolerance in *Solanum lycopersicum* is limited and inadequate. The stress tolerant tomato germplasm includes accessions of *S. cheesmanii*, *S. chilense*, *S. lycopersicum* var. *cerasiforme*, *S. pennellii*, *S. peruvianum* and *S. pimpinellifolium*. Drought tests show that *S. chilense* is five times more tolerant than cultivated tomato (Kumar et al., 2012).

iii. **Salinity tolerant lines:** Genetic variation for salt tolerance during seed germination in tomato has been identified within cultivated and wild species. A cross between a salt sensitive tomato line (UCT5) and a salt-tolerant *S. esculentum* accession (PI174263) showed that the ability of the tomato seed to germinate rapidly under salt stress is genetically controlled (Pena and Hughes, 2007). In pepper, salt stress significantly decreases germination, shoot height, root length, fresh and dry weight and yield. See details in Table 3.6.

Yildirim and Güvenç (2006) reported that pepper genotypes Demre, Ilica 250, 11-B-14, Bagci Carliston, Mini Aci Sivri, Yalova Carliston, and Yaglik-28 can be useful as sources of genes to develop pepper cultivars with improved germination under salt stress. In the fruit crops, rootstocks Bappakai, Kurrukan of mango; Rangpur lime, Cleopatra mandarin of citrus, and *Z. nummularia* of ber were found to be more tolerant to saline soil levels of up to 5.3 ds/m and saline water irrigations (Singh, 2010; Kumar et al., 2012).

4. **Climate proofing though Genomics and Biotechnology:** Increasing crop productivity in unfavorable environments will require advanced technologies to complement the traditional methods which are often unable to prevent yield losses due to environmental stresses. National and international institutes are retooling for plant molecular genetics research to enhance traditional plant breeding and benefit from the potential of genetic engineering to increase and sustain crop productivity (CGIAR, 2003).

i. **Quantitative trait loci (QTLs) and gene discovery for tolerance to stresses:** Genetic enhancement using molecular technologies has revolutionized plant breeding. The use of molecular markers as a selection tool provides the potential for increasing the efficiency of breeding programs

TABLE 3.6

List of Some Varieties and Advanced Line Tolerant to Abiotic Stresses

S. No.	Tolerant	Crop	Varieties/Rootstock	Advanced Line
1	Drought/ rainfed	Tomato	Arka Vikas, Arka Meghali	RF-4A
		Chilli	Arka Lohit	IIHR sel.-132
		Onion	Arka Kalyan	MST-42 & MST-46
		Aonla	Goma Aishwarya	Clonal sel. NA-7
		Annona	Arka Sahan	-
		Pomegranate	Ruby	-
		Fig	Deanna and Excel	
2	Salinity	Okra	Pusa Sawani	-
		Onion	Hisar-2	-
		Mango	Kurrukan, Bappakai (rootstock)	
		Citrus	Rangpur lime and Cleopatra mandarin (rootstock)	
		Grape	Dogridge (rootstock)	
3	Photo-insensitive	Lablab bean	Arka Jay, Arka Vijay	IIHR sel.-16-2
		Cowpea	Arka Garima, Arka Suman, Arka Mangala	-
		French bean	Arka Anoop, Arka Bold	IIHR-19-1
4	Frost	Onion	Arka Kalyan, PBO-1	-
		Tomato	Pusa Sheetal	-
5	High temperature	Capsicum	-	IIHR sel.-3
		Tomato	Pusa Hybrid-1	-
		Cauliflower	Pusa Meghna	IIHR-316-1 & IIHR-371-1
		Peas	-	IIHR-1 & IIHR-8
		Radish	Pusa Chetki	-
		Carrot	Pusa Vrishti, Pusa Kesar	-
		Cucumber	Pusa Barkha	-

Source: Rai and Yadav (2005), Yadav et al. (2012), Muthukumar and Selvakumar (2013).

by reducing environmental variability, facilitating earlier selection, and reducing subsequent population sizes for field testing. Martin et al. (1989) identified drought tolerance in tomato, and three QTLs were linked to water use efficiency in *Solanum pennellii* based on ^{13}C composition. An identified four QTLs were associated with tomato seed to germinate rapidly under the drought and salt tolerance which was contributed by *S. pimpinellifolium* (Foolad et al., 2003). Lin et al. (2006) identified random amplified polymorphic DNA (RAPD) markers linked to heat tolerance in The World Vegetable Center (AVRDC), tomato line CL5915. "Osmotin" gene in potato, tolerant to water stress condition, has also been developed.

ii. **Engineering stress tolerance:** Environmental stress tolerance is a complex trait and involves many genes (Wang et al., 2003). In response to stresses, both RNA and protein expression profiles change. Approximately 130 drought-responsive genes have been identified using microarrays (Reymond et al., 2000; Seki et al., 2001). These genes are involved with transcription modulation, ion-transport, and carbohydrate metabolism and transpiration control. *DREB1A*, *CBF* and *HSF* genes are transcription factors implicated in drought and heat response, respectively (Sung et al., 2003; Sakuma et al., 2002). The *CBF/DREB1* genes have been used successfully to engineer drought tolerance and increased stress tolerance without plant growth retardation in tomato and other crops (Hsieh et al., 2002; Kumar et al., 2012).

5. **Carbon trading:** This is a type of compensatory allowance in terms of money value for the generation of carbon sink. In India carbon trading remains neglected. It is well known and widely accepted that plants are a very potent source of utilization of carbon. This is the reason that plantation drives form massive campaigns across the globe. In continents where the created biomass base is considerably large, there is an opportunity of carbon trading. It has been observed that a mandarin (Citrus reticulata Blanco) plant at the age of 6 years had the potential to store 1.6 tc/ha (Mehta et al., 2016). Thus plantation drives, while lessening the CO_2 content in the environment, can be an opportunity for earning.

3.4 CONCLUSION

The demand for horticultural produce is increasing due to rising consumerism, and needs to be met in a sustainable manner. The succulent horticultural crops are highly sensitive to heat, radiation, drought, salinity, and flooding. Direct and indirect observations indicate increased levels of carbon dioxide and increase in temperature from 1–5°C by the end of this century with changing rainfall patterns and greater frequency of extreme events of droughts and floods (IPCC, 2007). Elevated CO_2 has a positive effect ranging from 24% to 51% on the productivity of crops like mango, citrus, grapes, guava, fig, annona, tomato, capsicum, onion, cucumber, and melons. However, increase in temperature affects the crop duration, flowering, fruiting, fruit size, quality, and fruit ripening with reduced productivity and economic yield. Therefore, the overall impact of climate change depends on the interaction effect of elevated CO_2 and rising temperature. Accurate impact analyses of global warming on horticultural crops are required to evolve adaptive measures and future strategies to cope up with climate change and global warming (Shetty et al., 2013; Pathak et al., 2012). Selection of adaptive crops and adoption of mitigation measures are in the way of sustainability in the production of horticultural crops.

REFERENCES

Afroza, B., Wani, K.P., Khan, S.H., Jabeen, N., Hussain, K., Mufti, S. and Amin, A. 2010. Various technological interventions to meet vegetable production challenges in view of climatic change. *Asian Journal of Horticulture*, 5(2), pp. 523–29.

AVRDC. 1979. *Annual Report*. Asian Vegetable Research and Development Center. Shanhua, Taiwan. p. 173.

AVRDC. 1981. *Annual Report.* Asian Vegetable Research and Development Center. Shanhua, Taiwan. p. 84.

AVRDC. 2005. *Annual Report.* AVRDC-The World Vegetable Center. Shanhua, Taiwan.

AVRDC. 2006. *Vegetables Matter.* AVRDC-The World Vegetable Center, Shanhua, Taiwan.

AVRDC. 2009. *Guide: Grafting Sweet Peppers for Production in the Hot-Wet Season.* Asian Vegetable Research and Development, Shanhua, Tainan, Taiwan, p. 8.

Bahadur, A., Chatterjee, A., Kumar, R., Singh, M. and Naik, P.S. 2011. Physiological and biochemical basis of drought tolerance in vegetables. *Journal of Vegetation Science,* 38(1), pp. 1–16.

Bates, B., Kundzewicz, Z.W., Wu, S. and Palutikof, J. 2008. *Climate Change and Water: IPCC Technical Paper VI.* United Nations Environment Programme. World Meteorological Organization, Geneva, pp. 210.

Bhardwaj, M.L. 2012. Challenges and opportunities of vegetable cultivation under changing climate scenario. A training manual on vegetable production under changing climate scenario, pp. 13–18.

Bindi, M., Fibbi, L., Lanini, M. and Miglietta, F., 2001. Free air CO_2 enrichment (FACE) of grapevine (Vitis vinifera L.): I. Development and testing of the system for CO_2 enrichment. *European Journal of Agronomy,* 14(2), pp. 135–143.

CGIAR. 2003. Applications of molecular biology and genomics to genetic enhancement of crop tolerance to abiotic stresses-a discussion document. Interim Science Council Secretariat, FAO.

Cheeseman, J.M. 1988. Mechanisms of salinity tolerance in plants. *Plant Physiology.* 87, pp. 57–550.

Davis, J.H.S. 1997. Phaseolous beans. In Wien, H,C. (Ed.), *The Physiology of Vegetable Crops.* CAB International, Wallingford, UK, pp. 409–428.

Deuter, P. 2008. Defining the impact of climate change on horticulture in Australia. Garnaut Climate Change Review, Department of Primary Industries and Fisheries, Queensland. pp. 1–23.

Drew, M.C. 1981. Plant responses to anaerobic conditions in soil and solution culture. *Commentaries in Plant Science,* 2, pp. 209–223.

Edelstein, M. 2004, March. Grafting vegetable-crop plants: pros and cons. *VII International Symposium on Protected Cultivation in Mild Winter Climates: Production, Pest Management and Global Competition,* 659, pp. 235–238.

Erickson, A.N. and Markhart, A.H. 2002. Flower developmental stage and organ sensitivity of bell pepper (Capsicum annuum L.) to elevated temperature. *Plant, Cell & Environment,* 25(1), pp. 123–130.

FAO. 2001. Climate variability and change: A challenge for sustainable agricultural production. Committee on Agriculture, Sixteenth Session Report, 26–30 March, 2001. Rome, Italy.

FAO. 2009. Global agriculture towards 2050 Issues Brief. High level expert forum. Rome, pp. 12–13.

Foolad, M.R. 2004. Recent advances in genetics of salt tolerance in tomato. *Plant Cell, Tissue and Organ Culture,* 76(2), pp. 101–119.

Foolad, M.R., Zhang, L.P. and Subbiah, P. 2003. Genetics of drought tolerance during seed germination in tomato: inheritance and QTL mapping. *Genome,* 46, pp. 536–545.

Hazra, P., Samsul, H.A., Sikder, D. and Peter, K.V. 2007. Breeding tomato (Lycopersicon esculentum Mill) resistant to high temperature stress. *International Journal of Plant Breeding and Genetics,* 1(1), pp. 31–40.

Hsieh, T.H., Lee, J.T., Charng, Y.Y. and Chan, M.T. 2002. Tomato plants ectopically expressing Arabidopsis CBF1 show enhanced resistance to water deficit stress. *Plant Physiology,* 130(2), pp. 618–626.

IMD, Annual Climate Summary. 2010. India Meteorological Department, Pune (MH). Government of India, Ministry of Earth Sciences, p. 27.

IPCC (Inter-governmental Panel on Climate Change). 2001. *Climate Change: Impacts, Adaptation and Vulnerability.* Inter-governmental Panel on Climate Change. New York, USA.

IPCC. 2007. The physical science basis. In Solomon, S., Qin, D., Manning, M., Chen, Z., Marquis, M., Averyt, K.B., Tignor, M., Miller, H.L. (Eds.), *Contribution of Working Group I to the Fourth Assessment Report of the IPCC.* Cambridge University Press, Cambridge, UK, p. 996.

Jones, R.A. 1984, December. The development of salt-tolerant tomatoes: breeding strategies. *Symposium on Tomato Production on Arid Land,* 190, pp. 101–114.

Kawase, M. 1981. Anatomical and morphological adaptation of plants to water-logging. *Horticultural Science,* 16, pp. 30–34.

Kimball, B.A. 1983. Carbon dioxide and agricultural yield: An assemblage and analysis of 430 prior observations. Agronomy Journal, 75, pp. 779–788.

Kricksen, P.J. 2008. Conceptualizing food systems for global environmental change research. Global Environmental Change, 18(1), pp. 234–245.

Kumar, R., Solankey, S.S. and Singh, M. 2012. Breeding for drought tolerance in vegetables. *Journal of Vegetation Science,* 39(1), pp. 1–15.

Kuo, D.G., Tsay, J.S., Chen, B.W. and Lin, P.Y. 1982. Screening for flooding tolerance in the genus Lycopersicon. *Horticultural Science,* 17(1), pp. 76–78.

Lin, K.H., Lo, H.F., Lee, S.P., George Kuo, C., Chen, J.T. and Yeh, W.L. 2006. RAPD markers for the identification of yield traits in tomatoes under heat stress via bulked segregant analysis. *Hereditas,* 143(2006), pp. 142–154.

Martin, B., Nienhuis, J., King, G. and Schaefer, A. 1989. Restriction fragment length polymorphisms associated with water use efficiency in tomato. *Science,* 243(4899), pp. 1725–1728.

Matsubara, S. 1989. Studies on salt tolerance of vegetables, 3: Salt tolerance of rootstocks. *Scientific Reports of the Faculty of Agriculture Okayama University.* 73, pp. 17–25.

Mehta, L.C., Singh, J., Chauhan, P.S., Singh, B. and Manhas, R.K. 2016. Biomass Accumulation and Carbon Storage in Six-Year-Old Citrus reticulata Blanco. Plantation. *Indian Forester,* 142(6), pp. 563–568.

Miglietta, F., Magliulo, V., Bindi, M., Cerio, L., Vaccari, F.P., Loduca, V. and Peressotti, A. 1998. Free air CO_2 enrichment of potato (Solanum tuberosum L.): development, growth and yield. *Global Change Biology,* 4(2), pp.163–172.

Muthukumar, P. and Selvakumar, R. 2013. *Glaustas Horticulture.* New Vishal Publications, New Delhi, India. p. 544.

Nederhoff, E.M. 1994. Effects of CO_2 concentration on photosynthesis, transpiration and production of greenhouse fruit vegetable crops. Nederhoff.

Okimura, M., Matsuo, S., Arai, K. and Okitsu, S. 1986. Influence of soil temperature on the growth of fruit vegetable grafted on different stocks. Bulletin of the Vegetable and Ornamental Crops Research Station. Series C. Kurume (Japan). C9, pp. 3–58.

Opena, R.T. and Lo, S.H. 1981. Breeding for heat tolerance in heading Chinese cabbage. In: Proceedings of the 1st International Symposium on Chinese Cabbage. Talekar, N.S., Griggs, T.D., AVRDC, Shanhua, Taiwan.

Opena, R.T., Green, S.K., Talekar, N.S. and Chen, J.T. 1989. Genetic improvement of tomato adaptability to the tropics: Progress and future prospects. In: Green, S.K. (Ed.), Tomato and Pepper Production in the Tropics. AVRDC, Shanhua, Taiwan, pp. 70–85.

Pandita, M.L. and Singh, N. 1992. Vegetable production under water stress conditions in rainfed areas. In: Kuo, C.G. (Ed.), *Adaptation of Food Crops to Temperature and Water Stress.* AVRDC, Shanhua, Taiwan, pp. 467–472.

Pathak, H., Aggarwal, P.K. and Singh, S.D. 2012. *Climate Change Impact, Adaptation and Mitigation in Agriculture: Methodology for Assessment and Applications.* Indian Agricultural Research Institute, New Delhi, India.

Pena, R.D. and Hughes, J. 2007. Improving vegetable productivity in a variable and changing climate. *An Open Access Journal published by ICRISAT*, 4(1), pp. 1–22.

Phene, C.J. 1989. Water management of tomatoes in the Tropics Tomato and pepper production in the Tropics. In International Symposium on Integrated Management Practices 21–26 Mar 1988 Tainan (Taiwán). Asian Vegetable Research and Development Center, Shanhua (Taiwán).

Rai, N. and Yadav, D.S. 2005. *Advances in Vegetable Production*. Researcher Book Center, New Delhi, India.

Rao, G.G.S.N., Rao, A.V.M.S. and Rao, V.U.M. 2009. *Trends in Rainfall and Temperature in Rainfed India in Previous Century. Global Climate Change and Indian Agriculture Case Studies from ICAR Network Project*. ICAR Publication, New Delhi, India, pp. 71–73.

Reymond, P., Weber, H., Damond, M. and Farmer, E.E. 2000. Differential gene expression in response to mechanical wounding and insect feeding in Arabidopsis. Plant Cell, 12, pp. 707–720.

Romero, L., Belakbir, A., Ragala, L. and Ruiz, J.M. 1997. Response of plant yield and leaf pigments to saline conditions: effectiveness of different rootstocks in melon plants (Cucumis melo L.). *Soil Science and Plant Nutrition*, 43(4), pp. 855–862.

Sakuma, Y., Liu, Q., Dubouzet, J.G., Abe, H., Shinozaki, K. and Yamaguchi-Shinozaki, K. 2002. DNA-binding specificity of the ERF/AP2 domain of Arabidopsis DREBs, transcription factors involved in dehydration-and cold-inducible gene expression. *Biochemical and biophysical research communications*, 290(3), pp. 998–1009.

Schwarz, D., Rouphael, Y., Colla, G. and Venema, J.H. 2010. Grafting as a tool to improve tolerance of vegetables to abiotic stresses: thermal stress, water stress and organic pollutants. *Scientia Horticulturae*, 127(2), pp. 162–171.

Seki, M., Narusaka, M., Abe, H., Kasuga, M., Yamaguchi-Shinozaki, K., Carninci, P., Hayashizaki, Y. and Shinozaki, K. 2001. Monitoring the expression pattern of 1300 Arabidopsis genes under drought and cold stresses by using a full-length cDNA microarray. *The Plant Cell*, 13(1), pp. 61–72.

Sharma, I.M. 2012. Changing disease scenario in apple orchards: Perspective, challenges and management strategies. In Proceedings of the National Symposium on Blending Conventional and Modern Plant Pathology for Sustainable Agriculture, Indian Institute of Horticultural Research, Bengaluru, p. 123.

Shetty, P.K., Ayyappan, S. and Swaminathan, M.S. 2013. *Climate Change and Sustainable Food Security*. National Institute of Advanced Studies, Bangalore and Indian Council of Agricultural Research, New Delhi, India. ISBN: 978-81-87663-76-8.

Singh, B. 2005. *Protected Cultivation of Vegetable Crops*. III edition, Kalyani publishers, New Delhi, India, p. 161.

Singh, H.P. 2010. Ongoing research in abiotic stress due to climate change in horticulture. Curtain Raiser Meet on Research Needs Arising due to Global Climate Scenario, Baramati, Maharashtra, October, 29–30. pp. 1–23

Singh, H.P., Singh, J.P. and Lal, S.S. 2010. *Challenges on Climate Change-Indian Horticulture*, Westville Publishing House, New Delhi, India, p. 224.

Spaldon, S., Samnotra, R.K. and Chopra, S. 2015. Climate resilient technologies to meet the challenges in vegetable production. International Journal of Current Research and Academic Review, 3(2), pp. 28–47.

Srinivasa Rao, N.K. and Bhatt, R.M. 1992. Responses of tomato to moisture stress: Plant water balance and yield. *Plant Physiology and Biochemistry-New Delhi*, 19, pp. 36–36.

Stevens, M.A. 1978. Genetic potential for overcoming physiological limitations on adaptability, yield, and quality in the tomato. *Horticultural Science*, 13, pp. 673–678.

Sung, D.Y., Kaplan, F., Lee, K.J. and Guy, C.L. 2003. Acquired tolerance to temperature extremes. *Trends in Plant Science*, 8(4), pp. 179–187.

Wang, W., Vinocur, B. and Altman, A. 2003. Plant responses to drought, salinity and extreme temperatures: towards genetic engineering for stress tolerance. *Planta*, 218(1), pp. 1–14.

Weis, E. and Berry, J.A. 1988. Plants and high temperature stress. *Symposia of the Society for Experimental Biology*, 42, pp. 329–346.

Yadav, R.K., Kalia, P., Singh, S.D. and Varshney, R. 2012. Selection of genotypes of vegetables for climate change adaptation. Climate change impact, adaptation and mitigation in agriculture: Methodology for assessment and application. *Division of Environmental Sciences Indian Agricultural Research Institute, New Delhi*, 110, p. 012.

Yamaguchi, T. and Blumwald, E. 2005. Developing salt-tolerant crop plants: challenges and opportunities. *Trends in Plant Science*, 10(12), pp. 615–620.

Yildirim, E. and Güvenç, İ. 2006. Salt tolerance of pepper cultivars during germination and seedling growth. *Turkish Journal of Agriculture and Forestry*, 30(5), pp. 347–353.

4 Climate Change on Disease Scenario in Crops and Management Strategies

Harender Raj Gautam and I. M. Sharma
Dr. Y.S. Parmar University of Horticulture
and Forestry, Solan

CONTENTS

4.1 INTRODUCTION

Climate change is the biggest threat of the present century and it is already contributing to the deaths of nearly 400,000 people a year and costing the world more than US $1.2 trillion, thus wiping out 1.6% annually from the global GDP. The latest report of the Inter-Governmental Panel on Climate Change (IPCC) has reported an increase of 0.85°C (0.65°C–1.06°C) in the temperature between 1880 and 2012 (IPCC, 2014). Furthermore, the CO_2 concentration has increased from 280 ppm (pre-industrial value) to 401 ppm in 2015. The possible changes in temperature, precipitation, and CO_2 concentration are expected to have significant impact on crop growth. According to the 'Climate change and food security: risks and responses' report of 2016 of the Food and Agriculture Organization (FAO) of the United Nations, climate change is going to seriously impact the food security across different regions of the world. The International Fund for Agricultural Development (IFAD) states that at least 70% of the very poor live in rural areas, most of them depending partly or completely on agriculture for their livelihoods. The projected change in the global average temperature will likely be from 0.3°C to 0.7°C for the period 2016–2035,

63

relative to the reference period 1986–2005. There will be more frequent hot temperature extreme episodes over most land areas. Average precipitation will very likely
increase in the high- and parts of the mid-latitudes, and the frequency and intensity
of heavy precipitation will also likely increase on an average. An increase in temperature will trigger an increased demand for water for evapotranspiration by crops
and natural vegetation and will lead to more rapid depletion of soil moisture. There is
evidence that climate change has already negatively affected wheat and maize yields
in many regions and also at global level. IPCC in its 5th report (IPCC AR5) has
identified that four out of the eight key risks have close relations with or direct consequences to food security. Thus climate change can result in loss of rural livelihoods
and income, loss of marine and coastal ecosystems and livelihoods, loss of terrestrial
and inland water ecosystems and livelihoods, and food insecurity and breakdown of
food systems. Climate change will adversely affect agriculture globally (FAO, 2016;
Li et al., 2016). This will have a serious impact on food security all over the world.
All the studies indicate an adverse effect on our foodgrain production. Climate
change could result in a variety of impacts on agriculture. Changes in production
patterns will occur due to higher temperatures and changing precipitation patterns.
Agricultural productivity will also be affected due to increased carbon dioxide in
the atmosphere. Leading international agencies like IPCC (2007) and Universal
Ecological Fund (2011) have indicated the affect of climate change on agriculture
globally. According to these reports, there will be a 14% deficit in the global wheat
production, 11% deficit in rice, and 9% deficit in maize by 2020. The research findings coming from different parts of the world indicate that climate change will affect
many crops. According to the findings of the Australian scientists, climate change is
causing the early ripening of grapes. These findings are based on the harvesting data
of the last 64 years. Scientists attribute the fruit's ripening to climate warming and a
decline in soil water content, based on a comparison of decades of vineyard records.
Like other regions in the world, climate change will have an impact on Indian agriculture (Francesca et al., 2006; Garrett et al., 2006; Gautam, 2009a, b; Gautam and
Kuniyal, 2014). A warming of 0.85°C has occurred during 1882–2012. Variation in
climate will have a direct impact on the majority of the livelihood of the people. In
India, cereal productivity is projected to decrease by 10%–40% by 2100. The losses
in cereal production will be more pronounced in Rabi crops.

Climate change is affecting the crop production system in different ways, by
directly affecting the growth of the crops and/or by affecting all those factors which
affect the crops. Now, there are a number of studies which indicate that climate
change will affect pathogen development and survival rates, modify host susceptibility, and result in changes in the effects of diseases on crops (Harvell et al.,
2002). Plant diseases are one of the important factors which have a direct impact
on global agricultural productivity and climate change will further aggravate the
situation (IPCC, 2007). Combined infestation of pests and diseases in plants could
result in up to 82% losses in the attainable yield in case of cotton, and over 50%
losses for other major crops and if we combine these losses with post-harvest spoil-
age and deterioration in quality, then these losses become critical particularly for
resource-poor regions of the world (Oerke, 2006). Further, plant diseases are estimated to cause a yield reduction of almost 20% in the principal food and cash crops

worldwide (Thind, 2012). In the last 40 years, effective management of pests and diseases has played a key role in doubling the food production, but pathogens still claim 10%–16% of the global harvest (Chakraborty and Newton, 2011). It covered eight crops that together occupy half the world's cropland. In Asia, 14.2% of the potential production costing about US\$ 43.8 billion is lost due to diseases (Oerke et al., 2012). Climate models predict a gradual rise in CO_2 concentration and temperature all over the world, but are not precise in predicting future changes in local weather conditions. Local weather conditions such as rain, temperature, sunshine, and wind in combination with locally adapted plant varieties, cropping systems, and soil conditions can maximize food production as long as the plant diseases can be controlled. Currently, we are able to secure food supplies under these varying conditions. However, all climate models predict that there will be more extreme weather conditions, with more droughts, heavy rainfall, and storms in agricultural production regions. Such extreme weather events will influence where and when diseases will occur, and therefore impose severe risks on crop failure.

4.2 EFFECT OF INCREASED CO_2 CONCENTRATIONS ON PATHOGENS

The concentration of CO_2 in the atmosphere reached 379 ppm in 2005, which exceeds the natural range of values of the past 650,000 years (IPCC, 2007). An increase in CO_2 levels may encourage the production of plant biomass; however, productivity is regulated by water and nutrients' availability, competition against weeds, and damage by pests and diseases. Consequently, a high concentration of carbohydrates in the host tissue promotes the development of biotrophic fungi such as rust (Chakraborty et al., 2002). Thus, an increase in biomass can modify the microclimate and affect the risk of infection. In general, increased plant density will tend to increase the leaf surface wetness duration and regulate temperature, and so make infection by foliar pathogens more likely. Experimental research on the effects of high atmospheric CO_2 concentrations on plant–pathogen interactions has received little attention, and conflicting results have been published. Elevated levels of CO_2 can directly affect the growth of pathogens. Chakraborty et al. (2002) reported that the growth of the germ tube, appressorium, and conidium of *C. gloeosporioides* fungi is slower at high concentrations of CO_2 (700 ppm). Germination rates of conidia on leaves were lower at CO_2 concentrations of 700 ppm than those observed at 350 ppm. However, once the pathogen infects the plant, the fungus quickly develops and achieves sporulation. In contrast, the rate of germination sporulation was greater at high concentrations of CO_2 (700 ppm). Another study by Hibberd et al. (1996) evaluated powdery mildew in barley, and found that an acclimation of photosynthesis at elevated CO_2 and an infection-induced reduction in net photosynthesis caused larger reductions in plant growth at elevated CO_2; also, the percentage of conidia that progressed to produce colonies was lower in plants grown in high CO_2 (700 ppm) concentrations than in low CO_2 (350 ppm). Also, the percentage of conidia producing hyphae was lower in 700 ppm CO_2 due to a higher proportion of the spores being arrested at the appressorium stage. Tiedemann and Firsching (2000) analyzed the direct effects of elevated ozone and carbon dioxide on spring wheat infected with *Puccinia recondita*

f. sp. *tritici* and reported that ozone damage to leaves is largely dependent on both carbon dioxide concentrations as well as disease. Models can then be used to extrapolate, predict, and validate potential impacts. Some authors suggest that elevated CO_2 concentrations and climate change may accelerate plant–pathogen evolution, which can affect virulence. Research studies in this direction have been carried out. In this regard, Mulherin et al. (2000), evaluated the response of tobacco grown under elevated CO_2 to inoculation with tobacco mosaic virus (TMV) in two concentration of CO_2 (360 and 720 ppm) and found that plants grown at 720 ppm CO_2 produced fewer TMV lesions per leaf versus plants grown at 360 ppm CO_2. Eastburn et al. (2010) evaluated the effects of elevated CO_2 and O_3 on three soybean diseases, namely downy mildew (*Peronospora manshurica*), Septoria (*Septoria glycines*), and sudden death syndrome (*Fusarium virguliforme*) and reported that changes in the composition of the atmosphere altered the expression of the disease, and plant responses to the diseases varied considerably. The severity of downy mildew damage was significantly reduced at high levels of CO_2. In contrast, high levels of CO_2, alone or in combination with high concentrations of O_3, increased the severity of *Septoria glycines*. The concentrations of CO_2 and O_3 did not have an effect on sudden death syndrome. The authors concluded that high levels of CO_2 and O_3 induced changes in the soybean canopy density and leaf age, and likely contributed to disease expression modification. Kobayashi et al. (2006) evaluated the effects of elevated CO_2 concentrations on the interactions between rice, *Pyricularia oryzae,* and *Rhizoctonia solani* and found that rice plants were more susceptible to injury. Increased levels of CO_2 under changing climate have been reported to decrease the plant defense and increase susceptibility in coffee against *Colletotrichum gloeosporioides* that causes brown blight disease in tea because elevated CO_2 reduced the endogenous caffeine content in tea leaves. Furthermore, exogenous caffeine could induce jasmonic acid (JA) content under both CO_2 conditions in the absence of fungal infection; however, in the presence of fungal infection, caffeine increased JA content only under elevated CO_2.

Thus, the authors concluded that rice cultivated at sites with high concentrations of CO_2 may have an increased risk of infection by the above-mentioned pathogens. Overall, the effects of elevated CO_2 concentration on plant disease can be positive or negative, although in a majority of the examples reviewed by Chakraborty et al. (2000), disease severity increased. Similarly, the effects of ozone and other abiotic stresses can be in either direction and the effects of many stresses are interactive, so it is not possible to generalize.

4.3   EFFECT OF INCREASE IN TEMPERATURE ON PATHOGENS

Changes in temperature and precipitation regimes due to climate change may alter the growth stage, development rate and pathogenicity of infectious agents, and the physiology and resistance of the host plant (Chakraborty et al., 1998; Chakraborty and Dutta, 2003). A change in temperature could directly affect the spread of infectious diseases and their survival between seasons. There are indications of increased aggressiveness at higher temperatures of stripe rust isolates (*Puccinia striiformis*), suggesting that rust fungi can adapt to and benefit from higher temperatures (Mboup et al., 2012). Climate change is also reported to cause a shift in the geographical distribution of host–pathogens (Mina and Sinha, 2008). Temperature is one of the most important factors affecting the occurrence of bacterial diseases such as *Ralstonia solanacearum, Acidovorax avenae,* and *Burkholderia glumea.* Thus, bacteria could proliferate in the areas where temperature-dependent diseases have not been previously observed (Kudela, 2009). As the temperature increases, the duration of winter and the rate of growth and reproduction of pathogens may be modified (Ladányi and Horváth, 2010). Similarly, the incidence of vector-borne diseases will be altered.

Based on the published observations of the distribution of 612 crop pests collected over the past 50 years, scientists have reported that crop pests have been spreading north and south a little less than 2 miles (3.2 km) a year since 1960, though there's a lot of variety within individual species and this corresponds to increased temperatures during that period (Bebber et al., 2013). In their observation, the scientists comprehend that if pests continue to march polewards as the Earth warms, the combined effects of a growing world population and the increased loss of crops to pests will pose a serious threat to global food security. A change in temperature may favor the development of different dormant pathogens, which could induce an epidemic. Increase in temperature with sufficient soil moisture may increase evapotranspiration resulting in humid microclimate in crops and may lead to incidence of diseases favored under these conditions (Mcelrone et al., 2005). Diseases such as common bunt (*Tilletia caries*) and Karnal bunt (*Tilletia indica*) in wheat can be of importance under changing climatic conditions in regions with low productivity if proper seed treatment is not followed in this crop (Oerke, 2006). In India, in the last decade the disease scenario of chickpea and pigeon pea has changed drastically; dry root rot (*Rhizoctonia bataticola*) of chickpea and Phytophthora blight (*Phytophthora drechsleri* f. sp. *cajani*) of pigeon pea have emerged as a potential threat to the production of these pulses (Pande and Sharma, 2010). Higher risk of dry root rot has been reported in Fusarium wilt chickpea-resistant varieties in those years when the temperature exceeded 33°C (Dixon, 2012). In North America, needle blight (*Dothistroma septosporum*) is reported to be spreading northwards with increasing temperature and precipitation (Madden et al., 2007). In general, an increase in temperature would significantly raise the severity and spread of plant diseases but the quantity of precipitation could act as a regulator in deciding the increase or decrease in disease severity and spread (Woods et al., 2005). Temperature is one of the most important factors affecting the occurrence of bacterial diseases such as *Ralstonia solanacearum, Acidovorax avenae,* and *Burkholderia glumea.* Thus, bacteria could proliferate in areas where temperature-dependent diseases have not been previously observed (Kudela, 2009). Similarly, the incidence of most of the virus and other vector-borne diseases will be altered. This is because climate can substantially influence the development and distribution of vectors. Genetic changes in the virus through mutation and recombination, changes in the vector populations and long-distance transportation of plant material or vector insects due to trade of vegetables and ornamental plants have resulted in the emergence of tomato yellow leaf curl disease, African cassava mosaic disease, diseases caused by bipartite begomoviruses in Latin America, Ipomovirus diseases of cucurbits, tomato chlorosis caused by criniviruses, and the torrado-like diseases of tomato (Navas-Castillo et al., 2011).

Temperature can also affect disease resistance in plants, thus affecting the incidence and severity of the diseases. Temperature sensitivity to resistance has been reported for leaf rust (*Puccinia recondita*) in wheat, broomrape (*Orobanche Cumana*) in sunflower, black shank (*Phytophthora nicotianae*) in tobacco, and bacterial blight (BB) (*Xanthomonas oryzae* pv. *oryzae*) in rice (Gregory et al, 2009). Out of the 19 *Oryza glaberrima* accessions and *O. sativa* rice variety SUPA evaluated for rice BB caused by *Xanthomonas oryzae* pv. *oryzae* resistance under high temperature (35°C and 31°C in day and night, respectively), most accessions showed a reduction in the BB disease, whereas accession TOG5620 showed disease reduction from all the *X. oryzae* pv. *oryzae* strains under high temperature (Dossa et al., 2016).

4.4   EFFECT OF CHANGED MOISTURE REGIME ON THE DISEASE SCENARIO

Moisture can impact both host plants and pathogens in various ways. Some pathogens such as apple scab, late blight, and several vegetable root pathogens are more likely to infect plants with increased moisture content because forecast models for these diseases are based on leaf wetness, relative humidity, and precipitation measurements. Other pathogens like the powdery mildew species tend to thrive under conditions with lower (but not low) moisture (Colhoun, 1973; Coakley et al., 1999). A condition of drought is also expected to lead to increased frequency of tree pathogens due to indirect effects on host physiology (Desprez-Loustau et al., 2007). In Italy, the invasive exotic species *Heterobasidion irregulare* appears to be well-adapted to dispersal in the Mediterranean climate than the native *H. annosum* species (Garbelotto et al., 2010). Drought stress has been found to affect the incidence and severity of viruses such as Maize dwarf mosaic virus and Beet yellows virus (Clover et al., 1999; Olsen et al., 1990).

More frequent and extreme precipitation events that are predicted by some climate change models could result in longer periods with favorable pathogen environments. Host crops with canopy size limited by lack of moisture might no longer be so limited and may produce canopies that hold moisture in the form of leaf wetness or high-canopy relative humidity for longer periods, thus increasing the risk of pathogen infection (Coakley et al., 1999). Salinari et al. (2006) used two climate-change models to simulate future scenarios of downy mildew on grapevine (*Plasmopara viticola*). These empirical models predicted an increase in the disease pressure in each decade and more severe epidemics were a direct consequence of more favorable air temperature and rainfall reduction conditions during May and June. The simulation analysis suggests that the impact of increased temperature on enhancing disease pressure exceeded the limiting effect of reduced rainfall. From a biological point of view, this result can be explained by considering that temperature and wetness act together on the pathogen. Thus, the production of grapes in northwestern Italy would decrease. Some climate change models predict higher atmospheric water vapor concentrations with increased temperature and this would also favor pathogen and disease development. While physiological changes in host plants may result in higher disease resistance under climate change scenarios, host resistance to disease may be overcome quickly by more rapid disease cycles, resulting in a greater chance of pathogens evolving to overcome host-plant resistance. Some modeling studies predict changes in incidence and severity with rising temperature and other weather variables for important crop pathogens such as those for black sigatoka (*Mycosphaerella fijiensis*) in banana, grapevine downy mildew, and phoma stem canker (*Leptosphaeria maculans*) on oilseed rape (Evans et al., 2008; Ghini et al., 2008; Kocmánková et al., 2009). Similarly, a model was developed for the risk assessment of early outbreak or increases in the intensity of potato late blight (*Phytophthora infestans*) under climate change in central Europe (Kocmánková et al., 2009).

4.5 EFFECT OF CLIMATE CHANGE ON PLANT DISEASES

The climate influences the incidence as well as the temporal and spatial distribution of plant diseases (Gautam et al., 2013). The main factors that control the growth and development of diseases are temperature, light, humidity, and water. Changes in rainfall patterns and temperature can induce severe epidemics in plants because some types of pathogens will tend to favor others. Moreover, if these changes cause unfavorable conditions for pathogens, the diseases could be reduced or may not occur. In the presence of susceptible hosts, pathogens with short life cycles, high reproduction rates, and effective dispersion mechanisms respond quickly to climate change, resulting in faster adaptation to climatic conditions (Coakley et al., 1999). Harvell et al. (2002) demonstrated that warm winters with high night temperatures facilitate the survival of pathogens, accelerate the life cycles of vectors and fungi, and increase sporulation and aerial fungal infection. Climate change will also modify host physiology and resistance, and alter the stages and rates of the development of pathogens. Temperature and moisture govern the rate of reproduction of many pathogens (Caffarra et al., 2012). The longer growing seasons that will result from global warming will extend the amount of time available for pathogen reproduction and dissemination. Climate change may also influence the sexual reproduction of pathogens thereby increasing the evolutionary potential of individual populations. In the wheat belt of India in Punjab, while changes in temperature and humidity will reduce the importance of yellow rust (*P. striiformis*) and Karnal bunt (*T. indica*); the importance of leaf rust, foliar blights, *Fusarium* head blight and stem rust may increase in the future, particularly in the absence of resistance in wheat cultivars (Kaur et al., 2008). In plant diseases, there are some pathogens of major crops which have a huge potential to cause losses in those crops.

In wheat, *Sr31* stem rust resistance has been effective in cultivars for over 30 years, which can be overcome by new races of *Puccinia graminis* f. sp. *tritici* like Ug99 (Duveiller et al., 2007). According to estimates, Ug99 race of the stem rust can result in up to 10% yield loss in Asia alone, amounting to US$ 1–2 billion/year. The effect of changing climate has also been projected to increase the risk of serious Fusarium ear blight (*Fusarium graminearum* and *F. asiaticum*) epidemics on winter wheat in Central China by the middle of this century (2020–2050) (Zhang et al., 2014). Similar conclusions were reached in the UK also, where climate change models are predicting warmer, wetter winters for the country, resulting in greater incidence of Fusarium ear blight on wheat crops.

Similarly, banana wilt caused by *Xanthomonas* affects the food security of 70 million people in Uganda. In potato, economic production is often impossible without the application of pesticides. Coffee rust epidemics caused by *Hemileia vastrix* is another indicator of climate change with effects on a number of countries including Columbia, Central Amecica, Mexico, Peru, and Ecuador with higher intensities of the diseases between 2008 and 2013 (Avelino et al., 2015).

Late blight of potato caused by *P. infestans*, is considered to be the most economically important disease of potato worldwide. The disease can destroy a potato crop within a few weeks. Estimates of losses to late blight in the developing countries vary between US$ 3 and US$ 10 billion each year, and about US$ 750 million is spent on pesticides alone. In the temperate Indian hills which occupy about 20% of the acreage, a severe epiphytotic (epidemic) of late blight recurs every year resulting in 40%–85% yield loss. The disease now appears earlier in the northern part (November) and later in the eastern part (February) and within a wider temperature range i.e., 14°C–27.5°C than at 10°C–25°C recorded in earlier years (Luck et al., 2012). The impact of climate change on potato productivity in Uttar Pradesh was assessed using WOFOST crop growth model and it projected a decline in the productivity of Kufri Badshah, Kufri Bahar, and Kufri Pukhraj to the tune of 5.5%, 6.1%, and 7.0%, respectively, by 2020 and 9.4%, 10.9%, and 13.4%, respectively by 2055, without adaptation. In effective disease management strategy in potato, the pesticide usage may increase if the changing crop physiology interferes with the uptake and translocation of pesticides or changes in other climatic factors (e.g., more frequent rainfall, washing away residues of contact pesticides) indicate that there is a need for more frequent applications. Faster crop development at increased temperature could also increase the need for application of pesticides. In such situations, diseases caused by the pathogens that infect through stomata such as *Phyllosticta minima* (*Phyllosticta* leaf spot of maple) may be reduced. In soil-borne pathogens, increase in disease development for autumn and winter-infecting root and stem pathogens has been predicted due to the increased thermal time. In Tanzania, bacterial leaf blight (*Xanthomonas oryzae* pv. *oryzae*) of rice is predicted to increase and cause greater losses, with losses due to leaf blast (*Magnaporthe oryzae*) declining (Duku et al., 2016).

The results of this study indicate that the effects of climate change on plant disease can not only be expected to be uneven across diseases but also across geographies, as in some geographic areas losses increase but decrease in others for the same disease. Pathogens such as *Sclerotinia sclerotiorum*, which causes stem or white rot of oilseed rape and a wide range of vegetable crops, are likely to release spores in synchrony with earlier flowering of crops like oilseed rape. Excess moisture, on the other hand, favors some dreaded soil-borne diseases caused by *Phytophthora*, *Pythium*, *R. solani,* and *Sclerotium rolfsii*, especially in pulses. Viral and mycoplasma diseases are enhanced in crops due to increased vector activity. Different climate change variables will have different effects on different soil microorganisms and the associated biological processes as the soil is a highly complex ecosystem. Furthermore, such changes are highly dependent on the particular soil conditions and few generalizations attributable to climate change can be made.

Plant diseases are a major problem not only for food production but also for the quality and safety of the crop produce. Mycotoxins and pesticide residues are among the top food safety concerns and climate change can increase the toxicity levels of these toxicants. Changes in both temperature conditions and atmospheric composition may influence the severity of outbreaks of *Fusarium* head blight and the production of mycotoxins, which can affect the entire wheat value chain (Chakraborty and Newton, 2011).

Biological control is an integral part of integrated disease management strategy and microorganisms deployed or in the natural environment in air and soil are bound to be affected with the climate change. But, there is very little information on the effects of climate change on the biological control of plant disease (Ghini et al., 2011). Biocontrol of the fungal foliar disease gray mold (*Botrytis cinerea*) in greenhouse cucumber crops by the biocontrol agent *Trichoderma harzianum* T39 is more pronounced at higher temperatures and lower Relative Humidity (RH) levels.

In growth chambers, the duration of the wet period also increased the severity of late blight (*Phytophthora infestans*). In a biocontrol trial involving tomato plants inoculated with *P. infestans* sporangia, it has been reported that when environmental conditions are favorable for late blight development, the examined biocontrol agents are less effective (Elad and Pertot, 2014). Gray mold is promoted under conditions of moderate temperatures, high humidity, and water on the surface of plants and can be suppressed by rendering the crop's environment less conducive to disease development. Similarly, effects of climate change on commercialized biocontrol agents of plant diseases in Brazil indicated that *Bacillus subtilis* and *Trichoderma* spp. will be less affected than some other biocontrol agents, whereas the efficacy of *Coniothyrium minitans* and *Clonostachys rosea* may be reduced in some areas (Ghini et al., 2011).

Fungicide and bactericide efficacy may change with increased CO_2, moisture, and temperature (Schepers, 1996). The more frequent rainfall events predicted by climate change models could result in farmers finding it difficult to keep residues of contact fungicides on plants, triggering more frequent applications. Systemic fungicides could be affected negatively by physiological changes that slow down the uptake rates, such as smaller stomatal opening or thicker epicuticular waxes in crop plants grown under higher temperatures. The same fungicides could be affected positively by increased plant metabolic rates that could increase the fungicide uptake.

4.6 CHANGING DISEASE SCENARIO IN FRUITS AND VEGETABLES

Apple (*Malus domestica* Borkh.) is a major fruit crop grown in the hilly states of the country comprising Jammu & Kashmir, Himachal Pradesh, Uttarakhand, and Arunachal Pradesh. The economy of the people of these hill states largely depends upon this golden crop. Although the overall apple production has increased, yet the productivity per unit area is still quite low (7–11 t/ha) in comparison (25–35 t/ha) to other apple producing countries including New Zealand, USA, Germany etc. (Thakur et al., 2000). Out of the various factors responsible for low productivity, disease is one of the main contributing factors. Amongst the different diseases affecting apple, scab, premature leaf fall, powdery mildew, *Alternaria* leaf spot and blight, sooty mold, fly speck, moldy core, fruit rots (core rot, white rot, brown rot, black rot, bull's eye rot), white root rot, collar rot, seedling blight and cankers (stem brown, smoky blight, pink canker, *Cryptosporiopsis corticola* canker, *Fusicoccum* canker, nail head) are causing huge economic losses to both the orchardists and nursery-men, besides resulting in the untimely decline of the orchards. In the last two decades, changing climatic conditions have rapidly changed the disease scenario in Himachal Pradesh and other apple growing regions of the country.

The disease severity in any crop is dependent on the availability of primary inoculum, prevalence of congenial environmental conditions, and the cultivation of susceptible genotypes. Under the present situation in Himachal Pradesh, the cultivar Royal Delicious and its strains (Vance Delicious, Top Red) susceptible to almost all the diseases are covering about 90% of the total area under apple cultivation (Jindal et al., 2001). Secondly, the availability of primary inoculum, initiation, and the subsequent spread of disease in any crop is largely determined by the prevalent environmental conditions. It is, therefore, the climatic conditions, which play an important role in deciding the status of diseases in plants. Hence, in the present review, efforts have been made to find the effect of changing climatic [environmental data obtained from Hazard Ranking System (HRS) – Seobagh (Kullu) Meteorology Laboratory] factors (1993–2014) on the occurrence and severity of apple diseases in the Kullu district of Himachal Pradesh in particular and other apple growing districts of the state in general. Secondly, the indiscriminate use of fungicides has also played a significant role in changing disease scenario in apple.

The scab disease, which remained economically very important for about two decades (1978–1998) since its first appearance during 1977 and caused epiphytotics in the early eighties and mid-nineties, has now almost disappeared since the crop season of 1999 from the majority of apple orchards of Himachal Pradesh. The major reason for its sudden disappearance is the occurrence of extreme weather conditions during the year 1999 i.e., (i) prolonged dry weather conditions (6.9 mm in 4 rainy days and RH 54.8%) between the period of ascospore maturity to their complete exhaustion from pseudothecia (March 10–May 2, 1999) did not allow the manifestation of primary infection, as the leaves did not remain wet for the minimum period of 9 hours, a pre-requisite for the infection to occur, (ii) less rainfall (35.2 mm in 4 rainy days) during February 1999 (condition required for the initiation of asci and ascospore formation) have also contributed toward the lesser availability of primary inoculum for infection during 1999. Secondly, the appearance of premature defoliation disease in epidemic form has eclipsed the scab (Sharma, 2003a). Presently, the disease is confined in a few stray locations like Barsaini in Manikran valley of district Kullu, Janjelli in district Mandi, and Nichar and Sangla in district Kinnaur. Similar reports have been appearing from the Kashmir valley. Premature defoliation (*Marssonina coronaria*) first appeared in a small proportion in the Kullu Valley as well as in the apple orchards of Kotkai (Shimla district) in Himachal Pradesh during 1994 (Sharma and Bhardwaj, 1997).

Its appearance is mainly ascribed to the indiscriminate use of Evidence-based interventions (EBI) fungicides, particularly; hexaconazole (more effective against scab) since 1991 onwards gave sufficient time and chance for Marssonina to build up its inoculum to appear in severe form in the following years. It assumed severe to epiphytotic proportions in all the apple growing areas of the state during 1995–2001 (Sharma, 2003b). The appearance and subsequent spread of this disease during these years was largely dependent upon the occurrence of moderate to heavy rainfall during mid-June to mid-September. During 2002, the occurrence of drought-like conditions during mid-June to mid-August (43.4 mm in 7 rainy days) suppressed this disease to a great extent (diseases index 0%–12%) in the apple orchards of Kullu district and that of Himachal Pradesh in general. During 2006 and 2007, the prevalence of sufficient rains (122.3–138.6 mm) during July and August has again led to the appearance of this disease in severe form. Barring 2002–2004 (low incidence) and 2009, this disease continued to appear in moderate to severe form till date. The changing climate has contributed to make some minor diseases to become major ones warranting a separate disease management practice.

It has been observed that the occurrence of low disease severity of premature defoliations disease during 2002–2004 and non-spray of EBI fungicides in the absence of scab disease from 1999 onward, coupled with the occurrence of more rains during the late rainy seasons of 2002–2006, except 2005, resulted in appropriate conditions for *Alternaria* leaf spot disease to build up. Thereafter, it has been appearing in moderate to severe form till date. The Alternaria leaf spot and blight has emerged as one of the major diseases in the apple orchards of Kashmir valley. Furthermore, *Cryptosporiopsis corticola* canker appeared in severe forms in the apple orchards of Kullu district in particular and Himachal Pradesh in general during 2002 and is mainly attributed to the prevalence of low rainfall during the month of May (9.2 mm in 2 rainy days) and June (30.2 mm in 4 rainy days), accompanied with slightly higher temperature. It spread very fast during the month of July with the prevalence of slightly higher average temperature (26.9°C), low humidity (52.1%), and scanty rains (2 mm in 1 rainy day) and took on epiphytotic proportions. The canker phase of this disease did not flare up in the following years (2003–2007) and is occurring in low intensities in the various apple orchards of the state but its fruit rot (bull's eye rot) phase has been appearing in moderate form since 2003 due to the sufficient availability of pathogen inoculum since 2002 onward and occurrence of adequate rains during rainy season.

Among other diseases viz., moldy core and core rot mainly caused by *Alternaria* species, *Trichothecium roseum* fungi, occurring in extremely low frequencies earlier, suddenly assumed alarming proportions in the apple orchards of Himachal Pradesh in 2005. This disease again appeared in high frequencies from 2007 onward and resulting in huge economic losses. It has mainly been attributed to the appearance of *Alternaria* leaf spot and blight disease in apple orchards since 2002 which has possibly increased the availability of the pathogen's inoculum. Secondly, the occurrence of extremely low rainfall (15.7 mm in 3 rainy days and 5.2 mm in a single rainy day during the month of April in 2005 and 2007, respectively) at the flowering and fruit set stage in the month of April has resulted in successful establishment of infection by these pathogens of moldy core and core rot disease. Occurrence of low rainfall during April–May during 1999, 2002, 2005, 2007, 2010, and 2012, accompanied with low to moderate temperatures along with fewer sprays of EBI fungicides due to the disappearance of scab, has helped in the appearance of powdery mildew disease in higher severities since 1999. It also led to higher severity of fruit russeting during these years.

Changes in the rainfall pattern since 2002 onward (more rains in late rainy season) and use of imbalance fertilizers (high P) have aggravated the problem of collar rot, root rot, and pink cankers. Viral diseases (mosaic, chlorotic leaf roll, stem pitting, and grooving) are also on the rise due to the use of uncertified bud wood of pollinizer varieties for top-working, and have presently assumed a serious proportion. During the crop season of 2007, the environmental related disorder viz., sun scalding (noticed first time) and russeting in apple fruit were noticed to a greater extent. It is attributed to the rise in temperature during summer and rainy season (June–August).

4.7 INITIATIVES TO MITIGATE THE EFFECT OF CLIMATE CHANGE ON CROPS

To mitigate the impacts of climate change, the Government of India has launched the National Action Plan on Climate Change (NAPCC) in 2008. The Central Government has announced its intent to reduce the emissions intensity of its GDP by 20%–25% between 2005 and 2020, thus making a major contribution to mitigating climate change. The government has also formed an Expert Group on Low Carbon Strategy for Inclusive Growth under the Planning Commission to develop a roadmap for low-carbon development. It has also launched the Indian Network for Climate Change Assessment (INCCA), in October 2009, as a network-based program with the broad objectives of measuring, modeling, and monitoring the changes due to climate change. It brings together over 120 institutions and over 220 scientists from across the country. The agriculture mission under the NAPCC alone is to spend upwards of Rs 1 lakh crore over 5 years to make the primary sector more resilient to the inevitable changes in climate change. The National Mission for Sustainable Agriculture (NMSA) has also been formulated which derives its mandate from Sustainable Agriculture Mission which is one of the eight Missions outlined under NAPCC.

According to NAPCC, it is predicted that a loss of 10%–40% in production may occur by 2100 due to climate change. In recognition of the growing problem of climate change, India has declared a voluntary goal of reducing the emissions intensity of its GDP by 20%–25%, over 2005 levels, by 2020, despite having no binding mitigation obligations as per the Convention. A slew of policy measures were launched to achieve this goal. As a result, the emission intensity of India's GDP has decreased by 12% between 2005 and 2010. It is a matter of satisfaction that the United Nations Environment Programme (UNEP) in its Emission Gap Report 2014 has recognized India as one of the countries on course to achieving its voluntary goal. The National Agricultural Research System (NARS) has been carrying out comprehensive research and technology demonstration activities on coping with climate change under the flagship program of ICAR on National Initiative on Climate Resilient Agriculture (NICRA). NICRA was launched in 2011 and the project aims to enhance the resilience of Indian agriculture to climate change and climate vulnerability through strategic research and technology demonstration. The findings of the study indicate that an increase in the temperature by 1.5°C along with increase in rainfall up to 50% during soybean growth can reduce the soybean yield to the tune of 5%–10%. Furthermore, increase in the CO_2 concentration (360–720 ppm) favored soybean growth when the temperature was reduced by 1°C during 1980–2010. However, with increase in CO_2, the yield is masked by the adverse impact of increase in temperature on crop growth.

The Paris Agreement on climate change is a way forward in this direction. The agreement aims to limit the increase in global average temperatures to "well below 2°C above pre-industrial levels". It also aims to "pursue efforts to limit the temperature increase to 1.5°C above pre-industrial levels, recognizing that this would significantly reduce the risks and impacts of climate change". The agreement requires that the developed nations will continue to help the developing countries with the costs of going green, and the costs of coping with the effects of climate change. The target set for developed countries are obliged to 'mobilize' $100 bn a year of public and private finance to help the developing countries by 2020 in Copenhagen in 2009 continue their existing collective mobilization goal through 2025. We need a comprehensive approach to reduce the impacts of climate change – an approach that decreases emissions across all sectors.

4.8 NEED FOR ADOPTION OF NOVEL APPROACHES

Changing disease scenario due to climate change has highlighted the need for better agricultural practices and the use of eco-friendly methods in disease management for sustainable crop production. Owing to the changing climate and shift in the seasons, the choice of crop management practices based on the prevailing situation is important. In such scenarios, weather-based disease monitoring, inoculums monitoring, especially for soil-borne diseases and rapid diagnostics would play a significant role. There is a need to adopt novel approaches to counter the resurgence of diseases under changed climatic scenario. Integrated disease management strategies should be developed to decrease the dependency on fungicides. Other multipronged approaches include healthy seeds with innate forms of broad and durable disease resistance, and intercropping systems that foster refuges for natural bio-control organisms. In addition, monitoring and early warning systems for forecasting disease epidemics should be developed for important host–pathogens which have a direct bearing on the earnings of the farmers and food security at large. Such a diversified crop protection strategy has been highlighted in a comprehensive study on an integrated approach to control all foliar diseases in barley. Use of botanical pesticides and plant-derived soil amendments such as neem oil, neem cake, and karanja seed extract also help in the mitigation of climate change because it helps in the reduction of nitrous oxide emission by nitrification inhibitors such as nitrapyrin and dicyandiamide.

REFERENCES

Avelino, J., Cristancho, M., Georgiou, S., Imbach, P., Aguilar, L., Bornemann, G., Läderach, P., Anzueto, F., Hruska, A.J. and Morales, C. 2015. The coffee rust crises in Colombia and Central America (2008–2013): Impacts, plausible causes and proposed solutions. *Food Security*, 7(2), pp. 303–321.

Bebber, D.P., Ramotowski, M.A. and Gurr, S.J. 2013. Crop pests and pathogens move polewards in a warming world. *Nature Climate Change*, 3(11), p. 985.

Caffarra, A., Rinaldi, M., Eccel, E., Rossi, V. and Pertot, I. 2012. Modelling the impact of climate change on the interaction between grapevine and its pests and pathogens: European grapevine moth and powdery mildew. *Agriculture, Ecosystems & Environment*, 148, pp. 89–101.

Chakraborty, S. and Datta, S. 2003. How will plant pathogens adapt to host plant resistance at elevated CO_2 under a changing climate? *New Phytologist*, 159(3), pp. 733–742.

Chakraborty, S., Murray, G.M., Magarey, P.A., Yonow, T., O'Brien, R.G., Coft, B.J., Barbetti, M.J., Sivasithamparam, K., Old, K.M., Dudzinski, M.J., Sutherst, R.W., J., Archer, C and Emmett, R.W. 1998. Potential impact of climate change on plant diseases of economic significance to Australia. *Australasian Plant Pathology*, 27(1), pp. 15–35.

Chakraborty, S., Murray, G. and White, N. 2002. Potential impact of climate change on plant diseases of economic significance to Australia. *Australasian Plant Pathology*, 27, pp. 15–35.

Chakraborty, S. and Newton, A.C. 2011. Climate change, plant diseases and food security: an overview. *Plant Pathology*, 60(1), pp.2–14.

Chakraborty, S., Tiedemann, A.V. and Teng, P.S. 2000. Climate change: potential impact on plant diseases. *Environmental Pollution*, 108(3), pp. 317–326.

Clover, G.R.G., Smith, H.G., Azam-Ali, S.N. and Jaggard, K.W., 1999. The effects of drought on sugar beet growth in isolation and in combination with beet yellows virus infection. *The Journal of Agricultural Science*, 133(3), pp. 251–261.

Coakley, S.M., Scherm, H. and Chakraborty, S. 1999. Climate change and plant disease management. *Annual Review of Phytopathology*, 37(1), pp. 399–426.

Colhoun, J. 1973. Effects of environmental factors on plant disease. *Annual Review of Phytopathology*, 11(1), pp. 343–364.

Desprez-Loustau, M.L., Robin, C., Reynaud, G., Déqué, M., Badeau, V., Piou, D., Husson, C. and Marçais, B. 2007. Simulating the effects of a climate-change scenario on the geographical range and activity of forest-pathogenic fungi. *Canadian Journal of Plant Pathology*, 29(2), pp. 101–120.

Dixon, G.R. 2012. Climate change–impact on crop growth and food production, and plant pathogens. *Canadian Journal of Plant Pathology*, 34(3), pp. 362–379.

Dossa, G.S., Oliva, R., Maiss, E., Vera Cruz, C. and Wydra, K. 2016. High temperature enhances the resistance of cultivated African rice, Oryza glaberrima, to bacterial blight. *Plant Disease*, 100(2), pp. 380–387.

Duku, C., Sparks, A.H. and Zwart, S.J. 2016. Spatial modelling of rice yield losses in Tanzania due to bacterial leaf blight and leaf blast in a changing climate. *Climatic change*, 135(3–4), pp. 569–583.

Duveiller, E., Singh, R.P. and Nicol, J.M. 2007. The challenges of maintaining wheat productivity: pests, diseases, and potential epidemics. *Euphytica*, 157(3), pp. 417–430.

Eastburn, D.M., Degennaro, M.M., Delucia, E.H., Dermody, O. and McElrone, A.J. 2010. Elevated atmospheric carbon dioxide and ozone alter soybean diseases at SoyFACE. *Global Change Biology*, 16(1), pp. 320–330.

Elad, Y. and Pertot, I. 2014. Climate change impacts on plant pathogens and plant diseases. *Journal of Crop Improvement*, 28(1), pp. 99–139.

Evans, N., Baierl, A., Semenov, M.A., Gladders, P. and Fitt, B.D. 2008. Range and severity of a plant disease increased by global warming. *Journal of the Royal Society Interface*, 5(22), pp. 525–531.

FAO. 2016. Climate change and food security: Risks and responses, Rome 98 P.

Francesca, S., Simona, G., Francesco Nicola, T., Andrea, R., Vittorio, R., Federico, S., Cynthia, R. and Maria Lodovica, G. 2006. Downy mildew (Plasmopara viticola) epidemics on grapevine under climate change. *Global Change Biology*, 12(7), pp. 1299–1307.

Garbelotto, M., Linzer, R., Nicolotti, G. and Gonthier, P. 2010. Comparing the influences of ecological and evolutionary factors on the successful invasion of a fungal forest pathogen. *Biological Invasions*, 12(4), pp. 943–957.

Garrett, K.A., Dendy, S.P., Frank, E.E., Rouse, M.N. and Travers, S.E. 2006. Climate change effects on plant disease: Genomes to ecosystems. *Annual Review of Phytopathology*, 44, pp. 489–509.

Gautam, H.R. 2009a. Challenges of climate change and bio-energy to world food security. *Open Learning*, (July–December), pp. 49–51.

Gautam, H.R., 2009b. Effect of climate change on rural India. *Kurukshetra*, 57(9), pp. 3–5.

Gautam, H.R., Bhardwaj, M.L. and Kumar, R. 2013. Climate change and its impact on plant diseases. *Current Science*, pp. 1685–1691.

Gautam, H.R. and Kuniyal, C.P. 2014. Climate change is affecting apple cultivation in Himachal Pradesh. *Current Science*, 106(4), p. 498.

Ghini, R., Bettiol, W. and Hamada, E. 2011. Diseases in tropical and plantation crops as affected by climate changes: current knowledge and perspectives. *Plant Pathology*, 60(1), pp. 122–132.

Ghini, R., Hamada, E., Gonçalves, R.R.V., Gasparotto, L. and Pereira, J.C.R. 2008. Risk analysis of climate change on black sigatoka in Brazil. J. *Plant Pathology*, 90, p. 105.

Gregory, P.J., Johnson, S.N., Newton, A.C. and Ingram, J.S. 2009. Integrating pests and pathogens into the climate change/food security debate. *Journal of Experimental Botany*, 60(10), pp. 2827–2838.

Harvell, C.D., Mitchell, C.E., Ward, J.R., Altizer, S., Dobson, A.P., Ostfeld, R.S. and Samuel, M.D. 2002. Climate warming and disease risks for terrestrial and marine biota. *Science*, 296(5576), pp. 2158–2162.

Hibberd, J.M., Whitbread, R. and Farrar, J.F. 1996. Effect of elevated concentrations of CO_2 on infection of barley by Erysiphe graminis. *Physiological and Molecular Plant Pathology*, 48(1), pp. 37–53.

IPCC. 2007. *Intergovernmental Panel on Climate Change, The Fourth IPCC Assessment Report*. Cambridge University Press, Cambridge, UK.

IPCC. 2014. Climate Change 2014: Synthesis Report. Contribution of Working Groups I, II and III to the Fifth Assessment Report of the Intergovernmental Panel on Climate Change [Core Writing Team, R.K. Pachauri and L.A. Meyer (eds.)]. IPCC, Geneva, Switzerland, p. 151.

Jindal, K.K., Chauhan, P.S. and Mankotia, M.S. 2001. Apple productivity in relations to environmental components. In: *Productivity of Temperate Fruits*. Edited by K.K. Jindal and D.R. Gautam, Dr YS Parmar University of Horticulture and Forestry, Solan, pp. 12–20.

Kaur, S., Dhaliwal, L. and Kaur, P. 2008. Impact of climate change on wheat disease scenario in Punjab. *Journal of Research*, 45(3&4), pp. 161–170.

Kobayashi, T., Ishiguro, K., Nakajima, T., Kim, H.Y., Okada, M. and Kobayashi, K. 2006. Effects of elevated atmospheric CO_2 concentration on the infection of rice blast and sheath blight. *Phytopathology*, 96(4), pp. 425–431.

Kocmánková, E., Trnka, M., Juroch, J., Dubrovský, M., Semerádová, D., Možný, M. and Žalud, Z. 2009. Impact of climate change on the occurrence and activity of harmful organisms. *Plant Protection Science*, 45(Special Issue), pp. 48–52.

Kudela, V. 2009. Potential impact of climate change on geographic distribution of plant pathogenic bacteria in Central Europe. *Plant Protection Science*, 45(Special Issue), pp. 27–32.

Ladányi, M. and Horváth, L. 2010. A review of the potential climate change impact on insect populations-general and agricultural aspects. *Applied Ecology and Environmental Research*, 8(2), pp. 143–152.

Li, X., Ahammed, G.J., Li, Z., Tang, M., Yan, P. and Han, W. 2016. Decreased biosynthesis of jasmonic acid via lipoxygenase pathway compromised caffeine-induced resistance to Colletotrichum gloeosporioides under elevated CO_2 in tea seedlings. *Phytopathology*, 106(11), pp. 1270–1277.

Luck, J. et al. 2012. The effects of climate change on pests and diseases of major food crops in the Asia-Pacific region. Final Report for APN (Asia-Pacific Network for Global Change Research) Project, p. 73.

Madden, L.V., Hughes, G. and Bosch, F. 2007. The study of plant disease epidemics. American Phytopathological Society (APS), Saint Paul, MN.

Mboup, M., Bahri, B., Leconte, M., De Vallavieille-Pope, C., Kaltz, O. and Enjalbert, J. 2012. Genetic structure and local adaptation of European wheat yellow rust populations: the role of temperature-specific adaptation. *Evolutionary Applications*, 5(4), pp. 341–352.

Mcelrone, A.J., Reid, C.D., Hoye, K.A., Hart, E. and Jackson, R.B. 2005. Elevated CO_2 reduces disease incidence and severity of a red maple fungal pathogen via changes in host physiology and leaf chemistry. *Global Change Biology*, 11(10), pp. 1828–1836.

Mina, U. and Sinha, P. 2008. Effects of climate change on plant pathogens. *Environmental News*, 14(4), pp. 6–10.

Mulherin, K.M., Karowe, D.N. and Enyedi, A.J. 2000. Effects of elevated carbon dioxide on plant-pathogen interactions. In plant Biology 2000. Plant Biology Meeting. Symposium Elevated CO_2. Abstract number 368. San Diego, CA. USA.

Navas-Castillo, J., Fiallo-Olivé, E. and Sánchez-Campos, S. 2011. Emerging virus diseases transmitted by whiteflies. *Annual Review of Phytopathology*, 49, pp. 219–248.

Oerke, E. C. (2006). Crop losses to pests. *The Journal of Agricultural Science*, 144, pp. 31–43.

Oerke, E.C., Dehne, H.W., Schönbeck, F. and Weber, A. 2012. Crop production and crop protection: Estimated losses in major food and cash crops. Elsevier.

Olsen, A.J., Pataky, J.K., D'arcy, C.J. and Ford, R.E. 1990. Effects of drought stress and infection by maize dwarf mosaic virus in sweet corn. *Plant Disease*, 74, pp. 147–151.

Pande, S. and Sharma, M. 2010. Climate change: Potential impact on chickpea and pigeonpea diseases in the rainfed semi-arid tropics (SAT). In Proceedings of the 5th International Food Legumes Research Conference (IFLRC V) and 7th European Conference on Grain Legumes (AEP VII), Antalya, Turkey.

Salinari, F. et al.2006. Downy mildew (Plasmopara viticola) epidemics on grapevine under climate change. Global Change Biology, 12, pp. 1299–1307.

Schepers, H.T.A.M. 1996. Effect of rain on efficacy of fungicides on potato against Phytophthora infestans. *Potato Research*, 39, pp. 541–550.

Sharma, I.M. 2003a. Probable reasons for zero level of apple scab incidence in Kullu valley of Himachal Pradesh. *Journal of Mycology and Plant Pathology*, 33, pp. 96–100.

Sharma, I.M. 2003b. Influence of environmental factors on the development of pre-mature defoliation disease caused by Marssonina coronaria in apple and its management. *Journal of Mycology and Plant Pathology*, 33(1), pp. 89–95.

Sharma, I.M. and Bhardwaj, S.S. 1997. Diagnosis and management of premature defoliation diseases of apple in Kullu valley of Himachal Pradesh. In Proceeding International Conference on Integrated Plant Disease Management for Sustainable Agriculture organized by IPS and ICAR at IARI New Delhi w.e.f. 10 Nov., to 15 Nov., 1997. p. 329.

Thakur, J.R., Kumar, J. and Sharma, I.M. 2000. Present and Future of Horticulture in Himachal Pradesh. In: *New Options for Hill Agriculture*. Edited by D.S. Thakur, CSK HPKV RRS, Bajaura, pp. 64–73.

Thind, T. S. 2012, Fungicides in crop health security. *Indian Phytopathology*, 65(2), pp. 109–115.

Tiedemann, A.V. and Firsching, K.H. 2000. Interactive effects of elevated ozone and carbon dioxide on growth and yield of leaf rust-infected versus non-infected wheat. *Environmental Pollution*, 108(3), pp. 357–363.

Woods, A., Coates, K.D. and Hamann, A. 2005. Is an unprecedented Dothistroma needle blight epidemic related to climate change? *AIBS Bulletin*, 55(9), pp. 761–769.

Zhang, X., Halder, J., White, R.P., Hughes, D.J., Ye, Z., Wang, C., Xu, R., Gan, B. and Fitt, B.D. 2014. Climate change increases risk of fusarium ear blight on wheat in central China. *Annals of Applied Biology*, 164(3), pp. 384–395.

5 Solar Thermal Modeling of Microclimatic Parameters of Agricultural Greenhouse

Dilip Jain
ICAR-Central Arid Zone Research Institute

CONTENTS

5.1 INTRODUCTION TO MODELING OF AGRICULTURAL GREENHOUSE

A greenhouse is essentially an enclosed structure, which traps the short wavelength (0.38–0.78 μm) solar radiation and stores the long wavelength (0.78–3 μm) thermal radiation to create favorable microclimate for higher productivity. Although this trapped thermal energy supplies heating during the day, at night the temperature falls to undesirable levels that can affect those plants sensitive to low temperature in the winter. Thus, this enclosure is known as controlled environment greenhouse. At the same time, due to greenhouse effect, the temperature of the air in a greenhouse is greater than that outside, and can be used to modify the microclimate for crop cultivation. The greenhouse climate may be used for crop drying, distillation, biogas plant heating, space conditioning, and aquaculture.

A mathematical model is an abstract model that uses mathematical language to describe the behavior of a system. As the purpose of modeling is to increase our understanding of the world, the validity of a model rests not only on its fit to empirical observations, but also on its ability to extrapolate to situations or data beyond those originally described in the model. Thermal modeling of the greenhouse is essential not only to predict the covered environment, but also to help in optimizing the dimensional and operational parameters. The essentials of greenhouse modeling can be defined as (Tiwari 2002):

5.1.1   THE PURPOSE OF MODELING OF GREENHOUSE ENVIRONMENT

   i. To get the prehand environment data before going for true construction.
  ii. To optimize the various constructional parameters.
 iii. To optimize the operational parameters.
  iv. To ascertain the suitability of structure for a particular season and crop.
   v. To ascertain the maximum production of a crop from the greenhouse for a given thermal capacity.

5.1.2   THE COMPONENTS OF GREENHOUSE CAN BE CONSIDERED AS

   i. Canopy of greenhouse cover
  ii. Enclosed air of greenhouse
 iii. Floor under the greenhouse
  iv. Storage capacity inside the greenhouse
   v. North wall of the greenhouse (in case of having north wall in a greenhouse in the northern hemisphere of the earth)
  vi. Ground air collector for heating in winter
 vii. Earth heat exchanger
viii. Cooling pad for summer cooling
  ix. Aquaculture pond water (in case of aquaculture greenhouse) etc.

5.2   STEPS OF GREENHOUSE MODELING

5.2.1   THERMAL MODELING

The thermal modeling of agricultural greenhouse can be well understood based on the energy balance based mathematical model. Following steps should be considered for the identification of the problem to design the greenhouse:

Step I: Location (latitude) for greenhouse design, orientation of greenhouse, mode of greenhouse for heating, cooling, drying, aquaculture, or maybe a combination of these.

Step II: Energy balance equation for different components should be written in terms of solar fraction, solar radiation, ambient temperature, wind velocity, and various heat transfer coefficients in the quasi-steady state condition.

Step III: If possible sensitive or insensitive parameters should be identified before modeling.

Step IV: Energy balance equation should be solved to obtain the greenhouse room air temperature for the given climatic and design parameters.

Step V: Effect of various sensitive design parameters on the greenhouse room air temperature should also be studied for the given climatic parameters.

Step VI: For known optimum greenhouse room air temperature, the design parameters can be adopted for the erection of greenhouse.

5.2.2 Some Basic Assumptions

The energy balance equations for the different components of the greenhouse system can be written on the following assumptions:

  i. Greenhouse is east-west oriented (best orientation for northern hemisphere),
 ii. Absorptivity and heat capacity of air is negligible in comparison with plants and other materials,
iii. Storage capacity of the covering materials is negligible,
 iv. Heat flow is one dimensional and in a quasi-steady state condition,
  v. Radiative exchange within the walls and roofs of greenhouse is neglected,
 vi. Thermal properties of plants in the greenhouse are same as of water, and
vii. Radiation exchange between the floor and the north wall has not been considered due to small temperature difference between the floor and the wall.

5.3 UNDERSTAND THE DISTRIBUTION OF SOLAR ENERGY INSIDE THE GREENHOUSE (TIWARI, 2002; TAKI ET AL., 2018)

The solar radiation received from all the transparent walls of the greenhouse is trapped by the cover material and absorbed by various components of the greenhouse structure such as plants, north wall, and ground surface. The distribution is explained in Figure 5.1.

where;

$\sum I_i A_i \tau_i$ = Transmitted solar radiation inside the greenhouse

$F_n \sum I_i A_i \tau_i$ = Fraction of transmitted radiation falling on the north wall

$\rho_n F_n \sum I_i A_i \tau_i$ = Reflected radiation from the north wall

$(1 - \rho_n) F_n \sum I_i A_i \tau_i$ = Available solar radiation on the north wall

$\alpha_n (1 - \rho_n) F_n \sum I_i A_i \tau_i$ = Absorbed radiation of the north wall

Example of the energy balance on different components of the greenhouse (Figure 5.1)

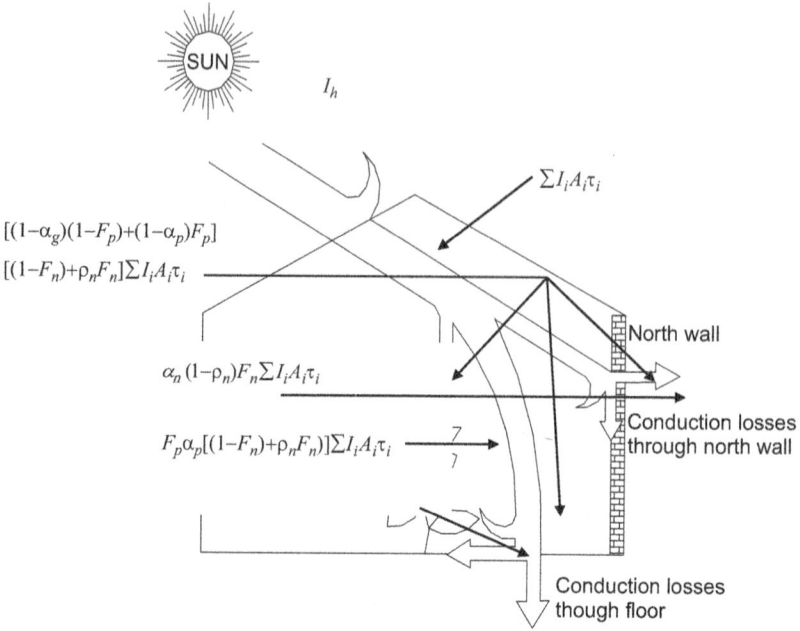

FIGURE 5.1 Distribution of solar energy inside the greenhouse having the north wall and plants.

Greenhouse is connected with ground air collector (Figure 5.2a).

a. For the north wall

$$\alpha_n(1-\rho_n)F_n\Sigma I_i A_i \tau_i = [h_{nr}(T\,|_{z=0} - T_r) + h_{na}(T\,|_{z=0} - T_a)]A_n \qquad (5.1)$$

<u>Rate of thermal energy absorbed by north wall after reflection of radiation from north wall</u>   <u>Rate of thermal energy transferred from north wall to greenhouse (q_{nr})</u>   <u>Rate of thermal energy conducted to outside air by north wall (q_{na})</u>

b. For plants

$$F_p\alpha_p\big[(1-F_n)+\rho_n F_n\big]\Sigma I_i A_i \tau_i = M_p C_p \frac{dT_p}{dt} + h_{pr}\big(T_p - T_r\big)A_p \qquad (5.2)$$

<u>Rate of thermal energy absorbed by the plant in the greenhouse</u>   <u>Rate of thermal energy stored in plants</u>   <u>Rate of thermal energy transferred from plant to greenhouse (q_{pr})</u>

c. For the floor

$$\alpha_g(1-F_p)\big[(1-F_n)+\rho_n F_n\big]\Sigma I_i A_i \tau_i = \left[\, h_{gr}\big(T\,|_{y=0} - T_r\big) + h_{g\infty}\big(T\,|_{y=0} - T_\infty\big)\,\right]A_g \qquad (5.3)$$

<u>Rate of thermal energy absorbed by the greenhouse floor</u>   <u>Rate of thermal energy transferred from ground surface to greenhouse (q_{gr})</u>   <u>Rate of thermal energy conducted inside the ground $(q_{g\infty})$</u>

(a)

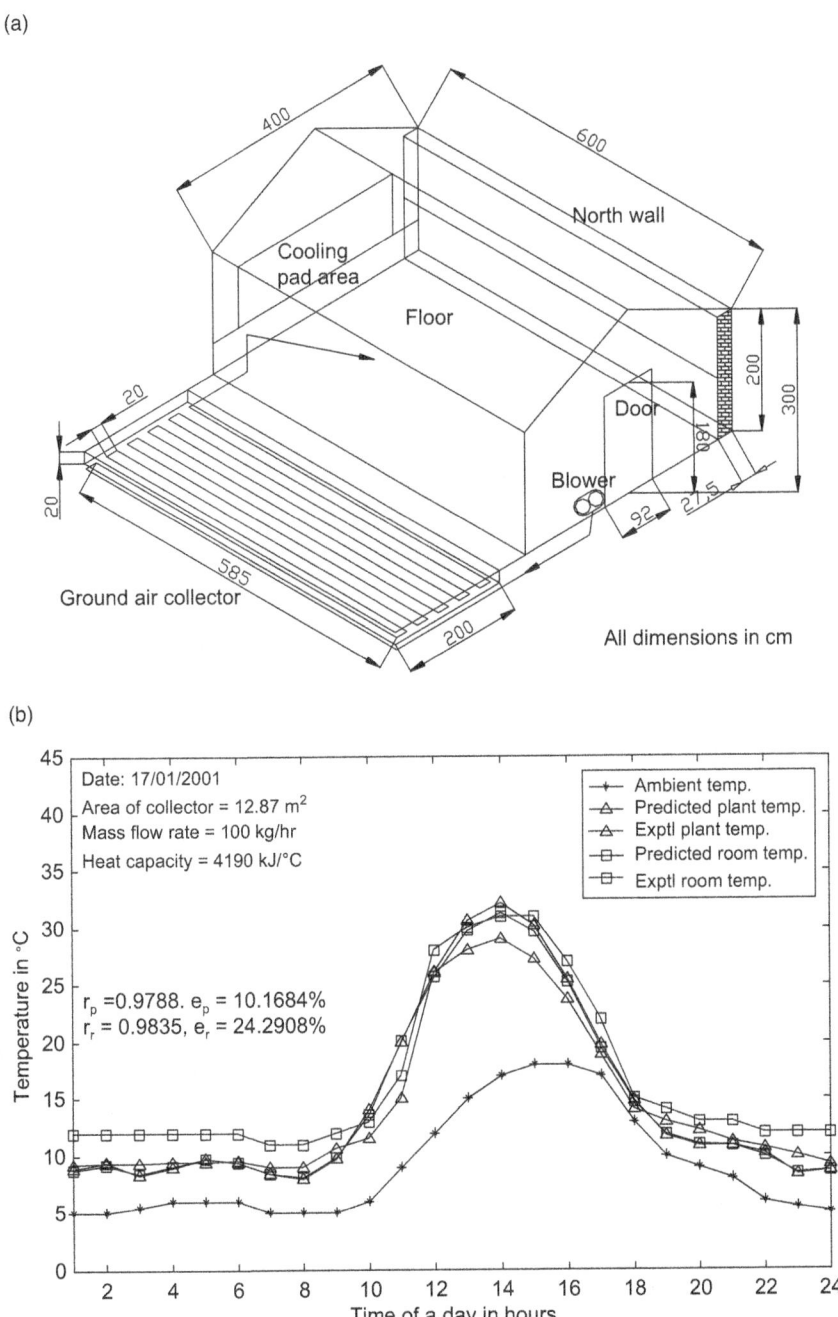

(b)

FIGURE 5.2 (a) Greenhouse attached with ground air collector and north wall. (b) Hourly variation of plant and room temperature in greenhouse with ground air collector.

d. For air

$$\left[\left(1-\alpha_g\right)\left(1-F_p\right)+\left(1-\alpha_p\right)F_p\right]\left[\left(1-F_n\right)+\rho_n F_n\right]\sum I_i A_i \tau_i + h_{nr}\left(T\big|_{z=0}-T_r\right)A_n$$

Rate of thermal energy retained by air of greenhouse (remaining part of energy after absorption by north wall, plants and floor)

Rate of thermal energy transferred from north wall to greenhouse $\left(q_{nr}\right)$

$$+h_{gr}\left(T\big|_{y=0}-T_r\right)+h_{pr}\left(T_p-T_r\right)A_p+\dot{Q}_u = \sum U_i A_i\left(T_r-T_a\right)+0.33NV\left(T_r-T_a\right)$$

Rate of thermal energy transferred from ground surface to greenhouse $\left(q_{gr}\right)$

Rate of thermal energy transferred from plant to greenhouse $\left(q_{pr}\right)$

Rate of energy gain from air collector

Rate of overall loss from greenhouse cover

Rate of thermal energy loss through ventilation

$$(5.4)$$

Similarly the energy balance equations can be written on various components of the greenhouse in any configuration.

5.4   EXAMPLES OF APPLICATION OF GREENHOUSE MODEL

5.4.1   A GREENHOUSE ATTACHED WITH GROUND AIR COLLECTOR FOR HEATING IN THE WINTER (JAIN AND TIWARI, 2002, 2003)

A ground air collector with efficient design can be effectively used for heating the greenhouse in winter season (Figure 5.2a). The optimum area of ground air collector, mass flow rate, and heat capacity were obtained as $17.55\,m^2$, 200 kg/h, and 20,950 kJ/°C, respectively, for the given size and shape of the greenhouse and climatic condition. The stored thermal energy of ground collector gives the increase in temperature of 6°C–7°C from the ambient during night (Figure 5.2b).

5.4.2   A GREENHOUSE WITH EVAPORATIVE COOLING PAD FOR COOLING IN THE SUMMER (JAIN AND TIWARI, 2003)

A thermal model was developed for a greenhouse attached with a cooling pad. The greenhouse was considered in two zones i.e., above the cooling pad level and cooling pad level representing in zone-I and zone-II, respectively (Figure 5.3a). The cooling system parameters like length of the greenhouse (for effect of cooling length), height of the cooling pad, and mass flow rate were optimized against the maximum temperature and thermal load leveling in the zone-I and zone-II. The average temperature in zone-II is 4°C–5°C less than the ambient temperature (Figure 5.3b). The temperature in the greenhouse increased along the length of the greenhouse due to the receiving solar incidence. The optimum parameters of cooling system were (i) length of the greenhouse as 6 m, (ii) mass flow rate as 0.6 kg/s, and (iii) height of the cooling pad as 1.75 m (with width of 3 m) for the given size and shape of the greenhouse and climatic condition.

(a)

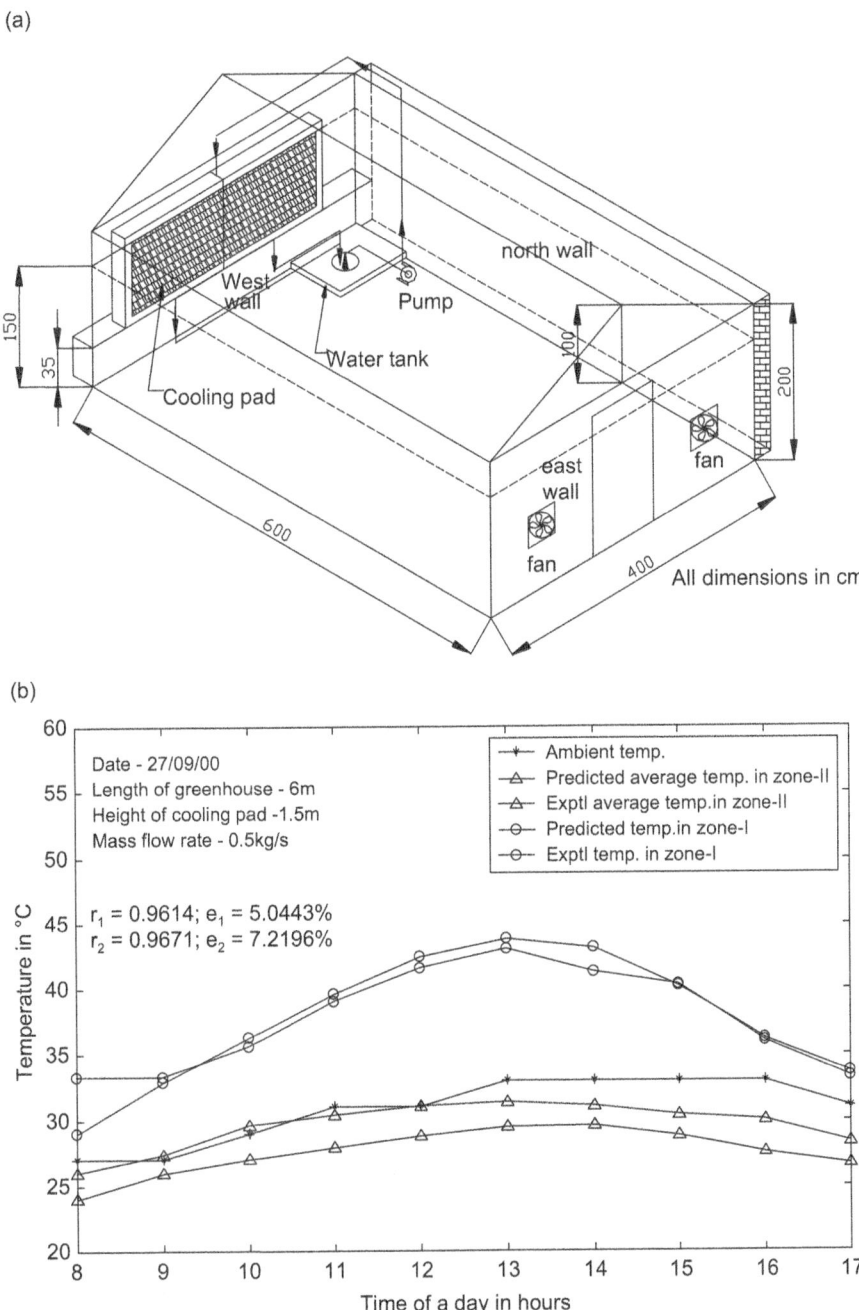

FIGURE 5.3 (a) Greenhouse attached with cooling pad and fan arrangement. (b) Hourly variation of room and ambient air temperature after evaporative cooling.

5.4.3   GREENHOUSE ATTACHED WITH TRAY-TYPE
CROP DRYING SYSTEM (JAIN, 2005)

The performance of an even shape greenhouse with a packed bed and crop dryer was evaluated for drying of onions (Figure 5.4a). The parametric study involved the effect of length and breadth of the greenhouse and mass flow rate of air on the temperatures of the crop (Figure 5.4b). It has been observed that the crop moisture

(a)

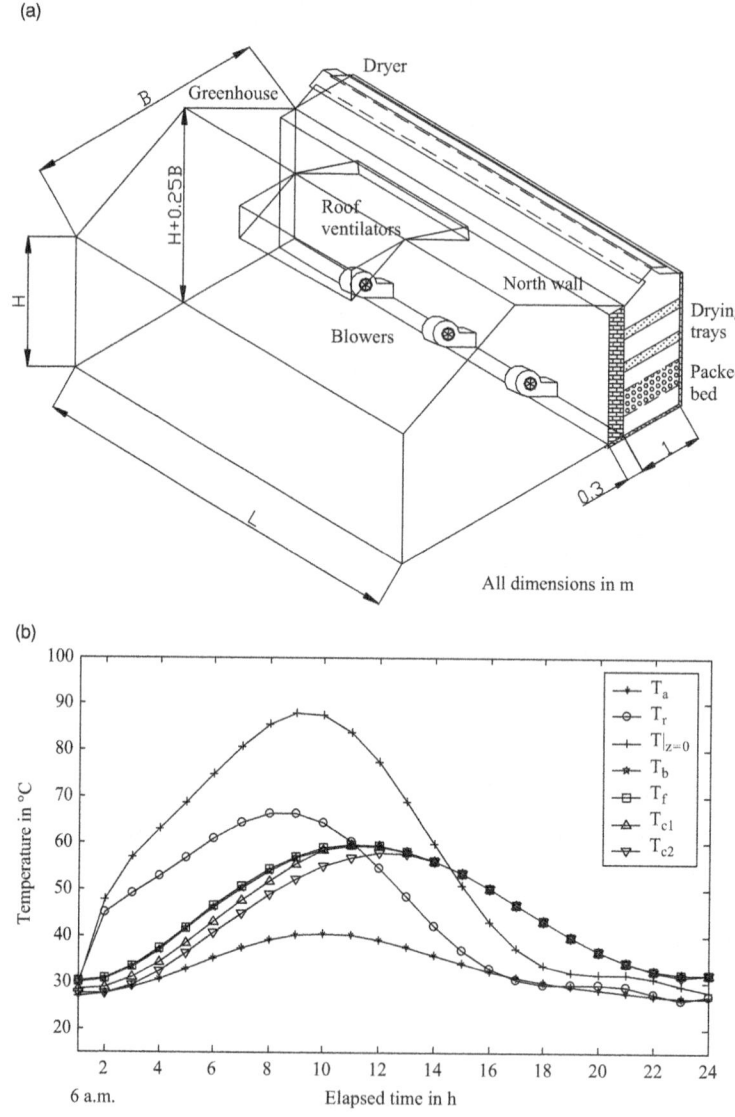

(b)

FIGURE 5.4   (a) Greenhouse attached with north wall, tray-dryer, and packed bed. (b) Hourly variation of temperature of various functional components and different stages of air of the greenhouse-based crop drying system for $L = 6\,\text{m}$, $B = 4\,\text{m}$, $\dot{m}_a = 0.167\,\text{kg/s}$.

content and drying rate decreases with the drying time of the day. A greenhouse of 6 m length, 4 m breadth, with a 0.278 kg/s air mass flow rate with a 0.25 m height of packed bed could dry 2,280 kg of onion from a moisture content of 6.14 to 0.21 kg water/kg of dry matter in a 24 h drying period.

5.5 AN AQUACULTURE GREENHOUSE SYSTEM (JAIN, 2007)

A simple trapezoidal design of aquaculture pond was proposed to attach with greenhouse (Figure 5.5a). Parametric studies involved the effects of length, breadth, depth, inclination of lining of fishpond, and air change in the greenhouse on the water

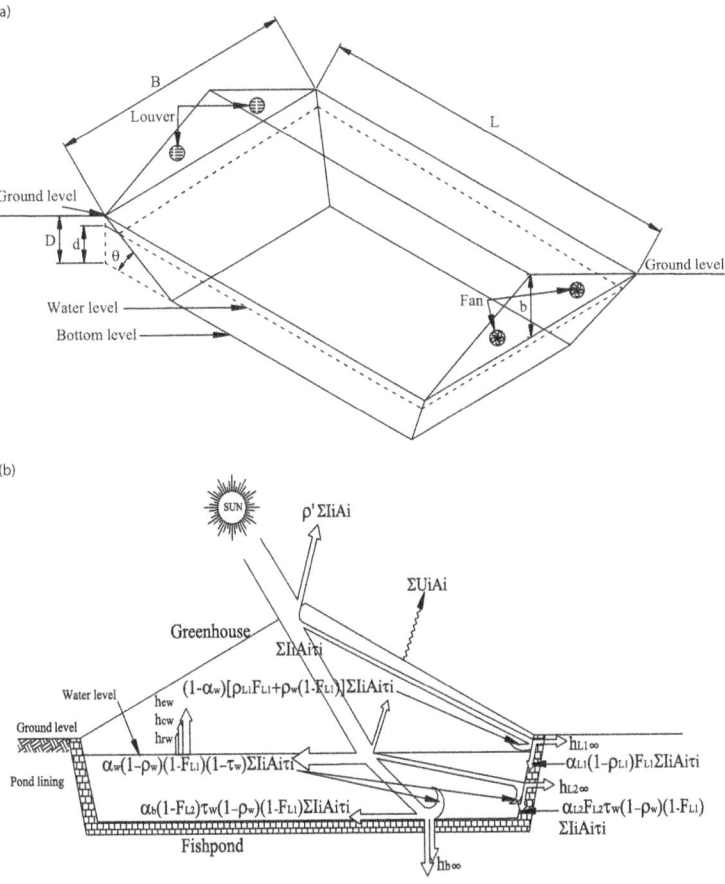

FIGURE 5.5 (a) Isometric view of fishpond-greenhouse system showing the various parameters of optimization. (b) Fishpond heating with greenhouse showing distribution of solar radiation and various heat transfer coefficients. (c) Hourly variation of air and water temperature on optimum parameters of fishpond-greenhouse system at $L = 30$ m, $B = 16$ m, $D = 1.25$ m, $d_w = 1$ m, $\theta = 75°$ and $N = 8$.

(Continued)

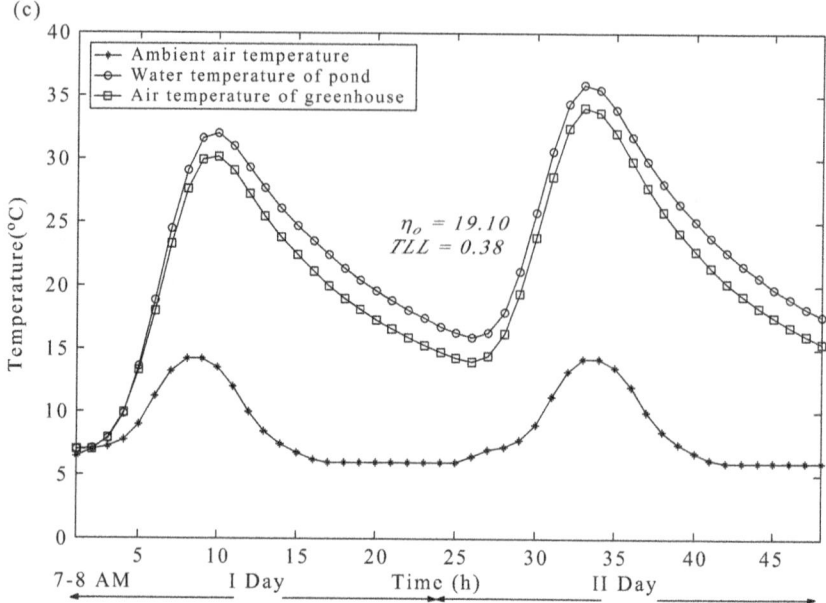

FIGURE 5.5 (CONTINUED)   (a) Isometric view of fishpond-greenhouse system showing the various parameters of optimization. (b) Fishpond heating with greenhouse showing distribution of solar radiation and various heat transfer coefficients. (c) Hourly variation of air and water temperature on optimum parameters of fishpond-greenhouse system at $L = 30\,\text{m}$, $B = 16\,\text{m}$, $D = 1.25\,\text{m}$, $d_w = 1\,\text{m}$, $\theta = 75°$ and $N = 8$.

heating in the fishpond. The performance of the fishpond was assessed in terms of temperature gain, mean thermal efficiency, and thermal load leveling. The optimum parameters for the fishpond were 30 m length, 16 m breadth, 1.25 m depth, 1.0 m water depth, 75° lining inclination, and 8 number of air changes per hour for maximum temperature gain, maximum thermal efficiency, and minimum thermal load leveling (Figures 5.5a and b). A 20°C increase in water temperature could be achieved during the day and 11°C in the night in the month of January (Figure 5.5c). The fishpond with greenhouse system can provide the favorable water temperature from 16°C to 35°C against 5°C–15°C temperature of ambient air for prawn fish farming during the winter season. The parametric studies revealed that 1 m depth of water with 0.25 m depth of freeboard is enough to get the optimum water temperature. The thermal efficiency of the fishpond-greenhouse system can be obtained as 19.1%.

5.6   CONCLUSION

Thermal modeling of agricultural greenhouse is an effective tool to predict the greenhouse environment and design the operative parameters for various applications. The research area in the greenhouse modeling for Indian scenario is still virgin in many aspects and needs to be properly studied for variable environment and regions, since it has a great scope in the country.

REFERENCES

Jain, D. and Tiwari, G.N. 2002. Modeling and optimal design of evaporative cooling system in controlled environment greenhouse. *Energy Conversion and Management*, 43(16), pp. 2235–2250.

Jain, D. and Tiwari, G.N. 2003. Modeling and optimal design of ground air collector for heating in controlled environment greenhouse. *Energy Conversion and Management*, 44(8), pp. 1357–1372.

Jain, D. 2005. Modeling the performance of greenhouse with packed bed thermal storage on crop drying application. *Journal of Food Engineering*, 71(2), pp. 170–178.

Jain, D. 2007. Modeling the thermal performance of an aquaculture pond heating with greenhouse. *Building and Environment*, 42(2), pp. 557–565.

Taki, M., Rohani, A. and Rahmati-Joneidabad, M. 2018. Solar thermal simulation and applications in greenhouse. *Information Processing in Agriculture*, 5, pp. 83–113.

Tiwari, G.N., 2002. Solar energy: fundamentals, design, modelling and applications. Alpha Science Int'l Ltd.

6 Development of Agriculture under Climate and Environmental Changes in the Brazilian Semiarid

Juliana Espada Lichston, Emile Rocha de Lima, and Magda Maria Guilhermino
Federal University of Rio Grande do Norte

Renato Dantas Alencar
Federal Institute of Rio Grande do Norte

CONTENTS

6.1 CHARACTERIZATION OF THE BRAZILIAN NORTHEAST SEMIARID REGION

The main ecosystem of the Brazilian Northeast is the semiarid biome, Caatinga, which occupies the equivalent of 11% of the national territory, encompassing the states of Alagoas, Bahia, Ceará, Maranhão, Pernambuco, Paraíba, Rio Grande do Norte, Piauí, Sergipe, and the North of Minas Gerais (Figure 6.1), totaling about 844.453 square kilometers of territory (Brasil, 2018).

The Northeastern Semiarid region has, like its main feature, the frequent droughts that can be characterized by scarcity and high spatial and temporal variability of

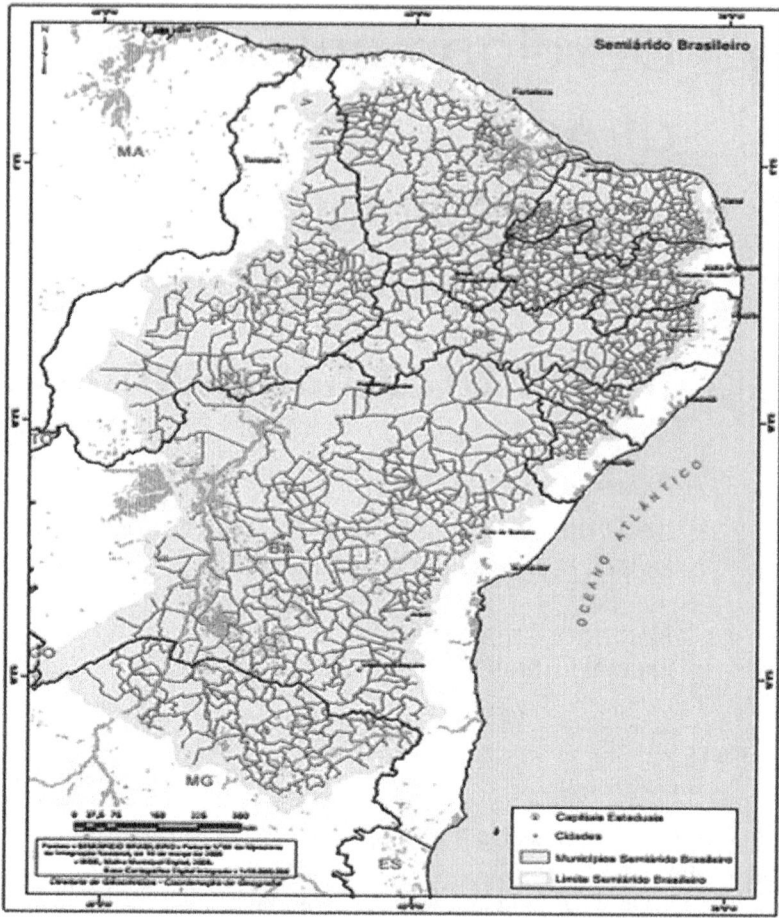

FIGURE 6.1   Delimitation of the Brazilian semiarid.

rainfall. The succession of years with drought is not rare (Sudene, 2018). Brazil is among the top ten countries in the world with regard to water availability; however, the distribution of these water resources is not uniform. The Northeast region has only 3% of the total volume of available water in the country (Montenegro and Ragab, 2012), facing water deficit during at least 70% of the year. In the northern parts of the Brazil Northeast region, events such as El Niño still inhibit cloud formation and reduce precipitation, causing even more severe droughts (Marengo et al., 2018). The climate in this region called the "Polygon of Droughts" is semiarid, hot, with low rainfall (between 250 and 800 mm annually), marked by two well-defined seasons, one dry and hot, and one dry and humid, with high evaporation, strong sunshine, and shallow and stony soil (Giulietti et al., 2004). In addition to high temperatures and low rainfall, the Brazilian semiarid region is still characterized by low phytomass production and poor soils with low regeneration rate (Sousa et al., 2012). The local vegetation can be described as arboreal or shrub forest, mainly containing low shrubs, many of them with spines, small leaves, and some other xerophytic characteristics (Leal et al., 2003). Rich in biodiversity, the biome has 178 mammal species, 591 birds, 177 reptiles, 79 amphibian species, 241 fish, and 221 bees. The Caatinga has an immense potential for the conservation of environmental services, sustainable use, and bioprospection that, if it's well explored, will be decisive for the development of the region and the country. About 22.5 million people currently live in the original area of the Caatinga, and most of their original ecosystems have been altered already, mainly through deforestation and burning, and despite being an exclusively Brazilian biome, it is one of the most threatened in the country (Marengo et al., 2018).

Among the economic activities developed in the region, considering the value of gross domestic products (GDP) in the semiarid region, the agricultural sector is one of the most outstanding sectors (INSA, 2014). In total, 82.6% of the rural labor force is equivalent to family farming. Droughts in Brazil, especially in the semiarid region, are the phenomenon that mostly affects costs due to its impacts on agriculture. An example of this was the drought of 2012/2013, which resulted in losses of US$ 1.6 billion to the economy, with the eight most important crops affected (beans, rice, corn, cotton, banana, sugar cane, soybean, and coffee) (Cunha et al., 2015).

6.2   AGRICULTURE IN THE BRAZILIAN NORTHEASTERN IN FRONT OF CLIMATE CHANGE AND SOIL USAGE

6.2.1   Diagnosis of Agriculture in the Region and its Adversities

The Northeast region of Brazil is the largest national banana producer, accounting for 34% of the total production. It also manages the production of cassava, with 34.7% of the total, and is the second largest producer of rice. It also ranks second in terms of fruit production, with about 27% of the national production in 2008. However, the share of Northeastern agricultural production in the whole country is still low. In 1995, of the total amount of Brazilian agribusiness, the region contributed only 13.6%. Sugarcane is the main agricultural product of the region, and cotton, soya, cashew, grape, mango, melon, and other fruits are also important for domestic consumption and export. In the Northeastern backwoods, subsistence agriculture predominates, sometimes affected by drought (IPEA, 2012).

Much of the agricultural activities in the Brazilian Northeast are developed on a fragile ecosystem, with limitations in edaphoclimatic order. In addition to climatic adversity affecting the agriculture, the anthropogenic actions do not help to mitigate the negative effects on production, especially in years with water scarcity (IPEA, 2012). Over time, there was no concern with the establishment of an infrastructure to support agroforestry activities and living with the semiarid climate (de Oliveira et al., 2009).

Agriculture itself is an activity that impacts the environment, whether on a large or small scale. This will depend on the techniques and practices that are used to cultivate the land. In many semiarid localities, the degradation of the environment begins with inefficient agricultural practices that remove the original vegetation covering the soil, leaving it vulnerable to erosive processes (Brasileiro, 2009).

The progress of environmental degradation in the region is due to several factors, including: inadequate agricultural practices, deforestation, soil infertility and compaction, erosion processes, salinization, cutting and burning, in addition to overgrazing, which reduces soil porosity and decreases water infiltration rates (Brasileiro, 2009; Sousa et al., 2012).

Soils in the biome are mostly nutrient-poor and have little physical structure to support agricultural activities. Soil degradation through a series of physical, chemical, and hydrological processes causes the destruction of the biological potential of the land (IPEA, 2012). Continuous cultivation, with the withdrawal of agricultural products and without the replacement of nutrients, leads to the loss of fertility. In irrigated areas, the use of water with high levels of salts, the absence of drainage, and inadequate irrigation, lead to salinization. Failure to leach due to lack of rainfall further intensifies the accumulation of salts (Leal et al., 2003). However, erosion is considered the main factor in the loss of arable land, low crop yields, and silting of rivers and reservoirs, with severe damage to crop productivity and the environment (Perez-Marin et al., 2013).

As a consequence of this scenario, in the last decades large areas of the semiarid region have been degraded and in many cases, affected by an advanced stage of desertification (Sousa et al., 2012) (Figure 6.2).

Susceptible areas to desertification in Brazil cover the semiarid tropic, dry sub-humid, and surrounding areas, occupying around 1,340,000 km² and directly reaching 30 million people. Of this total, about 180,000 km² are already in serious and very serious desertification, concentrated mainly in the states of the Northeast, which have 55.25% of their territory reached in different degrees of environmental deterioration.

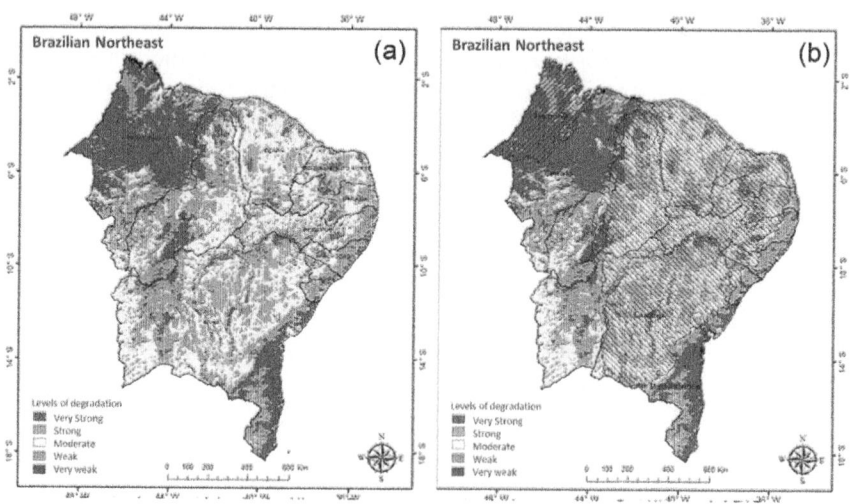

FIGURE 6.2 Desertification in the Brazilian Northeast. (a) Distribution in the states; (b) prevailing biomes. (Laboratory of Analysis and Processing of Satellite Images, LAPIS – adapted.)

Among the regions affected by desertification, four were characterized as high risk to desertification (Figure 6.3), and became known as Desertification Nucleus of Gilbués (PI), Irauçuba (CE), Seridó (PB), and Cabrobó (PE).

In these four areas, it was verified that the anthropogenic factor for the intense degradation, in general, was the replacement of the Caatinga by agriculture, livestock, and logging practices for the production of firewood and coal.

Desertification is accelerated by the human actions, through the use of inappropriate practices, bringing harmful consequences to the land and to those who take away their livelihood, exploiting the land intensively until the exhaustion of their natural fertility, without any practice of restoring this fertility and soil preservation (de Oliveira et al., 2009). This process frequently begins with the deforestation and the replacement of native vegetation by another cultivation, of different size and/or life cycle. Thus, the predominant semiarid vegetation of the Caatinga is replaced by the herbaceous grasses or short cycle crops (Perez-Marin et al., 2013).

With the advance of the desertification process, every year farmers lose their arable land together with a drought that has plagued the Seridó region for more than 4 years, increasing the rural exodus. The plantations that still can be made are located in the floodplain areas or even in the beds of temporary rivers silted and without riparian vegetation. Perennial rivers still lend themselves to irrigation by inefficient methods of crops unsuitable for the semiarid climate, such as elephant-grass for feeding dairy cows benefiting only those on the banks of the rivers.

Combating desertification and soil degradation are based on some axes and include preservation, conservation, and sustainable management of natural resources (de Oliveira et al., 2009), however, most agricultural establishments in the region

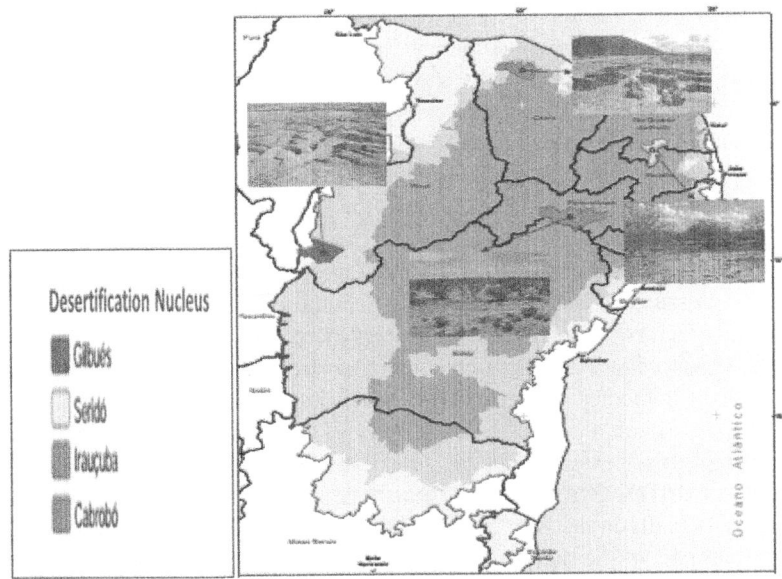

FIGURE 6.3 Areas with high risk of desertification in the northeast.

do not use any recommended agricultural practices to contain land degradation. Of the approximately 2,454,006 Northeastern agricultural establishments, almost half do not use any of the recommended practices and 25% use fires that do not help the local environment (IPEA, 2012).

6.2.2 FUTURE SCENARIOS

The Northeast region is naturally characterized like having high potential for water evaporation due to the enormous availability of solar energy and high temperatures. Increases in temperature associated with climate change due to global warming, regardless of what might occur with rainfall, would already be enough to cause greater evaporation of lakes, reservoirs, and greater evaporative demand of plants, that is, water will become a scarce commodity (Marengo et al., 2012).

Climate change prediction data and the South American behavior from Intergovernmental Panel on Climate Change (IPCC) (IPCC, 2001) predict a temperature increase for the entire continent and the projections indicate desertification of the Brazilian Northeastern semiarid in the scenarios for 2091–2100, resulting in significant losses of biodiversity (Pellegrino et al., 2007).

Temperatures may rise from 2°C to 5°C and precipitation may increase slightly by the end of the 21st century. Despite the positive anomalies of rainfall, it is expected that the "aridization" of the Northeast will occur due to the high temperatures causing an increase in evapotranspiration and a decrease in water availability in the region. Groundwater recharge is expected to decline by more than 70% by 2050 as a consequence of the changes in the climate (de Araújo Filho et al., 2002; Brasil, 2018; Montenegro and Ragab, 2012). The Caatinga can become a vegetation of arid zones, irreversibly, with predominance of xerophytes.

Climate change in Brazil threatens to intensify the difficulties of access to water. The combination of changes in climate in the form of lack of rainfall or low rainfall, accompanied by high temperatures and high rates of evaporation and competition for water resources, can lead to a potentially catastrophic crisis (de Medeiros et al., 2010; Santana, 2007; Marengo et al., 2018).

Agriculture is one of the economic activities that will be most affected by the impacts of climate change. Areas suitable for subsistence crops will be reduced, irrigated crops will need more water, space for agribusiness will be reduced, and crop productivity in general will be affected. Desertified areas will be abandoned, increasing pressure on marginal lands (II Seminário Sobre Mudanças climáticas, 2008). Other factors expected under the stress created in agricultural systems are a reduction of water flow and irrigation potential, increased incidence of pests and diseases, changes in biomes, and decrease in biodiversity of animals and plants (World Bank, 2013).

The preliminary assessment of the International Fund for Agricultural Development (FIDA, 2016) presents future cassava production as one of the main concerns, especially in the Northeast, where culture is a fundamental "anchor" and an important part of the regional culture. Several analyzed crops are vulnerable to dramatic temperature and precipitation oscillations. However, traditional crops may offer alternatives to compensate for the losses of other crops.

Despite the impact of climate change on the whole agricultural sector, family farmers are the most threatened agents and will need more support to adapt. Although it's a country with a high concentration of land, in Brazil, family agriculture is responsible for much of the national production. As a result, the losses caused by these events will not only affect the food security of the farmers directly dependent on this activity, but also the other consumers indirectly dependent on the cultivated products (FIDA, 2016).

To maintain the level of production, farmers commonly adjust planting dates, crop variety, stock rates, and water use, among other factors, in response to small climatic variations. In addition, as climate change becomes more severe, long-term adaptations should increase in importance. These adaptations not only include those directly relevant to agriculture: new technologies and management practices, increased use efficiency and water distribution systems, change in practices (fertilization, farming methods, irrigation), but also include public investments, policies, and other strategies that encourage adaptations (LEE et al., 2014).

The recovery of lost and compacted soils in areas that should be permanently protected, such as riparian forests, legal, and/or productive reserve areas and that are unproductive and degraded, is urgent. It is a sine qua non condition to recover the degraded areas, to recover the riparian forests, avoiding the silting of the water bodies and allowing the Caatinga to recover if we want future generations to inhabit it.

Climate change and its impacts will not occur immediately, although some effects may already be perceived. It is necessary to admit this reality to the donation of practices, habits, technologies, and investments in several sectors, and in particular in agriculture, one of the main bases of the economy in the Northeastern region of Brazil. With prior planning and innovative techniques, it is possible to reduce vulnerabilities and build resilience. In order to mitigate the effects of climate change, replanting and reforestation actions and the intensive fight against deforestation will be necessary, together with the adoption of agroecology techniques for economic production and biome recovery.

Investing in the sustainable development of family farming is an essential way to build a safer and fairer planet. The United Nations (UN) and Food and Agriculture Organization (FAO) in the International Year of Family Agriculture in 2014 emphasized the importance of family and small-scale agriculture for global food security and for the preservation of traditional foods, the adoption of agroecology, and the protection of agrobiodiversity, consequently, its role in mitigating global warming and climate change.

6.3 FAMILY FARMING AND THE USE OF TECHNIQUES AND PRACTICES FOR AGROECOLOGICAL DEVELOPMENT

The definition of family farming around the world varies according to the understanding and valorization of each country, with a broader range and categories of public policies offered to family agriculture in countries that base their agricultural economy on family production, such as European countries, the United States, and New Zealand. The differentiation of family farmers is associated with the formation of groups throughout history, their varied cultural heritages, their particular professional and life experience, the access to and the differentiated availability of a set of factors, including education and training, availability of natural resources, agricultural landscapes, human and social capital, differentiated access to markets, and the socioeconomic insertion of producers and their income and wealth generation capacities. The differentiated universe of family farmers is composed of groups with particular interests, their own survival and production strategies, which react differently to similar challenges, opportunities and constraints, and therefore demand treatment compatible with differences (Buainain et al., 2014).

In Brazil, family farming, until recently known as subsistence agriculture, was only recognized with the creation of the National Program for the Strengthening of Family Agriculture (PRONAF) in 1996, and only 10 years later in 2006, when the Family Agriculture Law was enacted, Law 11.326, July 24, 2006, which considers family farmers as those who practice activities in rural areas and who simultaneously meet the following requirements: (i) do not have an area greater than 4 fiscal modules; (ii) predominantly use family labor in the economic activities of its establishment; (iii) have family income predominantly from the economic activities linked to the establishment; (iv) run their establishment with their family. In Brazil, there are 4,366,267 family farming establishments, occupying 30.5% of the total area of rural establishments and representing 77% of jobs in the agricultural sector and producing 70% of the food consumed by Brazilians.

Family farming in the Northeast covers more than half of the family farming establishments in Brazil, with less than 35% of agricultural land in a semiarid environment.

Family agriculture plays a leading role in the adoption of agroecological production and conservation techniques for the recovery and preservation of the environment. The adoption of agroforestry systems is one of the strategies that must be adopted in a climate change scenario, and consists in the use of integrated farming systems with shrubs and trees (logging, fruit, honey, and medicinal), providing environmental, economic, and social advantages for the farmers and their properties (Arco-Verde and Amaro, 2012). These systems should encourage the creation of, for example, poultry, goats, sheep, pigs, and the breeding of bees and earthworms, which will assist in the cycling of nutrients and increase productivity in function of the environmental services that these animals promote. These systems aim to increase production in a continuous way and in the same area, using sustainable management practices compatible with the culture of the local population (ALTIERI, 2002), besides helping to conserve the environment, especially in areas subject to degradation.

In these agroforestry systems, the tree species used should preferably be native or naturalized species, which usually have multiple functions in the system, related to obtaining wood, fodder, fruit, medicinal, and honey, as well as having a craftsmanship and ornamental function. Pruning is an important practice for the implantation and management of these systems and the material resulting from pruning should be used in the area for the preparation of other techniques or as a cover to the environment of the plants.

Another set of technologies that can contribute to the development of family agriculture is the construction of low cost infrastructures for the collection and storage of rainwater, strategies for the greater infiltration of rainwater into soils, and the storage of these waters for use in the periods of drought, among them, we highlight: successive barrage of containment of sediments, subterranean dam, cauldron or tank of stone and cisterns.

The **successive barrage of sediment containment** are built with stones, roman arch shaped arrays lying down, in streams. The goal is the retention of the sediments generated by the leaching process, as a consequence of the misuse of the soil (de Oliveira et al. 2010).

The **underground dam** is a water structure that aims to intercept the flow of surface and underground water through an impermeable septum (plastic canvas, stone wall, or compacted clay, etc.), which serves as a technological alternative for the use of rainwater, avoiding to drain on the soil surface, where they can cause erosion. The rainwater in the underground dams is stored in the soil profile, in order to allow the creation or elevation of the existing water table, allowing the exploitation of an effluent agriculture.

The **cauldron or stone tank** is a technology that stores rainwater in the crevices of rocky outcrops. These cavities are cleaned to form a water reservoir. The trench-shaped cauldrons, with a small width, and great length and depth, present better conditions and efficiency of water storage for longer periods, due to the reduced area of evaporation (Brito, 2006).

The **cisterns** are rainwater reservoirs, which can be supplied with roofing water for human supply and were widespread in Brazil. The cisterns called "sidewalk cisterns" can also be constructed with the objective of obtaining water for agricultural and animal production, which in these cases have larger sizes and the waters are captured by means of a cement board built on the ground (cisterns boardwalk) or in the course of small streams (cistern-runoff).

6.3.1 Applications of Agroecological Techniques – Case Study

Agroecological technologies for the reclamation of degraded soils (frogs, bushes, vegetation structures from pruning and/or thinning and lowering of trees) are efficient and effective techniques in the fight against desertification, of very low cost, and can be done manually and in a collective way. In order to contribute to the sustainable development of family agriculture and the recovery of the Caatinga biome, the EMATER Terra Viva project in line with the desertification mitigation policy proposed by the Rio Grande do Norte Technical Assistance and Extension Institute (EMATER-RN), in partnership with the Ministry of Science and Technology, in 2007. In order to reverse the desertification situation and the rural exodus, it was proposed in a collective effort among the family farmers of the Seridó region, the implantation of agroecological technologies (roasting and bale bushes, vegetation enleiraments, thinning and rearing of Caatinga) allow the recuperation and recomposition of the soils for the agrosystems' recomposition.

The family farmers participated in joint efforts to implement the agroecological actions recommended and motivated by EMATER. The descriptions adopted here on frogs and banding bars followed the recommendations in Lira Santos et al. (2007) and the rearing and thinning techniques of the Caatinga followed the recommendations in de Araújo Filho et al. (2002).

Renques (Figure 6.4) are vegetated and/or stone contour strings in degraded areas or in erosion and desertification processes, in lands that were agricultural, in strategic places for soil containment and reduction of flood velocities.

The vegetation structure changes are made with material from pruning and/or retraction of Caatinga trees present in rural areas. This pruning aims to remove branches that cause energy damage to the plant, thus improving its nutrient supply in the time of water shortage. The material coming from this pruning is grouped in the form of cords in contour (renques) or placed in the "foot of the trees" to maintain the humidity and to improve the contribution of organic matter in the soil.

The bore bushings (Figure 6.5) are made in the beds of water bodies, streams of micro basins or paths of floods that had high slope and physical conditions suitable for the mooring of the ends of the bush as well as the nearby existence of a place with rocks of quality and quantity, preferably, loose. As for storage, the stones should be arranged so that they are well joined together, forming transverse walls in the river beds. The first layers placed at the base of the creek must be of larger stones and carefully arranged. The stones that make up the walls of the upstream and downstream slopes, must also be the largest available near the work, and the "core" of the bush should be filled with stones of smaller size.

The **thinning** (Figure 6.6) of the arboreal-shrub vegetation of the Caatinga according to Pereira Filho (2013), consists of the selective control of woody species so that with the reduction of the density of undesirable trees and shrubs, and also shading, around 30%–40%, allowing the penetration of the solar rays, so that when

FIGURE 6.4 Renques.

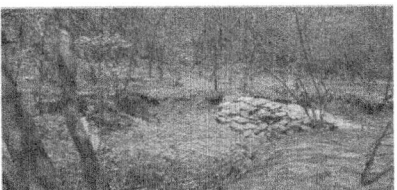

FIGURE 6.5   Examples of boring brushings.

FIGURE 6.6   Example of thinning – Trangola Settlement, in Seridó Region, Currais Novos, RN, Brazil. Photos by M. Guilhermino, 2015.

the rains begin, the seeds of the herbaceous plants germinate and they develop satisfactorily. Also according to the same author, the thinning transforms the biomass of shrubs and trees to the herbaceous stratum, increasing considerably the dry matter originating from grasses and herbaceous dicotyledons, which allows the creation of ruminants in a sustainable way, conserving the Caatinga. However, it is important to emphasize that the success of thinning will depend on the degree of occupation of the area by trees and shrubs, the stage time, rainfall conditions, and the topographic location of the ecological site.

The lowering of the woody plants (Figure 6.7) of the Caatinga is a way to cut the woody species at a height of 30–40 cm from the soil (popularly called Broca) in the final third of the dry season, so that in the beginning of the rains the plants

February 2016   September 2015   *M. Guilhermino, 2016*

FIGURE 6.7   Example of lowering.

can regrow, making available green mass of shrubs and trees which otherwise would be unavailable for the animals to reach, especially for the goats. In this way, the goats can graze without damaging the trees and without exhausting the vegetation.

In the following year from the implantation of the branches and bushes, it was observed in relation to the mitigation of the desertification process that most of the bore bushes were efficient in the retention of soil and organic matter, avoiding the silting of springs.

With the rains and the leaching process, there was also accumulation of organic matter rich in seeds, upstream of the rails and along the ravines, where spontaneous revegetation can be noticed due to the humidity and soil fixed there. Experience in the construction of these technologies allowed the farming families to develop a level of environmental awareness related to the issues of desertification and deforestation, recovery of the Caatinga biome, importance of trees, as well as issues related to the rational use of soil and water sources. But, above all, learning the collective work, transforming the bureaucratic figure of the "Rural Workers' Association", a first condition to participate in the TERRA VIVA project, in a living and effective experience, giving new hopes and directions to the rural communities leading them to paths such as the diversification of agricultural activities, collective actions in the production of sweets, cakes, and handicrafts, actions with children in the making of seedlings of native plants that will be used in the reforestation of the areas and in the organization of agroecological trails for rural tourism, which help to supplement the income of families.

The implementation of barbells and renques require monitoring and subsequent actions, such as, if necessary, increasing the wall of the bush and the rows so that the ground is retained and does not exceed the installations over time.

In areas where the woody vegetation has been manipulated, it is notorious to increase the litter and herbaceous stratum and treetops, generating a pleasant climate in these areas.

This work, despite its importance and its success in adopting technologies, did not go as expected because there was no continuity of the work, that is, there was neither any follow-up of the evolution of the herbaceous plants and bushes nor the construction of others in the degraded areas. This is because the biome recovery process takes time, and farmers have to leave their family units in order to work to make a better income for themselves and their families, since their land no longer bears economic agricultural activities due to the exhaustion of resources, desertification process, and timid public policies that do not meet the real demand of family agriculture.

Aware of this problem, the Multidisciplinary Group for Studies in Public Policies for Sustainable Development of Family Agriculture, Agroecology and Recovery of the Caatinga Biome (GEPARN from the Jundiaí Agricultural School – EAJ at the Federal University of Rio Grande do Norte – UFRN), in partnership with the Research Group CAATINGUEIROS (Agroecology and Sustainable Development of the Semiarid Region)), proposes the construction of a public policy (PP) entitled "Defeso da Caatinga". The PP was poposed based on Agenda 21, the recommendations of the Regional Conference on Sustainable Development of the Caatinga – Caatinga Rio + 20 and based on the CATINGUEIROS scientific knowledge and expertise in the various areas of knowledge. The "Defeso da Caatinga" provides subsidies for family farmers to work on the recovery of the Caatinga biome from the implantation of agroecological techniques of soil conservation, such as rows and banding, vegetation material dressing, Caatinga thinning, lowering and monitoring according to the recommendations of the EMATER and for the ecological restoration of the areas of environmental protection (APP) and legal reserve (RL). All the actions are according to the requirements of the New Forest Code, Federal Law No. 12.651, of May 25, 2012 of the Brazilian federal government.

As a result of the implementation of CAATINGA's DEFESO PP (Figure 6.8), so that the family units can start agroecological production systems for the formation and/or complementation of household income, act on the conservation of natural resources (those resources that are used for life and for the production of water, soil, air, and vegetation), and ensure the preservation of APP and RL areas, the second PP proposal entitled "**ECO-TEC – Ecological Technologies for the Production and Conservation of Natural Resources**" follows. These technologies will be able to leverage production systems through techniques that optimize the use of natural resources and waste production.

The proposals presented here are based on the following assumptions, essential information for the development of educational actions, research and extension and construction of public policies aimed at the sustainable development of the rural

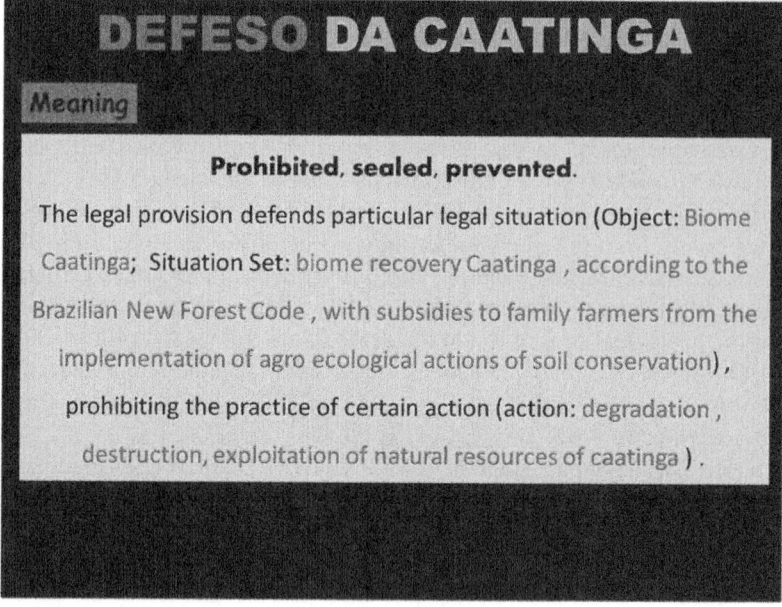

FIGURE 6.8   Public policy booklets Defeso da Caatinga.

family society, the recovery of the Caatinga biome and its rational use, ecological restoration of the areas of permanent protection and RL, are: Collectivity and Education – the incentive to collective actions is in the form of groupings, associativism, and cooperativism through education, training, and digital inclusion with the objective of empowerment and autonomy of the rural family society. These actions favor the collective strength and liberation through knowledge, education, and access to the local and distant regions and markets, characteristics that will favor the autonomy of the farmers, who are still very less articulated, often isolated in their rural area, far from the urban possibilities and still forgotten by the society in general. Recovery of the Caatinga Biome, Natural Resource Conservation, Rational Use of the Caatinga and Agroecology – the recovery of the Caatinga where the productive units of family agriculture are inserted is essential to the survival of the countryside and the city.

Ensuring the recovery of the soil, water, vegetation, fauna and, consequently, the self-esteem and love of the land of the farmer is a sine qua non for the development of family farming. The conservation of natural resources means the possibility of using soils, water, and vegetation through their rational use with agroecological techniques in function of the production and economic value of the products. Pluriactivity, slow food, non-agricultural rural activities – multiproduction that is able to guarantee biodiversity respecting the seasons of the year according to the oscillations of nature and consumer market are the characteristics that fit the climatic reality of the Caatinga and the condition of size and peculiarities of the family unit. Activities of different natures (agricultural and non-agricultural activities, such as rural tourism in its different forms as well as artezanato, through the valorization of local landscape features and biotic and abiotic elements) can contribute to the reward of uncertain production scale and for the composition of a decent income to the family farmer.

The recovery of the Caatinga Biome is urgent, and it is a necessity for all if we want future generations to inherit a decent planet. For this, family farmers are essential pieces in this reconstruction together with researchers and extensionists so that we can contribute and focus our efforts on the answers and needs of the recovery and ecological restoration work of the Caatinga Biome.

REFERENCES

Altieri, M. 2002. *Bases científicas para uma agricultura sustentável*. Guaíba: Agropecuária, 1 ed. p. 592.

de Araújo Filho, J.A., de Carvalho, F.C., Garcia, R. and de Sousa, R.A. 2002. Efeitos da manipulação da vegetação lenhosa sobre a produção e compartimentalização da fitomassa pastável de uma caatinga sucessional. Embrapa Caprinos e Ovinos-Artigo em periódico indexado (ALICE).

Arco-Verde, M.F. and Amaro, G. 2011. Cálculo de indicadores financeiros para sistemas agroflorestais. Embrapa Roraima.

Brasil. 2018. *Ministério do Meio Ambiente (MMA)*. Biomas: Caatinga. Brasília, 2012. Disponível em: http://www.mma.gov.br/biomas/caatinga. Accessed in: 05/09/2018.

Brasileiro, R.S. 2009. Alternativas de desenvolvimento sustentável no semiárido nordestino: da degradação à conservação. *Scientia Plena*, 5(5), pp. 1–12.

Brito, D. 2006. Metodologia para Seleção de Alternativas de Sistemas de Drenagem. 117f (Doctoral dissertation, Dissertação de Mestrado. Universidade Federal de Brasília), p. 117.

Buainain, A.M., ALVES, E.D.A., da Silveira, J.M. and Navarro, Z. 2014. *O mundo rural no Brasil do século 21: a formação de um novo padrão agrário e agrícola*. Brasília, DF: Embrapa.

Cunha, A.P.M., Alvalá, R.C., Nobre, C.A. and Carvalho, M.A. 2015. Monitoring vegetative drought dynamics in the Brazilian semiarid region. *Agricultural and Forest Meteorology*, 214, pp. 494–505.

FIDA, Fundo Internacional de Desenvolvimento Agrícola. 2016. Mudança do clima e os impactos na agricultura familiar no Norte e Nordeste do Brasil. Centro Internacional de Políticas para o Crescimento Inclusivo (IPC-IG) Programa das Nações Unidas para o Desenvolvimento.

Giulietti, A.M., Bocage Neta, A.L., Castro, A.A.J.F., Gamarra-Rojas, C.F.L., Sampaio, E.V.S.B., Virgínio, J.F., Queiroz, L.P., Figueiredo, M.A., Rodal, M.J.N., Barbosa, M.R.V. and Harley, R.M. 2004. Diagnóstico da vegetação nativa do bioma Caatinga. Biodiversidade da Caatinga: áreas e ações prioritárias para a conservação, pp. 48–90.

INSA, Intituto Nacional Do Semiárido. 2014. Abastecimento urbano de água: panorama para o semiárido brasileiro. Campina Grande: INSA.

IPCC, Intergovernmental Panel on Cimate Change. 2001. *Climate Change 2001: Impacts, Adaptation, and Vulnerability. Contribution of Working Group II to the Third Assessment Report of the Intergovernmental Panel on Climate Change.* Cambridge: Cambridge University Press, p. 1032.

IPEA, Instituto de Pesquisa Econômica Aplicada. 2012. *An agricultura no nordeste brasileiro: oportunidades e limitações ao desenvolvimento. Texto para discussão.* Brasília: Rio de Janeiro.

Leal, I.R., Vicente, A. and Tabarelli, M. 2003. *Herbivoria por caprinos na Caatinga da região de Xingó: uma análise preliminar. Ecologia e conservação da caatinga.* Recife: Ed. Universitária da UFPE, pp. 695–715.

Lee, D.R., Edmeades, S., De Nys, E., McDonald, A. and Janssen, W. 2014. Developing local adaptation strategies for climate change in agriculture: A priority-setting approach with application to Latin America. *Global Environmental Change*, 29, pp. 78–91.

Marengo, J., Nobre, C.A., Chou, S.C., Tomasella, J., Sampaio, G., Alves, L.M., Obregón, G.O., Soares, W.R., Betts, R., Kay, G. 2018. Riscos das mudanças climáticas no Brasil: análise conjunta Brasil-Reino Unido sobre os impactos das mudanças climáticas e do desmatamento na Amazônia. INPE. 2011. Disponível em: http://www.ccst.inpe.br. Accessed in: 01/09/2018.

Marengo, J.A., Tomasella, J. and Soares, W.R. 2012. Extreme climate events in the Amazon basin. Theoretical and Applied Climatology, 107, pp. 73–85. doi: 10.1007/s00704-011-0465-1.

de Medeiros, J.F., de Mesquita, L.X., de Oliveira, F.D.A. and Maracajá, P.B. 2010. Tecnologias para contenção de solo e água subterrânea: uma experiência de extensão rural na região do seridó-rn. *Informativo Técnico do Semiárido*, 1(1), pp. 13–27.

Montenegro, S. and Ragab, R. 2012. Impact of possible climate and land use changes in the semi arid regions: A case study from North Eastern Brazil. *Journal of Hydrology*, 434, pp. 55–68.

de Oliveira, E.M., dos Santos, M.J., de Araújo, L.E. and da Silva, D.F. 2009. Desertificação e seus impactos na região semi-árida do Estado da Paraíba Desertification and its impacts upon the semi-arid area of the State of Paraíba. *Ambiência*, 5(1), pp. 67–79.

de Oliveira, J.B., Alves, J. and França, F. 2010. Barragens sucessivas de contenção de sedimentos. Cartilhas temáticas tecnologias e práticas hidroambientais para convivência com o Semiárido, v. 1.

Pellegrino, G.Q., Assad, E.D. and Marin, F.R. 2007. Mudanças climáticas globais e a agricultura no Brasil. *Revista Multiciência*, 8, pp. 139–162.

Perez-Marin, A.M., Cavalcante, A.D.M.B., Medeiros, S.S.D., Tinôco, L.B.D.M. and Salcedo, I.H. 2013. Núcleos de desertificação do semiárido brasileiro: ocorrência natural ou antrópica? *Parcerias Estratégicas*, 17(34), pp. 87–106.

Santana, M.O. 2007. *Atlas das áreas susceptíveis à desertificação do Brasil*. Brasília: MMA/ SRH/UFPB.

Santos, S.C.L., Medeiros, J.F., Mesquita, L.X., Oliveira, F.A., Maracajá, P.B. 2007. Tecnologias para contenção de solo e água subterrânea: uma experiência de extensão rural na região do Seridó-RN. Informativo Técnico do Semiárido Grupo Verde de Agricultura Alternativa. Mossoró – Brasil, 1(1), pp. 13–27.

II Seminário Sobre Mudanças climáticas. 2008. Implicações Para O Nordeste. Carta de Fortaleza.

Sousa, F.P., Ferreira, T.O., Mendonça, E.S., Romero, R.E. and Oliveira, J.G.B., 2012. Carbon and nitrogen in degraded Brazilian semi-arid soils undergoing desertification. *Agriculture, Ecosystems & Environment*, 148, pp. 11–21.

Sudene Semiárido. 2018. Brasília, 2008. Disponível em: http://www.sudene.gov.br/ acessoainformacao/institucional/area-de-atuacao-da-sudene/semiarido. Accessed in: 05/08/2018.

World Bank. 2013. *Impacts of Climate Change on Brazilian Agriculture*. Washington, DC: World Bank Group.

7 Role of PGPR in Sustainable Agriculture under Changing Scenario of Climate Change

Gaurav Sood, Rajesh Kaushal,
and Shweta Gupta
Dr. Y.S. Parmar University of Horticulture and Forestry

CONTENTS

7.1 INTRODUCTION

The main problem hampering the crop growth and productivity in today's world is drought, which is the result of global climate change events and is estimated to have reduced cereal productivity by 9%–10% (Lesk et al., 2016). More than 50% of the cultivable lands by 2050 are going to face devastating drought consequences on plant growth (Vinocur & Altman, 2005). Plant and water relationships were affected by drought stress at both cellular and whole plant levels, resulting in various physiological complex processes and phenotypical responses in plants. Oxidative stress is generated within the sub-cellular compartments due to alleviated levels of reactive oxygen species (ROS). ROS consist of superoxide radical (O_2), hydrogen peroxide (H_2O_2), and the hydroxyl radical (OH), all of these affect the building components of the cell (lipids, proteins, carbohydrates, nucleic acids, etc.) and cause cell demise (Mittler, 2002). Therefore, there is an increased interest among the scientists in finding solutions to water-associated problems such as drought and its impacts on food security. Particularly, there is an utmost need to redress different solutions, which will improve the drought tolerance in crop plants, so as to satisfy the food requirement with the limited water resources in today's world (Mancosu et al., 2015). Crop productivity can be increased by inoculating the plants facing drought stress with plant growth promoting rhizobacteria (PGPR) (Ngumbi and Kloepper, 2016). The thin layer of soil immediately surrounding the plant roots is an extremely important and active area for root activity and metabolism and is known as rhizosphere. A large number of microorganisms such as bacteria, fungi, protozoa, and algae coexist in the rhizosphere. The microorganisms that colonize the rhizosphere can be classified according to their effects on plants and the way they interact with roots, some being pathogens whereas others trigger beneficial effects. The microorganisms inhabiting the rhizosphere and beneficial to plants are termed PGPR. Various species of bacteria like *Pseudomonas, Azospirillum, Azotobacter, Klebsiella, Enterobacter, Alcaligenes, Arthrobacter, Burkholderia, Bacillus,* and *Serratia* have been reported to enhance the plant growth. There are several PGPR inoculants currently commercialized that seem to promote growth through at least one mechanism; suppression of plant disease (Bioprotectants), improved nutrient acquisition (Biofertilizers), or phytohormone production (Biostimulants).

The biological control agents suppress plant disease through at least one mechanism; induction of systemic resistance and production of siderophores or antibiotics. Nitrogen fixing biofertilizers provide only a modest increase in the crop's nitrogen uptake. The elemental sulphur present in the soil must be transformed or oxidized into sulphate by the bacteria before it is available for plants. The inoculation of sulphur-oxidizing bacteria (*Thiobacillus*) onto the seeds of high S-demanding crops has proved to be quite successful in making sulphur more available for the plants (Vidyalakshmi et al., 2009). The rock phosphate is an approved source of phosphorus but its availability to plants is limited under most growing conditions. Phosphorus oxidizing bacteria help in making this phosphorus available to the plants. The phytohormones they produce include indole-acetic acid, cytokinins, gibberellins, and inhibitors of ethylene production. PGPR also helps in degrading the organic pollutants. *Azospirillum* sp. shows osmoadaptation and can survive under salinity/osmolarity due to the accumulation of compatible solutes. The bacteria like *P. fluorescens* can survive under dry conditions and hyperosmolarity. The use of PGPR offers an attractive way to supplement chemical fertilizer, pesticides, and supplements; most of the isolates result in a significant increase in the plant height, root length, and dry matter production of shoot and root of plants (Gupta et al., 2017; Sood et al., 2018a). PGPR acts as a component in integrated management systems in which reduced rates of agrochemicals and cultural control practices are also used as biocontrol agents. Such an integrated system could be used for transplanted vegetables to produce more vigorous transplants that would be tolerant to nematodes and other diseases for at least a few weeks after transplanting to the field. Thus, in the present scenario of climate change the effects of abiotic stress (drought stress) on crop plants can be reduced by the application of biofertilizers that can also act as bioprotectants and biostimulants.

7.2 HISTORICAL BACKGROUND OF PGPR

Inoculation of plants with beneficial bacteria can be traced back to centuries ago. From experience, the farmers knew that when they mixed the soil taken from a previous legume crop with the soil in which non-legumes were to be grown, yields often improved. By the end of the 19th century, the practice of mixing "naturally inoculated" soil with seeds became a recommended method of legume inoculation in the USA. A decade later, the first patent ("Nitragin") was registered for plant inoculation with *Rhizobium* sp. Eventually, the practice of legume inoculation with rhizobia became common. For almost 100 years, *Rhizobium* inoculants have been produced around the world, primarily by small companies. For the large majority of less developed countries in Asia, Africa, and Central and South America, inoculant technology has had no impact on the productivity of the family farm, because inoculants are not used or are of poor quality. Inoculation with non-symbiontic, associative rhizosphere bacteria, like *Azotobacter*, was used on a large scale in Russia in the 1930s and 1940s (Reddy, 2014). The practice had inconclusive results and was later abandoned. Interest in *Azotobacter* as an inoculant for agriculture has only recently been revived. Two major breakthroughs in plant inoculation technology occurred in the late 1970s:

i. *Azospirillum* was found to enhance non-legume plant growth by directly affecting plant metabolism.
ii. Biocontrol agents, mainly of the *Pseudomonas fluorescens* and *P. putida* groups, began to be intensively investigated.

In recent years, various other bacterial genera, such as *Bacillus*, *Flavobacterium*, *Acetobacter*, and several *Azospirillum* related microorganisms have also been evaluated. The immediate response to soil inoculation with associative, non-symbiotic plant growth promoting bacteria (PGPB) (but also for rhizobia) varies considerably depending on the bacteria, plant species, soil type, inoculant density, and environmental conditions.

7.3 MECHANISMS OF ACTION

A. Direct mechanisms:

It involves providing plants with the resources/nutrients that they lack, such as fixed Nitrogen, Iron, and Phosphorus, Manganese and production of Cytokinins, Gibberellins, Indoleacetic Acid, and Ethylene (Figures 7.1–7.3).

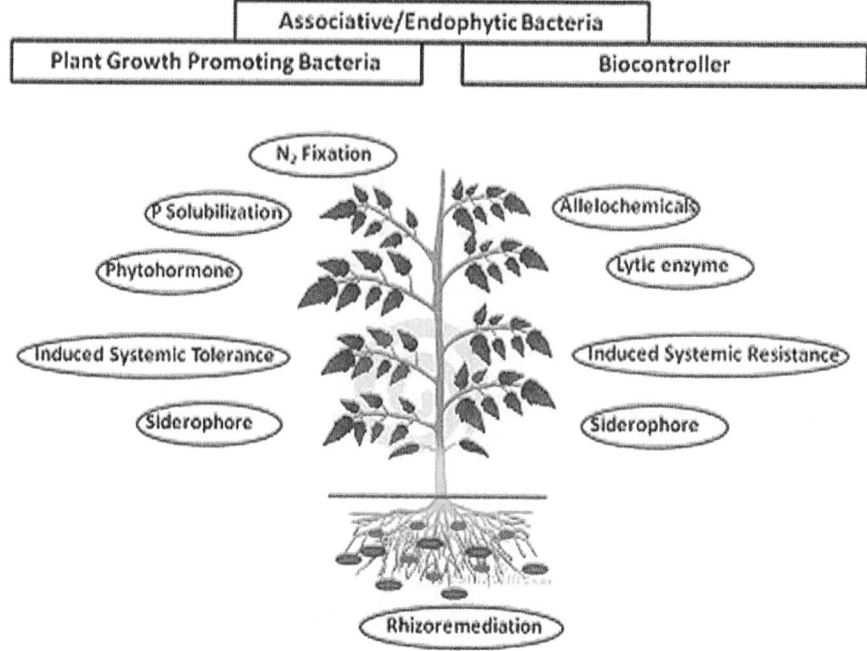

FIGURE 7.1   Mechanisms of PGPR action. (*Source*: Jha et al., 2013.)

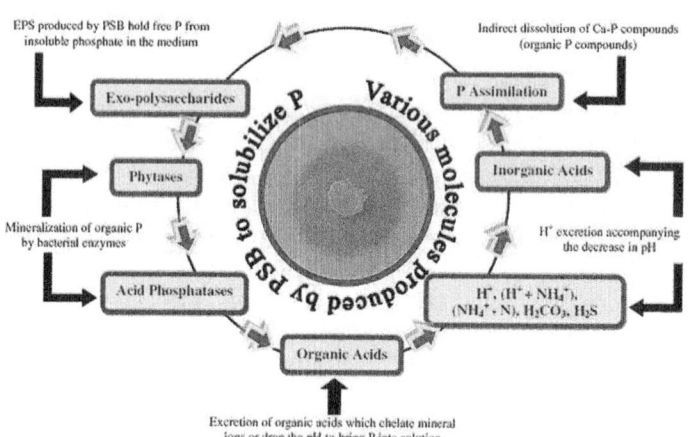

FIGURE 7.2   Phosphorus solubilization by PGPR. (*Source*: Ahemad and Kibret, 2014.)

B. Indirect mechanisms:

PGPR that indirectly enhance plant growth via suppression of phyto-pathogens do so by a variety of mechanisms (Glick, 2012). These include the ability to produce: antibiotics, lytic enzymes, siderophores (Figure 7.4), competition, and induced systemic resistance (ISR).

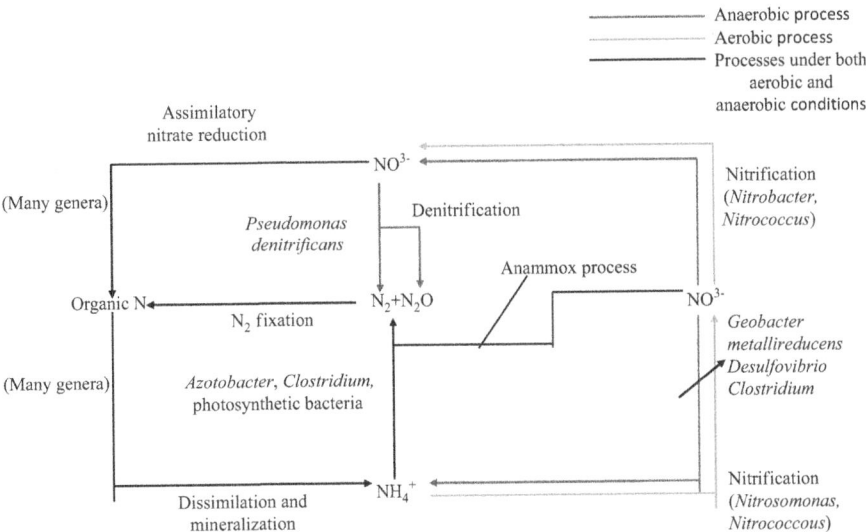

FIGURE 7.3 Nitrogen fixation by PGPR.

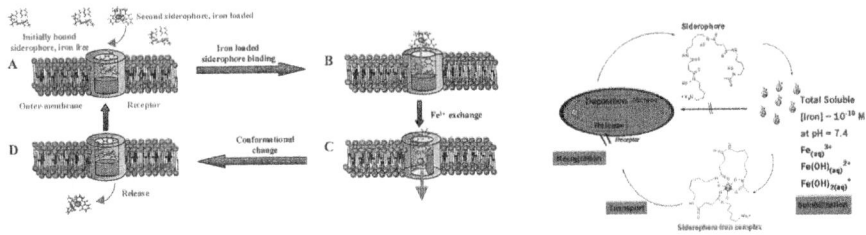

FIGURE 7.4 Siderophore production by PGPR. (*Source*: Stintzi et al., 2000.)

ISR: PGPB can trigger a phenomenon in plants known as ISR that is phenotypically similar to the systemic acquired resistance (SAR) that occurs when plants activate their defense mechanisms in response to infection by a pathogenic agent (Figure 7.5). ISR-positive plants are said to be "primed" so that they react faster and more strongly to a pathogen attack by inducing defense mechanisms.

ISR does not target specific pathogens. ISR involves jasmonate and ethylene signaling within the plant and these hormones stimulate the host plant's defense responses to a range of pathogens. ISR does not require any direct interaction between the resistance-inducing PGPB and the pathogen. Besides ethylene and jasmonate, other bacterial molecules such as the O-antigenic side chain of the bacterial outer membrane protein lipopolysaccharide, fagellar proteins, pyoverdine, chitin, glucans, cyclic lipopeptide surfactants, and salicylic acid have all been reported to act as signals for the induction of systemic resistance (Naz et al., 2007).

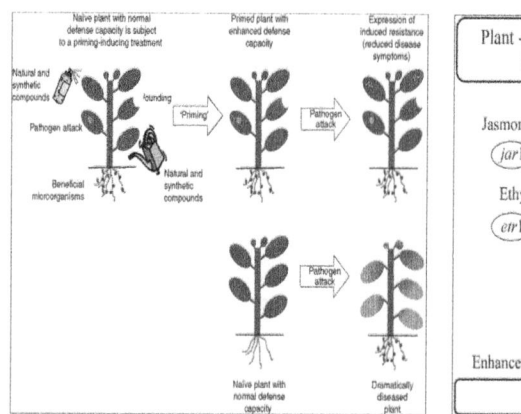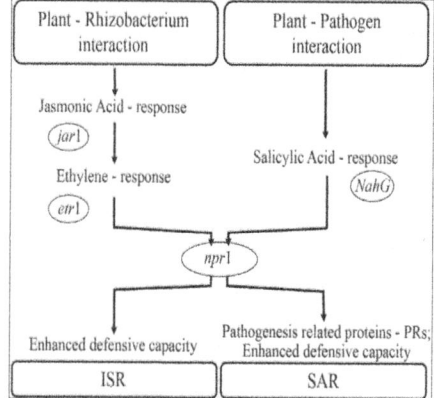

FIGURE 7.5   Induced systemic resistance by PGPR. (*Source*: Beneduzi et al., 2012.)

7.4   CHARACTERISTICS OF A SUCCESSFUL PGPR FOR FORMULATION DEVELOPMENT

To develop a successful PGPR formulation, the rhizobacteria should possess:

a. High rhizosphere competence	b. High competitive saprophytic ability
c. Enhanced plant growth	d. Ease for mass multiplication
e. Broad spectrum of action	f. Excellent and reliable control
g. Safe to environment	h. Compatible with other rhizobacteria
i. Should tolerate desiccation, heat, oxidizing agents, and UV radiations	

7.5   PGPR APPLICATION METHODS

PGPR-based formulations are of two types: solid-based and liquid. Liquid are easy to apply, applied at a rate of 250 mL/acre. These have very good shelf life (12 months from the date of manufacture), with cell count 1×10^9 cells protected with additives from biotic stress. Whereas, carrier-based formulations are applied at a rate of 5 kg/acre with cell count 1×10^7 but usually have a short life of 3–6 months. Cells are not protected and moisture loss is there. PGPR are delivered through several means based on survival nature and infection of the pathogen. They are delivered through seed, soil, foliage, rhizomes, or through a combination of several methods of delivery (Table 7.1 and Figure 7.6).

The details of worldwide studies on PGPR are summarized in Table 7.2.

Advantages
- Activate soil biologically, supplement chemical fertilizers by 25%–30%.
- Increase the yields by 10%–40%.
- Decompose plant residues, and stabilize C: N ratio of soil.
- Improved structure and water-holding capacity of soil.

TABLE 7.1

Various Methods and Modes of Application of PGPR

Sr. No.	Methods	Rate	Methods of Application
1.	Seed inoculation	500–750 g/ha of seed	Inoculant is mixed with 20% jaggery and slurry is prepared and poured on the seeds to form a thin coating on the seeds.
2.	Root and seedling treatment	1 kg/10 L of water/ha	Dip the root portion of the seedlings in this suspension for 15–30 min and transplant immediately. Generally, the ratio of inoculant and water is 1:10.
3.	Soil application	5 kg/100 kg farm yard manure (FYM)/ha	5 kg *Azotobacter* and 5 kg phosphate solubilizing bacteria are mixed with 50–100 kg of well-decomposed cattle manure for an area of 1 ha. The mixture of bio-fertilizer and cattle manure sprinkled with water is kept for 24 h and then broadcasted into soil at the time of sowing.
4.	Sets or tuber treatment	1 kg/40–50 L water/ha	Prepare culture suspension by mixing 1 kg of inoculant in 40–50 L of water. The cut pieces of planting material required for sowing one acre are kept immersed in the suspension for 30 min. Bring out the cut pieces and dry them in shade for some time before planting. After planting, the field is irrigated within 24 h.

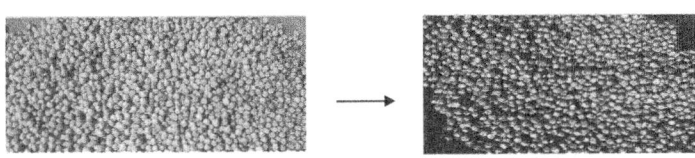

Seed treatment

Root and seedling dip

Soil treatment

FIGURE 7.6 Methods of application of PGPR.

- Stimulates plant growth by secreting growth hormones.
- Secrete fungistatic and antibiotic-like substances.
- Solubilize and mobilize nutrients.
- Increase tolerance to drought, salt stress.
- Eco-friendly, non-polluting, and cost-effective method.

TABLE 7.2

Worldwide Studies on PGPR

PGPR as Biofertilizers			
PGPR	**Plant**	**Result of Bacterial Addition**	**References**
Pseudomonas sp. PS1 (Pots)	Greengram (*Vigna radiata* (L.)	Increased plant dry weight, nodule numbers, total chlorophyll content, leghaemoglobin, root N, shoot N, root P, shoot P, seed yield, and seed protein	Ahemad and Khan (2012)
Bacillus subtilis (Field trials)	Caulifower (*Brassica oleracea var botrytis* L.)	Increased number of non-wrapper leaves, curd diameter, curd depth, curd weight, and yield	Kaushal and Kaushal (2013)
Rhizobium strain MRP1 (Pots)	Pea (*Pisum sativum*)	Increased growth, nodulation and leghaemoglobin content, amount of N and P nutrients in plant organs, seed yield, and seed protein of pea plants	Ahemad and Khan (2011)
Bacillus subtilis (both net house and field trials)	Wheat (*Triticum aestivum* L.)	Increased wheat yield, number of tillers per plant, grain number per spike, 1,000 grain weight, and biomass of wheat plants	Sood et al. (2018a)
Bacillus subtilis (both net house and field trials)	Maize (*Zea mays* L.)	Increased plant height, 1,000 seed weight, yield, number of cobs per plant, cob length, number of rows, seeds per line, and biomass of maize plants	Sood et al. (2018b)
PGPR under stress conditions			
Paenibacillus polymyxa	Pepper	Increased biomass of plants and elicited induced systemic resistance against bacterial spot pathogen *Xanthomonas axonopodis* pv. Vesicatoria untreated plants	Phi et al. (2010)
PGPR as biocontrol agent			
Ralstonia pickettii QL-A6	**Bacterial wilt of tomato** (*Ralstonia solanacearum*)		Wei et al. (2013)

7.6 CONSTRAINTS TO COMMERCIALIZATION

The success of these microbial inoculants depends on the availability of microbes as a product or formulation, which facilitate the technology to transfer from laboratory to land. The constraints include:

- Lack of suitable screening protocol for the selection of promising candidate of PGPR.
- Lack of sufficient knowledge on the microbial ecology of PGPR strains and plant pathogens.
- Optimization of fermentation technology and mass production of PGPR strains.

- Inconsistent performance and poor shelf life.
- Lack of patent protection and lack of multi-disciplinary approach.
- Determination of those traits that are most important for efficacious functioning and subsequent selection of PGPB strains with appropriate biological activities.
- Consistency among regulatory agencies in different countries regarding what strains can be released to the environment, and under what conditions genetically engineered strains are suitable for environmental use.
- Selection of PGPB strains that function optimally under specific environmental conditions as the responses usually depend on several environmental factors. (i) The type of soil and its characteristics. (ii) The inadequacy of organic matter especially for dry land. (iii) Soil water deficit and high temperature (hyper-thermia) are prominent abiotic factors that affect nitrogen fixation in dry land agriculture. (iv) Native microbial population opposes the inoculants. In general, predatory organisms often already present in the soil are more adapted to the environment and outcompete the inoculated population.
- Apart from environmental factors, deficiencies in handling procedure are a major cause of underperformance in real-life application. The high sensitivity to temperature and other external conditions of these 'living' inputs, calls for enormous caution at the stage of manufacture/culture, transportation/distribution, and application.
- Region-wise unavailability of soil-specific strains.
- Shelf life and limited demand.
- Unavailability of proper transportation and storage facilities.
- Lack of publicity, lack of coordination among researchers, extension workers, and farmers.

7.7   FUTURE THRUST

- Microbial strains which can compete with indigenous ones and work over a range of soils and agro-climatic conditions need to be isolated, multiplied, and made available to farmers.
- Development of ideal, cheaper, and easily available carriers for increasing longevity of bio-fertilizers.
- Research should be done with regard to suitability of biofertilizers against adverse conditions.
- Publicity and training programs.
- More understanding of root-microbe communication for effective rhizosphere competence is required.

REFERENCES

Ahemad, M. and Khan, M.S. 2011. Functional aspects of plant growth promoting rhizobacteria: recent advancements. *Insight Microbiology*, 1(3), pp. 39–54.
Ahemad, M. and Khan, M.S. 2012. Productivity of greengram in tebuconazole-stressed soil, by using a tolerant and plant growth-promoting Bradyrhizobium sp. MRM6 strain. *Acta Physiologiae Plantarum*, 34(1), pp. 245–254.

Ahemad, M. and Kibret, M. 2014. Mechanisms and applications of plant growth promoting rhizobacteria: current perspective. *Journal of King Saud University-Science*, 26(1), pp. 1–20.

Beneduzi, A., Ambrosini, A. and Passaglia, L.M. 2012. Plant growth-promoting rhizobacteria (PGPR): their potential as antagonists and biocontrol agents. *Genetics and Molecular Biology*, 35(4), pp. 1044–1051.

Glick, B.R. 2012. Plant growth-promoting bacteria: mechanisms and applications. *Scientifica*, 2012. doi:10.6064/2012/963401

Gupta, S., Kaushal, R., Spehia, R.S., Pathania, S.S. and Sharma, V. 2017. Productivity of capsicum influenced by conjoint application of isolated indigenous PGPR and chemical fertilizers. *Journal of Plant Nutrition*, 40(7), pp. 921–927. doi:10.1080/01904167.2015.1093139

Jha, P.N., Gupta, G., Jha, P. and Mehrotra, R. 2013. Association of rhizospheric/endophytic bacteria with plants: a potential gateway to sustainable agriculture. *Greener Journal of Agricultural Sciences*, 3(2), pp. 73–84.

Kaushal, M. and Kaushal, R. 2013. Plant growth promoting rhizobacteria-impacts on cauliflower yield and soil health. *The Bioscan*, 8(2), pp. 549–552.

Lesk, C., Rowhani, P. and Ramankutty, N. 2016. Influence of extreme weather disasters on global crop production. *Nature*, 529(7584), p. 84.

Mancosu, N., Snyder, R.L., Kyriakakis, G. and Spano, D. 2015. Water scarcity and future challenges for food production. *Water*, 7(3), pp. 975–992.

Mittler, R. 2002. Oxidative stress, antioxidants and stress tolerance. *Trends in Plant Science*, 7(9), pp. 405–410.

Naz, I., Ahmad, M., Alam, S., Tahir, M. and Raziq, F. 2007. Control of root and collar rot disease, a serious threat to chillies production in NWFP. *Sarhad Journal of Agriculture*, 23(2), p. 451.

Ngumbi, E. and Kloepper, J. 2016. Bacterial-mediated drought tolerance: current and future prospects. *Applied Soil Ecology*, 105, pp. 109–125.

Phi, Q.T., Park, Y.M., Seul, K.J., Ryu, C.M., Park, S.H., Kim, J.G. and Ghim, S.Y. 2010. Assessment of root-associated Paenibacillus polymyxa groups on growth promotion and induced systemic resistance in pepper. *Journal of Microbiology and Biotechnology*, 20(12), pp. 1605–1613.

Reddy, P.P. 2014. Plant Growth Promoting Rhizobacteria for Horticultural Crop Protection. Springer, New Delhi.

Sood, G., Kaushal, R., Chauhan, A. and Gupta, S. 2018a. Indigenous plant-growth-promoting rhizobacteria and chemical fertilisers: impact on wheat (Triticum aestivum) productivity and soil properties in North Western Himalayan region. *Crop and Pasture Science*, 69(5), pp. 460–468.

Sood, G., Kaushal, R., Chauhan, A. and Gupta, S. 2018b. Effect of conjoint application of indigenous PGPR and chemical fertilizers on productivity of maize (Zea mays L.) under mid hills of Himachal Pradesh. *Journal of Plant Nutrition*, 41(3), pp. 297–303.

Stintzi, A., Barnes, C., Xu, J. and Raymond, K.N. 2000. Microbial iron transport via a siderophore shuttle: a membrane ion transport paradigm. *Proceedings of the National Academy of Sciences*, 97(20), pp. 10691–10696.

Vidyalakshmi, R., Paranthaman, R. and Bhakyaraj, R. 2009. Sulphur Oxidizing Bacteria and Pulse Nutrition A Review. *World Journal of Agricultural Sciences*, 5(3), pp. 270–278.

Vinocur, B. and Altman, A., 2005. Recent advances in engineering plant tolerance to abiotic stress: achievements and limitations. *Current Opinion in Biotechnology*, 16(2), pp. 123–132.

Wei, Z., Huang, J., Tan, S., Mei, X., Shen, Q. and Xu, Y. 2013. The congeneric strain Ralstonia pickettii QL-A6 of Ralstonia solanacearum as an effective biocontrol agent for bacterial wilt of tomato. *Biological Control*, 65(2), pp. 278–285.

8 Status and Prospect of Precision Farming in India

Neelam Patel
Indian Agriculture Research Institute

Rohitashw Kumar and Dinesh Kumar Vishwakarma
Sher-e-Kashmir University of Agricultural Sciences and Technology of Kashmir (SKUAST-K)

CONTENTS

8.1 INTRODUCTION

Agriculture is the backbone of our country's economy, accounting for almost 13% of Gross Domestic Product (GDP) and employing 70% of the population. Though this is a rosy picture of our agriculture, how long can it continue to meet the growing demands of the ever-increasing population? This is a difficult question to be answered, if we depend only on traditional farming. Agriculture in India is climate restricted though it is having a large area under crop husbandry. About 48% of the geographical area of the country receives less than 1,000 mm rainfall and the rest about 1,000–2,500 mm. The difficulty is that the rainfall occurs in 3–4 months' duration, making rainwater storage imperative for crop irrigation purpose. However, the available water for irrigation cannot cover the net cultivated area and the water resources have assumed a declining trend in agriculture. Only 48% of the cultivated area is presently irrigated. Irrigation cover cannot be increased as the available 1,143 BCM of water would not extend further. By 2050 our water need (both irrigation and total need) would cross the availability level. To achieve the increased food production of 494 million tons by 2050, our irrigated area should increase from 79 million ha to 146 million ha. Similarly, the cultivated land area cannot increase beyond 2 million ha during the period 2010–2050. In future there is a serious situation where there is no possibility for increasing water, land, or energy for increased crop production in order to achieve the food security.

To meet the forthcoming demands and challenges we have to divert our attention towards new technologies for revolutionizing our agricultural productivity for sustainable feed security. In the post-green revolution period, agricultural production has become stagnant, and horizontal expansion of cultivable lands has become limited due to the burgeoning population and industrialization. In 1952, India had 0.33 ha of available land per capita, which is reduced to 0.15 ha at present. It is essential to develop eco-friendly technologies for maintaining crop productivity. Since long, it has been recognized that agro-climatic variability, crops, and soils are not uniform over area and time. Over the last decade, technical methods have been developed to utilize modern electronics to respond to agro climatic variability. Such methods are known as spatially variable crop production, geographic positioning system (GPS)-based agriculture, site-specific and precision farming (precision agriculture). The term "spatially variable crop production" seems to be more accurate and descriptive than the term Precision Agriculture (PA). The concept of PA avails the recent developments in micro irrigation, sensors networks, greenhouse, and protected agriculture structures. It is also certain that the availability of labor for agricultural activities is going to be in short supply in the future. The time has now arrived to exploit all the modern tools available by bringing information technology and agricultural science together for improved economic and environmentally sustainable crop production. PA is an integrated crop management system that attempts to match the kind and amount of inputs with the actual crop needs for small areas within a farm field. This goal is not new, but new technologies now available allow the concept of PA to be realized in a practical production setting and this is known as Precision Farming (PF).

PF is a package of technologies to enhance the input use efficiency with an aim to sustainably increase the productivity without causing any negative impact to the environment. In advanced countries, GPS, Remote sensing (RS), and geographical information system (GIS) technologies along with long-term yield monitoring of farms both at a macro and micro level, have been successfully adopted to assist in the decision-making at the farm level when embarking on PF. However, innovative use of such technologies has been a limitation in the Indian agricultural system. Considering the small land holding and subsistence agriculture in India, the technological adaption, especially for PF, has not been realized as yet. Several drawbacks have been listed to be the reasons, which include (i) *lack of simple and affordable technologies for the assessment of variation of the soil nutrient content in a field*; (ii) *lack of machinery and guiding instruments to affect variable rate technologies in small farm holdings*; (iii) *lack of perception about calendar to follow the farm operations timely and systematically*; and (iv) *lacunae in technology transfer methodologies*.

In a very *doable sense*, PF involves consideration to improve the yield, quality, and farm income through enhancing the input use efficiency (water and nutrients). Several of the technological innovations and crop management tools concerning the PF that warrant attention are: (i) location and crop-specific surface covered structure design; (ii) cost-effective, durable, and high-quality drip-fertigation systems design; (iii) energized or gravity drip systems design; (iv) establishing solar powered irrigation system feasibility wherever needed; (v) soil quality profiles over the cropping seasons across the crops and management strategies; (vi) sensor development; (vii) technical guidance on fertigation and irrigation scheduling specific to individual farms; (vii) agronomic support throughout the growing season; (ix). GIS, RS, and GPS technology for PF; (x) training of farmers and filed operators in the management of PF technologies including climate systems; and (xi) training in farm level post-harvest management and for good agricultural practices (GAP) certification.

Most of the current technologies used in PF agriculture to a limited extent, are applied from other countries with limited improvisation. Moreover, several of the knowledge gaps existing in India are yet to be established for PF agriculture.

8.2   DEFINITION OF PF

PF is defined as an information and technology–based farm management system to identify, analyze, and manage variability within fields by doing all practices of crop production in the right place, at the right time, and in the right way for optimum profitability, sustainability, and protection of the land resource. PA is a systems approach to farming for maximizing the effectiveness of crop inputs. PA is a management philosophy or approach to the farm and is not a definable prescriptive system. It identifies the critical factors where yield is limited by controllable factors, and determines intrinsic spatial variability. It is essentially more precise farm management made possible by modern technology. The variations occurring in crop or soil properties within a field are noted, mapped, and then management actions are taken as a consequence of continued assessment of the spatial variability within that field by adoption of site-specific management systems using RS, GPS, and GIS. PA requires special tools and resources to recognize the inherent spatial variability associated with soil characteristics, crop growth, and to prescribe the most appropriate management strategy on a site-specific basis. It offers a potential step change in the productive efficiency. The more suitable definition for PF in the context of the Indian farming scenario could be: the precise application of agricultural inputs based on soil, weather, and crop requirements to maximize sustainable productivity, quality, and profitability. Today, because of increasing input costs and decreasing commodity prices, the farmers are looking for new ways to increase efficiency and cut costs. PF technology would be a viable alternative to improve profitability and productivity.

The potential of PF for economical and environmental benefits could be visualized through the reduced use of water, energy, fertilizers, herbicides, and pesticides besides the farm equipments. Instead of managing an entire field based upon some hypothetical average condition, which may not exist anywhere in the field, a PF approach recognizes site-specific differences within fields and adjusts management actions accordingly. PA offers the potential to automate and simplify the collection and analysis of information. It allows management decisions to be made and quickly implemented on small areas within larger fields.

8.3   RATIONALE OF PF IN INDIA

The "Green revolution" of the 1960s has made our country self-sufficient in food production. In 1947, the country produced a little over 6 million tons of wheat; in 1999, our farmers harvested over 72 million tons, taking the country to the second position in the world in wheat production. In the last 5 decades, the production of food grains has increased more than threefold, and the yield during this period has increased more than twofold. All this has been possible due to high input application, like increase in fertilization, irrigation, pesticides, higher use of high yielding varieties (HYVs), increase in cropping intensity, and increase in mechanization of agriculture.

   i. Fatigue of Green Revolution

      The Green Revolution has, of course, contributed a lot. However, even with the spectacular growth in the agriculture, the productivity levels of many major crops are far below the expectation. We have not achieved even the lowest level of the potential productivity of Indian HYVs, whereas the world's highest productive country has crop yield levels significantly higher than the upper limit of the potential of Indian HYVs. Even the crop yields of India's agriculturally rich state like Punjab is far below the average yield of many high productive countries (Ray et al., 2001).

   ii. Natural Resource Degradation

The Green Revolution is also associated with negative ecological/environmental consequences. The status of the Indian environment shows that, in India, about 182 million ha of the country's total geographical area of 328.7 million ha is affected by land degradation. Of this, 141.33 million ha are due to water erosion, 11.50 million ha are due to wind erosion, and 12.63 and 13.24 million ha are due to water logging and chemical deterioration (salinization and loss of nutrients), respectively. On the other end, India shares 17% of the world's population, 1% of the gross world product, 4% of the world's carbon emission, 3.6% of CO_2 emission intensity, and 2% of the world's forest area. One of the major reasons for this status of the environment is the population growth of 2.2% in 1970–2000. The status of the Indian environment, although not alarming when compared to developed countries, gives an early warning.

In this context, there is a need to convert this green revolution into an evergreen revolution, which will be triggered by farming systems approach that can help to produce more from the available land, water, and labor resources, without either ecological or social harm (Swami Nathan, 2002). Since PF proposes to prescribe tailor-made management practices, it can help to serve this purpose.

8.4   NEED OF PA

i. **For assessing and managing field variability:** We know that our fields have variable yields across the landscape because of variations in management practices, soil properties, and/or environmental characteristics. One's mental information database about how to treat different areas in a field requires years of observation and implementation through trial-and-error. Today, that level of knowledge of field conditions is difficult to maintain because of the variable farm sizes and changes in areas farmed due to annual shifts in the leasing arrangements. PA offers the potential to automate and simplify the collection and analysis of information.

ii. **For doing the right thing in the right place at the right time:** After assessing the variability, PA allows management decisions to be made and implemented at the right time and in the right places on small areas within larger fields.

iii. **For higher productivity:** Since PF proposes to prescribe tailor-made management practices, it will definitely increase the yield per unit of land, provided nature's other uncontrollable factors are in favor.

iv. **For increasing the effectiveness of inputs:** Increased productivity per unit of input used indicates increased efficiency of the inputs.

v. **For maximum use of minimum land unit:** After knowing the land status, a farmer tries to improve each and every part of the land and uses it for production purpose.

8.5   PRECISION IN WATER MANAGEMENT

PF aims to manage production inputs over many small management zones rather than on large zones. It is difficult to manage inputs at extremely fine scales in irrigation systems. However, site-specific irrigation management can potentially improve the overall water management in comparison to irrigated areas of hundreds of hectares. Agricultural cropping systems depend on the use of water resources for survival, and water needs vary spatially in fields because of spatial soil variability (texture, topography, water holding capacity, and infiltration and drainage rate). Therefore, the need for irrigation may differ between the different zones of a particular field. While moving irrigation systems apply water at constant rates, some areas of the field may receive too much water and others not enough. Precision irrigation, an existing aspect of PA just beginning to be explored, means applying the right amount of water in the right place. The use of PA for irrigation water management worldwide is still in the development stage and requires a lot of investigation and experimental work to determine its feasibility and applicability.

The availability of some low-cost data gathering methods, positioning systems, and the developments in computer programming will help in regulating the depth of water within a field. So the next generation in irrigation scheduling is not just about when and how much, but about when, where, and how much to irrigate. A precision irrigation system expected to have the ability to apply the right amount of water directly where it is needed, therefore, is saving water through preventing excessive water runoff and leaching.

PF has the potential to increase the economic efficiencies by optimally matching the irrigation inputs to yields in each area of a field and thus reducing costs. The potential economic benefit of precision irrigation lies in reducing the cost of inputs or increasing yield for the same inputs. The notion of spatially varied irrigation is predicated on the hypothesis that the crop is non-uniform and the water requirements are similarly non-uniform, probably as a result of the differences in root zone conditions. It is also assumed that the yield will be maximized if each plant is supplied with water exactly matching its individual requirements. However, evidence to support these hypotheses is not readily found in the literature. The crop response to water has been studied extensively leading to the development of crop production functions for most crops. Also reasonably well-known is the spatial variation in crop performance largely as a result of unintended spatial variations in the depths of irrigation applied. Less well-studied is the variation in crop response to water across a field, that is, variation in the crop production function across the field. It is the presence of this variation that provides the justification for spatially varied irrigation.

8.6   STATUS OF PF GLOBALLY

Research efforts into precision irrigation were initiated in the USA in the early 1990s. Initially this work largely centered on the modification of center pivot and lateral move irrigation machines to give spatially varied applications of water and nitrogen with the system control based on stored databases of spatially referenced data. A range of methods for implementing valve control to achieve the desired application rate have been trialed, including programmable logic controllers and addressable solenoid valves. The variable rate water application systems employed include multiple sprinklers or groups of sprinklers for time-proportional pulsing, and a variable-aperture sprinkler with time-proportional control for a comprehensive review of research undertaken in the USA since the early 1990s on precision irrigation with moving irrigation systems. Interest by European researchers grew through the 2000s and the emphasis shifted to the purpose and performance of spatially varied irrigations. Other recent work at Washington State University has focused on the development and testing of digital control systems using onboard computer to implement radio-based transmitted instructions and the installation of both sprays and low-energy precision application (LEPA) on the same machine for plot research in Montana. Additional work has also been undertaken in Europe examining the yield response to non-uniform water applications under moving irrigation systems, and in New Zealand investigating the water savings and economic benefits of precision irrigation using center pivots. Clearly, the ease and consistency with which the location of moving irrigation machines can be determined, along with the large number of nozzles, and the presence of computer control, offer a ready means of differential irrigation.

8.6.1 Water Savings through PF

The primary goal of PF is to apply an optimum amount of irrigation throughout fields. No study was found to evaluate the site-specific or variable rate irrigation (VRI) as a necessary or essential component of precision irrigation. However, it is a possible or even desirable component. It is also seen by many as the most likely means of achieving significant water savings. While conditions could exist for which the aggregated optimum input for the entire field is greater than the amount usually applied in a conventional uniform application to the field, most researchers expect a reduction in water use on at least parts of fields, if not a reduction in the value aggregated over entire fields. They have identified that this has only recently been achieved for precision irrigation and then only in a few instances. They reviewed much of the work prior to that date and suggested that opportunities for water savings accrue by not irrigating non-cropped areas, by reducing irrigation applications to adapt to specific problems, and by optimizing the economic value of water applied through irrigation. Results from case studies of VRI reviewed showed water savings in individual years ranging from zero to 50%, and savings averaged over a number of years from 8% to 20%, depending on the previous irrigation management. They concluded that VRI could save 10%–15% of water used in conventional irrigation practice.

While there are no Australian studies pointing to the potential water savings from precision irrigation, water savings of about 25% are possible through improvements in application efficiency obtained by spatially varied irrigation applications adjusted to suit the available water holding capacity of the soils.

8.6.2   YIELD AND PROFIT THROUGH ADOPTION OF PF TECHNOLOGIES

Studies specifically evaluating the yield and profit potential of precision or VRI have involved both modeling and field experimental approaches. The yield of potatoes under a center pivot equipped for spatially varied from its applications. Yields per unit of water applied were more (4% and 6%) in two consecutive years over those for uniform irrigation management. Booker et al. (2007) analyzed yields and water use efficiency (WUE) for spatially varied irrigation in differet years. They concluded that cotton seems too unpredictable to manage with spatially varied irrigation. This result is supported by the work of Bronson et al. (2007) who concluded that management zones for upland cotton based on landscape position were not justified, and by Clouse (2006) who in a modeling study obtained conflicting results between variable rate and uniform irrigation of cotton depending on the scheduling strategy adopted. Crop modeling has been shown to be an important and effective means of determining the value (yield and profitability) of variable rate strategies in PF. The same is the case for spatially varied irrigation. Nijbroek et al. (2003) investigated the economics of irrigation management zones for soybeans in the south-eastern USA.

The model CROPGRO-Soybean was used to determine optimal irrigation strategies for each zone and the results compared to various uniform strategies applied to the whole field. Varying the irrigation strategies for the individual zones gave the highest return although the differences were small at US$ 16 per ha between the best and worst management. DeJonge and Kaleita (2006) and DeJonge et al. (2007) used the model CERES-Maize to explore the feasibility of irrigation of corn in Iowa, USA. They investigated that the benefits of spatially varied irrigation. Irrigation was shown to reduce both the spatial and temporal variability in yield and spatially varied irrigation gave higher yields than uniform irrigation. The results from the above studies showed that there is potential for yield improvements but that the benefits may not cover the costs of the technology required for spatially varied applications. Heermann et al. (2002) similarly concluded that site-specific irrigation management increased risk and the potential economic benefit from it is small when the farmer's tolerance for risk is low. Almas et al. (2003) suggested that caution is required and that the benefit of changing to VRI from uniform application methods needs to be assessed before adoption. Their results indicate that substantial field variability and high crop prices are required for VRI to be profitable. It also depends heavily on the useful life of the equipment. Yule et al. (2008) estimated payback periods ranging from 5 to 20 years for adoption of VRI in dairy and cropping in New Zealand. To date, there is a lack of significant evidence that investment in a precision irrigation system can provide sound financial returns to irrigators. It remains to be seen whether the costs can be reduced significantly or whether a simpler form of precision irrigation is needed that does not involve spatially varied applications.

8.7   STATUS OF PF IN INDIA

The concept of PF is not new in India. Farmers try their best to get the maximum possible yield with the information and technologies available to them, but unless and until the total information about their fields and advanced technologies are available, they cannot do PF in the perfect sense. In India, a major problem is the small field size. More than 58% of operational holdings in the country have size less than 1 ha. Only the states of Punjab, Rajasthan, Haryana, and Gujarat which are having more than 20% of agricultural lands have operational holding size of more than 4 ha. When contiguous fields with the same crop are considered, it is possible to obtain fields of over 15 ha area in which similar crop management techniques are followed. Such fields can be considered for the purpose of initiating the implementation of PF. There is a scope of implementing PA for horticultural crops having wider scope for PA in the cooperative farms. Nearly two-thirds of the arable land in India is rain-fed. The crop yields are very low ($\approx$1 t/ha) and very good potential exists for increasing the productivity of rain-fed cropping systems with PF.

The Indian Council of Agricultural Research, State Agricultural Universities, and National Committee on Plasticulture Applications in Horticulture are promoting research efforts for the encouragement of PF technology. The Ministry of Agriculture, Ministry of Water Resources, and different State governments have sponsored promotional activities for PF technologies. But its application at a commercial level was encouraged by the formation of a National Committee on the Use of Plastics in Agriculture (renamed as National Committee on Plasticulture Applications in Horticulture). The committee established Plasticulture Development Centres (Figure 8.1) (renamed as Precision Farming Development Centres) in different agro-climatic conditions of the country for conducting research on different PF technologies namely micro irrigation, surface covered cultivation (polyhouse, shade net, mulching, low tunnels), nursery raising, sensor development and automation, decision support system (DSS) development, water conservation through pond lining, low-cost – low-energy drip system, micro irrigation operated through solar energy, micro irrigation in canal command and fertigation to take the proven technologies to the farmers through demonstrations and capacity building, etc.

8.8   MICRO IRRIGATION STUDIES

Micro irrigation system has emerged as an appropriate water-saving and production augmenting technique for wide spaced crops, viz. grapes, banana, coconut, mango, pomegranate, citrus, guava, sapota, tea, coffee, cardamom, and also for commercial crops like cotton, tobacco, and sugarcane. The organized efforts in the sector of plasticulture were initiated way back in 1981 with the constitution of National Committee on Plasticulture Application (NCPA) under the then Ministry of Petroleum, Chemicals and Fertilizers (reconstituted as National Committee on Plasticulture Application in Horticulture (NCPAH) under the Ministry of Agriculture, Govt. of India. The Govt. of India announced a massive subsidy scheme of 250 crores during the Eighth Plan (1992–1997). During the Ninth Plan, plasticulture application got a further thrust with an outlay of Rs. 375 crores, i.e., 50% higher than the Eighth Plan. The government allocation kept increasing through the Ninth, Tenth, and Eleventh Plan periods. At present, the assistance under the "Centrally sponsored scheme on development of horticulture through plasticulture interventions" is available for all types of drip/ micro irrigation systems such as on line drip irrigation systems, in line systems,

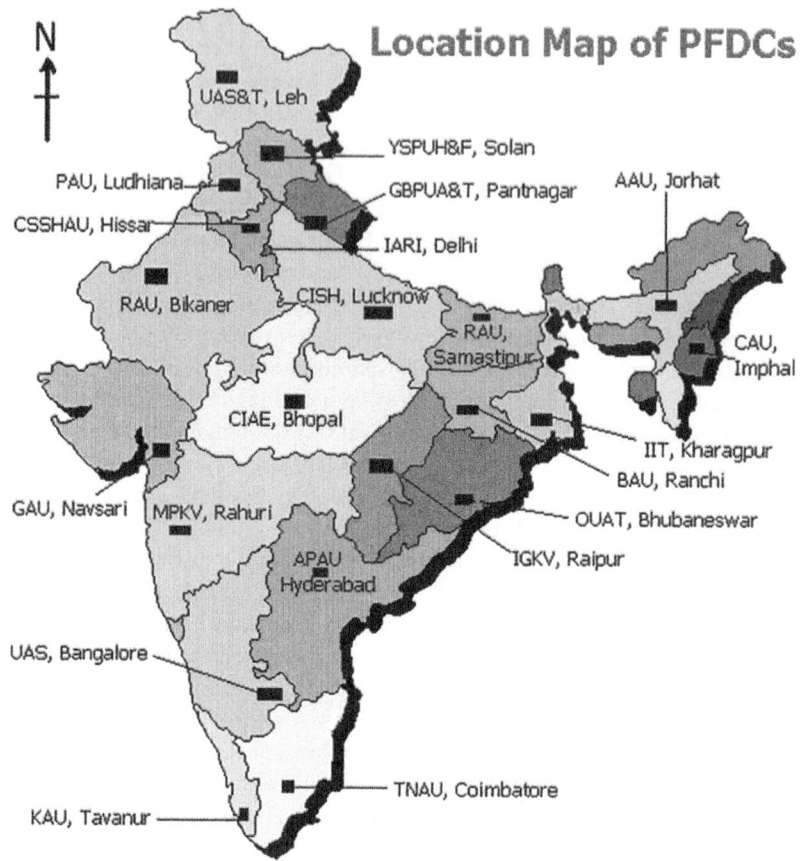

FIGURE 8.1   Location of Plasticulture Development Centres, NCPAH.

subsurface drip irrigation systems, micro tube, micro jets, fan-jets, micro sprinkler, mini sprinklers, misters, and similar other low discharge irrigation systems.

Drip irrigation through the trickle supply of water drops continuously keeping the soil moist in the rhizosphere has opened new vistas in the agricultural scenario, especially for the horticultural crops. Regular research work and the beginning of adoption of drip irrigation systems started in the mid-1970s. Drip irrigation wets a part of the land surface and water is supplied under the shaded area of plants. In drip irrigation, water is conveyed in closed conduits to almost each plant, therefore providing lesser opportunity of transporting fewer weeds to the field in comparison to surface irrigation methods. Thus the overall reduction is achieved in weed infestation in drip irrigated fields. Drip irrigation is the only method to irrigate the crops frequently as per their water requirement. Drip irrigation has been proved to be very effective in saving irrigation water and increasing the yield of vegetable crops with good quality of produce. Increased growth and yield with drip irrigation has been reported in several crops between 7% and 112% depending on the crops/varieties and the method of irrigation compared. The yield increase in drip irrigation compared to conventional irrigation methods varies from 20% to 100%, whereas the saving in water ranges from 40% to 70% besides 50%–60% saving in labor. The water and fertilizer savings through drip fertigation have been reported to be 40%–70% and 30%–50%, respectively (INCID, 1994; Narayanamoorthy, 1997a). Kadam et al. (1995) recorded higher WUE (374 kg/ha/cm) with drip irrigation over furrow irrigation (214 kg/ha/cm). The water and fertilizer savings through drip fertigation system have been reported to be 40%–70% and 30%–50%, respectively (Rekha and Mahavishnan, 2008).

Drip irrigation is the most effective way to supply water to okra crop, which not only saves water but also increases yield due to continuous maintenance of moisture content near field capacity (Gowda et al., 2001). Punamhoro et al. (2003b) studied the performance of okra under different irrigation methods, viz. drip irrigation with bucket kit and drum kit, micro sprinkler, overhead sprinkler irrigation, flood irrigation, check basin irrigation, and furrow irrigation. The results of the study revealed that the highest WUE of 2.52 q/ha/cm was recorded in drip irrigation with bucket kit, while the lowest WUE of 1.06 q/ha/cm was noticed with flood irrigation.

A study carried out on heavy soils and sub-humid climatic conditions of the South Gujarat region suggests that a large scale adoption of drip method of irrigation in sugarcane in the South Gujarat area can help to solve the problems of water logging and secondary salinization which are increasing in this region (Parikh et al., 1993). Muralidhara et al. (1994) also reported that B–C ratio was found greater for mulberry. Malik et al. (1994) recorded higher green pea yield when N-fertilizer was applied through drip compared to other methods. The highest yield of brinjal under drip irrigation was recorded at 100% irrigation requirement treatment on sandy loam soil in coastal Orissa (Mishra and Paul, 2009). Water saving to the extent of 20% in banana was reported while 50% saving under drip irrigation was achieved as compared to surface method of irrigation reported by (Pawar et al., 2001). Drip irrigation at 80% of evaporation replenishment could bring about nearly 25% saving in the irrigation water in banana (Srinivas, 1998). Comparative studies of drip versus surface method were conducted at Haryana Agriculture University (HAU), Hissar, on onion, sugar beet, and potato. The results revealed that yield and WUE were higher under drip irrigation than surface irrigation. Kumar and Khanna (1998) have studied the implications of low-cost bubbler irrigation system through a case study for Kinnow orchard.

Kumathe et al. (1998) compared the effects on yield through drip and ring basin methods of irrigation at the Regional Research Station, Raichur, for kagzi lime and banana. Malik et al. (1994) recorded higher green pea yield when the fertilizer N was applied through drip compared to other methods (Magar, 1995). Cost and return of structure of crops growth under drip and conventional system (per acre) had been given by Mane (1993). For the designing of drip irrigation system F value for the number of outlets is given by Mane et al. (2008). Economic feasibility of micro irrigation systems for various vegetables, viz. cabbage, potato, brinjal, chilli, cauliflower, and tomato were worked out for Nainital tarai region of Uttar Pradesh, India (Manjunatha et al. 2001). Between furrow and drip irrigations, drip irrigation produced significantly higher dry chilli yield with 42% higher WUE over furrow method (Veeranna et al., 2001). Drip irrigation for brinjal not only offers water economy, but also provides a high yield of the produce, which in turn gives higher net return than traditional furrow irrigation (Chauvan and Shukla, 1990). The highest WUE of 3,621 kg/ha cm was reported under drip irrigation, whereas it was 118.8 kg/ha cm in furrow irrigation in brinjal (Sivanappan and Padmakumari, 1980).

In close growing crops, the spacing between two drippers, spacing between two laterals, and the number of drippers per lateral are estimated by taking into consideration the movement of waterfront in the vertical as well as the horizontal direction in the soil (Murthy, 1998). The wetting soil surface of the two emitters behaved exactly as it would under two different single emitter systems till 2 h after the start of the system. But at 4 h after the start of the system, the wetting pattern of the soil profile was similar to one under a single emitter system (Patel and Rajput, 2001).

Micro sprinkler had a better economics than drip (Satyendra et al. 2007). For irrigating a 4 ha field, the sprinkler, flood, and drip irrigation may require 15 HP, 10 HP, and 1HP motors, respectively (Shantaraj 1983). Infection of apple by *Glomeralla cingulata* has been eliminated after the replacement of sprinkler systems of irrigation with drip system of irrigation (Subramanian et al., 1997).

A number of studies have also been carried out in the context of sugarcane using experimental data, which have found substantial water saving and productivity gains due to drip method of irrigation in sugarcane cultivation (Venugopal and Rajkumar, 1998). Saline water can be used with drip irrigation, unlike with any other method (Reddi et al. 2002). In fact, drip irrigation system is particularly advantageous when the water is saline (Mandal and Jana 1998). Before implementation of drip irrigation an analysis should be made regarding the costs, returns, cost-benefit ratio, and financial feasibility of the drip irrigation system under different crops (Patil and Angadi, 1997). The reduction in yield was 26%–52% in most of the crops under hexagonal geometry, while in tomato the yield reduction was marginal (Singh, 1978). The reduction in water consumption in micro irrigation also reduces the energy use (electricity) that is required to lift water from irrigation wells (Narayanamoorthy, 1997a, 2001).

Change in water productivity due to drip irrigation was calculated by Sivanappan (1994). The discounted value of the benefit-cost ratio (BCR) and the rate of return (RR) on drip investment are markedly higher in established than in newly established orchards (Sivanappan et al. 1987). Cost benefits and payback period of micro irrigation for various crops had been worked out by Sivanappan (1990, 1993). The pattern of movement and distribution of soil water resulting from drip sources can be quite different from those resulting from the more conventional modes of irrigation (Sivanappan, 1979). Till date, about 8.0 million ha area is under drip and sprinkler irrigation in India, covering different horticultural and grain crops. The area under micro irrigation is increasing gradually. The adoption of micro irrigation technology is more rapid in the western and southern states (Table 8.1). The area identified suitable for micro irrigation (drip and sprinkler) is given in Table 8.2.

TABLE 8.1
Area Covered (ha) under Drip and Sprinkler Irrigation as on March 31, 2012

S. No.	States	Drip	Sprinkler	Total
1	Andhra Pradesh	655,767	320,881	976,648
2	Arunachal Pradesh	613	0	613
3	Assam	116	129	245
4	Bihar	578	33,201	33,780
5	Chhattisgarh	8,798	133,961	142,759
6	Goa	849	1,544	1,544
7	Gujarat	299,279	246,852	546,131
8	Haryana	17,400	541,910	559,309
9	Himachal Pradesh	116	581	697
10	Jharkhand	1,273	8,842	10,115
11	Karnataka	293,593	385,675	679,268
12	Kerala	16,593	4,004	20,597
13	Madhya Pradesh	101,288	171,437	272,726
14	Maharashtra	779,295	346,600	1,125,895
15	Manipur	30	0	30
16	Mizoram	72	106	178
17	Nagaland	0	3,962	3,962
18	Orissa	12,109	43,958	56,067
19	Others	15,000	30,000	45,000
20	Punjab	26,758	11,533	38,291
21	Rajasthan	53,934	1,090,033	1,143,967
22	Sikkim	23,460	11,339	34,799
23	Tamil Nadu	201,396	28,196	22,592
24	Uttar Pradesh	13,963	16,832	30,794
25	Uttaranchal	38	6	44
26	West Bengal	538	150,576	151,114
Total		2,522,854	3,581,309	6,104,163

8.8.1 FERTIGATION STUDIES

The application of fertilizers through irrigation systems (fertigation) has become a common practice in modern irrigated agriculture. Increased yields, improvement in quality of product, irrigation and fertilizer use efficiencies, and protection of the soil environment are some of the main characteristics of this method, which made it very popular throughout the world. In some countries "Fertigation" is regarded as the second "Green Revolution". Fertilizers delivery systems may also be used for the application of herbicides, pesticides, and other chemicals (often referred to as "chemigation"). The reason why fertigation has become the state-of-the-art in vegetable cultivation is that nutrients can be applied to plants in the correct dosage and at the time appropriate for the specific stage of plant growth. Fertigation has increased dramatically in the past 15 years, particularly for sprinkler and drip systems.

TABLE 8.2
Area Suitable for Micro Irrigation in India

Crops	Cultivated Area (m ha)	Area Suitable for Drip Irrigation (m ha)
Sugarcane	4.10	4.10
Condiments & spices	2.19	1.40
Fruits	3.40	3.40
Vegetables	5.30	5.30
Coconut	1.90	1.90
Oil seeds	26.20	1.90
Cotton	9.00	9.00
Total	54.49	27.00

Source: H.P. Singh, Proceeding of International Conference on Micro and
Sprinkler Irrigation System, 2000.

For drip systems, the expansion is mostly in horticultural and high-value crops. With increased irrigation, a corresponding increase in fertigation has taken place. It will continue to grow since such systems result in less water usage and better uniformity and lend themselves to the technique much more readily than the less water-efficient and non-uniform furrow and flood systems being replaced. Fertigation uses either granular or liquid fertilizers, which are dissolved in water, and injected into the micro irrigation system. Nutrients can be applied through drip or spray systems, and can vary in concentration and composition. Fertigation provides uniform and relative ease of distribution of nutrients and can be fine-tuned to the nutritional requirements of a particular crop. In general, application of fertilizer with irrigation water gives a better crop response than either broadcast or foliar application. Fertigation is a must in order to realize the full potential and benefits of micro irrigation system.

Fertigation studies on onion with different fertilizer doses had been conducted by Patel and Rajput (2000). Earlier studies in arecanut demonstrated a yield increase of 45% with drip irrigation over basin method (Bhat and Sujatha, 2009) and improved nutrient movement of soil P and K in root zone with drip fertigation. An attempt has been made to study the effects of fertigation using commercially available granular fertilizers on okra, tomato, onion, and broccoli (Patel and Rajput, 2002a). More than 15% increase in the yield of okra was observed in fertigation when compared to the broadcasting method of fertilizer application at the same level of fertilizer application (Patel and Rajput, 2001). The limited root zone and the reduced amount of mineralization are the main reasons for the reduced nutrient availability to the plants with normal method of fertilizer application under drip irrigation.

A methodology was developed for using granular fertilizer for fertigation of broccoli (Patel and Rajput, 2002b). The yield response of pomegranate was compared when fertilizer was applied through fertigation and when fertilizer was broadcasted (Patel and Rajput, 2002c). Studies were undertaken to assess the effects of fertigation through drip on the growth, yield, and quality of banana (Pawar et al., 2001). A study was conducted to select the best fertilizer injector on the basis of its fixed cost, operating cost, nutrients concentration, head loss, and technical knowhow (Prasad 1997). A study on Bangalore Blue Grapes with seven treatments of normal and water-soluble fertilizer at various constraints level through drip fertigation was made by Shivashankar and Khan (1994). The quality of fruits like grapes and sapota were increased under fertigation (Shivshankar et al., 1998). As much as 50% of the N requirement of brinjal was reduced under drip irrigation over furrow irrigation. Drip irrigation with 75% nitrogen resulted in maximum plant height, leaf area, number of branches, and dry matter production over band placement of 100% nitrogen through furrow irrigation.

Water and fertilizer savings to the extent of 40%–70% and 30%–38%, respectively, with comparable yield levels were possible under the trickle fertigated crop as compared to the furrow-irrigated crop of potatoes (Chawla and Narda, 2001). On drip fertigation, cabbage and okra showed an increase in yield (46%–60%) as compared to conventional methods of irrigation and fertilizer application (Parikh et al., 1996). The application of herbicide through drip irrigation reduces the chance of human contact with the chemical and also minimizes the chances of leaching of the chemical (Patil, 1997). Lesser movement of K was attributed after fertigation due to large plant uptake of K. When nitrogen phosphorous and potassium (NPK) were applied through drip irrigation, higher tomato yield was obtained with 75% of the recommended dose (Singh et al., 2006). Highest yields of high-quality fruits of tomato were obtained with 50% trickle applied N + K grown on polyethylene mulched beds. In tomato, fertigation 1/2 N and K and black poly mulch was found to be good with respect to yield and growth parameters like yield of 121.3 t/ha, fruit weight (64.5 g), number of fruits/plant (62), yield/plant (4 kg), number of branches/plant (7.7), and number of clusters/plant (12.3). The fruit dry matter content (41.2%) was highest in the treatment 1/2 N and K fertigation through multi-K + black ploy mulch (Prabhakar et al., 2001). Drip fertigation of 80% recommended dose with water-soluble fertilizer registered 22.3% and 31.0% higher dry fruit yield over drip and furrow irrigation methods even with the same level and method of normal fertilizer application (Muralidhar et al., 1994). In tomato, there was considerable saving of fertilizers and water through fertigation using water-soluble fertilizers (Jeyabal et al., 2000). Application of 50% N and full dose of P and K as basal and remaining 50% N through fertigation at 15 days interval throughout the crop period significantly improved the yield and quality of tomato grown on coir pith mixed potting media absorbed more N than those conventionally.

In capsicum, 25% and 18% higher green fruit yield was recorded with fertigation and drip irrigation over furrow method (Gnanamurthy and Manickasundram, 2001). In chilli, fertigation using water-soluble fertilizers gave higher yield than soil application by about 15% and saved water by 40% (Jeyabal et al., 2000). Fertigation with water-soluble fertilizers in French beans saved 25% of fertilizers and recorded a yield increase of 28% over soil application of fertilizers (Jeyabal et al., 2000). But fruit yield of cucumber was maximum when N alone was fertigated through drip irrigation than fertigation of N along with K, NPK, and control (Rubeiz, 1990). In gherkins, fertigation with 100% NPK through poly feed and urea registered higher yield but considering the economics, 75% NPK through multi-K, mono-ammonium phosphate and urea was found to be the best (Jeyabal et al., 2000).

Fertigation at 300 kg N per ha provided the highest tuber yield (38.3 q/ha). Drip fertigation at 180 kg N per ha recorded tuber yield (30.6 q/ha) at par with furrow irrigation fertilized at 300 kg N/ha (30.5 q/ha) which indicate 40% nitrogen saving in potato (Patel and Patel, 2001). Potato crop fertilized by high frequency irrigation of fertigation techniques absorbed more N than those conventionally fertilized. In potato, four split nitrogen fertigation under drip irrigation resulted in higher WUE over furrow irrigation method (Keshvaiah and Kumaraswamy, 1993). Fertigation improved the yield of banana (Cv. robusta) and water economy (Mahalakshmi, 2000). The fertilizer savings through fertigation are presumably because fertilizer and water are applied to soil where active roots are concentrated (Srinivas, 1998). The increasing level of fertigation resulted in a significant increase in the height of the banana plant (Cv. Nendran). Maximum bunch weight of 9.82 kg was recorded in conventional crop geometry + 100% irrigation requirement + 100% N fertigation. Minimum of 7.5 kg bunch weight was recorded under normal % N fertigation (Pandey et al., 2013). Fertigation has proved to economize water and fertilizer with a corresponding lower expenditure in the cost of production and labor towards weeding, fertilization, and water application. An increase in the level of nitrogen and potassium fertigation improved the growth parameters of plants. However, differences beyond 100% were not significant. Furthermore, both levels and ratios of nitrogen and potassium fertigation influenced yield and yield attributes of banana robusta (Chandrakumar et al., 2001). A general reduction in the fruit quality was observed when irrigation was given in higher amounts at frequent intervals. TSS increased with increasing levels of fertilizer irrespective of the water levels (Hegde and Srinivas, 1990).

Crop duration was reduced by fertigation with 40 liters per day per pit + 75% of recommended N and K/pit in banana (robusta). Fertigation resulted in heavy bunches weighing' 36.5 kg with maximum number of hands (11) and number of fingers bunch (188.4). In banana, the same yield level obtained through conventional irrigation and fertilization practice can be achieved with half the fertilizer dose when applied through drip fertigation. Likewise, the WUE of 470 kg/ha mm in conventional irrigation was increased to 570 kg per ha per mm in subsurface drip fertigation system. In a 3 year old plantation of guava, fertigation at 75% recommended NK level with urea and multi-K gave 12.3% higher yield than soil application at 100% NK level, indicating a saving of 25% NK in addition to an improvement in productivity (Jeyabal et al., 2000).

Fertigation has been proved successful in a wide range of horticultural crop, particularly in fruit crops like citrus (Shrigure et al., 2000; Swietlik, 1992), grapes (Spayd et al., 1994) and date palm. Application of 10 liters of water/day + 13.5 9 urea and 10.5 g of muriate of potash/week through fertigation and soil application of super phosphate 278 g/plant in bimonthly intervals improved growth, yield, and quality characteristics in papaya. This might be due to the better physiological efficiency of the plant owing to supply of nutrients and water in splits through fertigation (Jeyakumar et al., 2001). In sapota, drip irrigation with fertigation (80% water-soluble fertilizer) increased fruit yield (5,800 kg/ha), whereas basin irrigation with recommended fertilizer application gave only 4,300 kg/ha (Gnanamurthy and Manicka Sundram, 2001).

In coconut, fertigation with water-soluble fertilizer at 80% recommended fertilizer improved the trunk girth (6%), number of fronds (18%), fruit bunches (21.5%), nut yield, and economized 20% fertilizer over control. In oil palm, fertigation with water-soluble fertilizer (80%) improved the trunk girth (18%), number of fronds (22%), and yield (83%) with a saving in fertilizer and water by 20% and 33% over control, respectively (Gnanamurthy and Manickasundram, 2001). In paprika, fertigation with urea and multi-K at 100% recommended NK level gave higher dry fruit yield of 63.8 q/ha which was 31.5% higher over the yield obtained with soil application of 100% NK and surface irrigation (48.5 q/ha) (Jeyabal et al., 2000). In okra, Kadam et al. (1993) also recorded higher WUE (374 kg/ha cm) under drip irrigation than furrow irrigation (214 kg/ha cm). Decreasing the fertilizer level by 20% than the recommended level, especially under fertigation conditions, may not affect the yield level in chilli because of *improved* fertilizer use efficiency. Fertigation under high density planting reduced the cost of production per kg of banana to as low as Rs. 0.83 with a possibility of economizing water and fertilizer with increase in productivity. The higher profit/rupee invested was realized with 150 g of N and K fertigation in 1:2 ratio (Chandrakumar et al., 2001). Trickle fertigation permits application of nutrients directly at the site of high concentration of active roots (Sivanappan et al., 1987). The cyclic regulation and continuous wetting of soil associated with drip irrigation maintained optimum moisture in the crop root zone, in turn, facilitating greater water and nutrient absorption (Rajput and Patel, 2002). Patel and Rajput (2003) also recorded higher agronomic efficiency in okra crop when drip fertigated with 60% recommended dose of fertilizer (RDF) than at 80% or 100% RDF. Ratoons help in extending the crushing schedule of sugar factories as they mature earlier than the plant crop due to early dehydration of tissues and flushing out of N. However, ratoon crop yields are lower than the plant crop yields due to soil compaction (Verma, 2002), decreased rate of soil fertility under continuous sugarcane cropping, and inefficient use of applied fertilizers (Sundara and Tripathi, 1999).

Sivakumar (2007) at Tamil Nadu Agriculture University (TNAU) conducted an experiment in mango cv. Ratna planted under high density plantation during 2005–2007 to study the influences of N and K nutrients applied through fertigation. Studies on uptake of nutrients in tissue culture banana cv. Robusta (AAA) also revealed that N, P, and K uptake increased linearly till the shooting stage (Nalina 2002). Kumar and Kumar (2007) recommend post shoot foliar spraying of standard operating procedure (SOP) 1.5% twice; first at the time of last hand opening and second 1 month from the first spraying to improve the bunch weight and quality of fruits in certain commercial varieties of banana. Mahalakshmi (2000a) revealed that under both the normal and high density system of planting, fertigation was effective in improving the yield and maintaining fruit quality. Nalina (2002) justified that the application of 150% of recommended NPK (i.e., 165:52.5:495 g) in four splits, viz. 2, 4, 6, and 8 months after planting, was found essential to increase the plant's growth and development, yield and quality in the plant and ratoon crops of technical committee banana. Improper and inadequate nutrition is one of the major causes of citrus decline in India. Studies on the decline of mandarins in Kerala showed that poor nutrient status of soil (Iyer and Iyengar, 1956) and neglect and lack of manuring are the main causal factors.

Results of a study conducted on 20-year-old seedling trees of sweet orange cv. Sathgudi revealed that the maximum yield and cost benefit ratio with better fruit quality could be obtained by the balanced nutrition through 400:150:300 g NPK per plant per year along with organic fertilizer (castor cake @ 7.5 kg) (Seshedri and Madhavi, 2001). Similarly, integrated nutrient management studies conducted at Akola revealed that application of chemical fertilizers along with neem cake significantly increased the yield with better quality fruits (Ingle et al., 2001). Ingle et al. (2003) found that in Nagpur mandarin, the number and weight of fruits and total soluble sugars were highest with application of 800 g N, 300 g P_2O_5, and 600 g K_2O along with 7.5 kg neem cake per plant per year. Tiwari et al. (1999) also obtained maximum yield of sweet orange with the application of 800 g N, 300 g P_2O_5, 600 g K_2O plus 15 kg neem cake/tree/year. Shirgure et al. (2001) found that fertigating Nagpur mandarin with NPK 50:140:70 kg/ha is good in improving the tree vigour, yield, and quality of fruits. Secondary and micronutrient deficiencies are common in almost all citrus species. Each state has its own recommendations which involve application of $ZnSO_4$ (0.3%–0.5%), $MgSO_4$ (0.2%–0.3%), $MnSO_4$ (0.1%–0.3%), $CuSO_4$ (0.3%), $FeSO_4$ (0.2%–0.3%), and Borax (0.05%–0.1%) two to three times on the new flushes to get good yield and quality fruits. The nutrient uptake studies conducted at the Tamil Nadu Agricultural University revealed that among the important nutrients, the demand for K was more, nearly twice as that of nitrogen in papaya (Veerannah and Selvaraj, 1984).

Trials conducted at the Tamil Nadu Agricultural University under the IPI collaboration project on the role of graded doses of K_2O (i.e., 0, 150, 300, and 450 g/plant) along with 300 g each of N and P_2O_5 per plant per year (Kumar et al., 2006) recorded the importance of potassium nutrition in significantly influencing the fruit weight, fruit number, and fruit yield per plant. According to Irulappan et al. (1984), six splits were found to be good while Ravichandrane et al. (2002) recommended 12 splits to get higher yield and quality of fruits in CO-2 papaya variety. Jeyakumar et al. (2001) scheduled the nutrient application under drip fertigation system at weekly intervals. Trials conducted at Assam (Mohan and Ahmed, 1987) and West Bengal (Sen, 1985) conditions stressed the importance of balanced nutrition to get higher yield of pineapple.

8.8.2 Protected Cultivation Studies

The continuous and rapid climate change and consequent biotic and abiotic stresses are posing a serious threat to the agricultural production system. Increasing cost of energy and inputs required for agricultural production are further adding to the woes of the farmer. In the above background, controlled environment agriculture in the form of protected cultivation and PF has emerged as potential alternative technology to combat such constraints. Protected cultivation technology as an integral component of PF has the ability and potential to enhance input use efficiency to a significant level so as to achieve sustainability in food production.

Besides sporadic efforts like Defense Research and Development Organization (DRDO) in Ladakh, protected cultivation technology in India for commercial production is hardly 3 decades old, whereas in the developed countries, namely Japan, Holland, Russia, UK, China, and others, it is about two centuries old. China started protected cultivation in the 1990s and today the area under protected cultivation in China is more than 2.5 m ha and 90% of the area is under vegetables. In China, low-cost protected technology, viz. plastic mulches, plastic low tunnels, and walk-in tunnels are being used on 80% of the total area under protected cultivation and perhaps this is the basic reason that today China is the largest producer of vegetables in the world. In recent years, Israel is one country which has taken big advantage of this technology by producing quality vegetables, flowers, fruits, etc. in water-deficit desert area for meeting not only its small domestic demand but also the huge export demands. Recently, India has established collaboration with Israel for demonstration of this technology at the Center for Protected Cultivation Technology (CPCT), Indian Agricultural Research Institute, New Delhi, which has paid good dividend and as a result we are discussing today the advances made in the application of this technology in India. Recently, two Centers of Excellence for Vegetables and Fruit Crop Nurseries established in Haryana are functioning very well at Karnal and Sirsa, respectively, in collaboration with Israel. Also Indo-Israel collaborative projects on protected cultivation in other states are doing well. The international associations with countries such as Israel are expected to act as a major boost to protected cultivation in India. The major challenges for the adoption of protected cultivation are lack of awareness, cost-effectiveness of the methods, and lack of market linkages. If these aspects are properly addressed by the industrial players and the government, then the protected cultivation industry will showcase remarkable growth in the next 5 years. India, at present, is the second largest producer of vegetables in the world. The area under all forms of protected cultivation is around 50,000 ha. Indo-American Hybrid Seeds (India), Bangalore, is the pioneer in India for making use of greenhouse technology since 1965 for the commercial productions of flower seeds.

The total area under protected cultivation in India is about 50,000 ha only. The state-wise area of protected cultivation under different crops is presented in Table 8.3. Maharashtra is the leading state in India with 15,000 ha area under protected cultivation. Mainly carnation, gerbera, rose, and capsicum crops are under practices in greenhouse. Karnataka is in second position with 10,000 ha area under protection cultivation followed by Himachal Pradesh which has an area of 5,000 ha. Other states which have areas under protected cultivation are Punjab, Uttarakhand, Haryana, U.P., Gujarat, Rajasthan, Jharkhand, J&K, Delhi, West

TABLE 8.3
Leading States in Protected Cultivation

S. No.	State	Area (ha)	Crops
1.	Maharashtra	15,000	Carnation, gerbera, rose, capsicum
2.	Karnataka	10,000	Roses, gerbera, carnation, seed, nursery
3.	Himachal Pradesh	5,000	Capsicum, carnation, gerbera, tuberose
4.	Punjab	4,000	Vegetable crops
5.	Uttarakhand	3,000	Gerbera, capsicum
6.	Tamil Nadu	2,100	Floricultural crops
7.	North-Eastern	2,000	Floricultural and vegetable crops

Other States – Haryana, Uttar Pradesh, Gujarat, Rajasthan, Jharkhand, J&K, Delhi, West Bengal, Orissa, Bihar, Madhya Pradesh

Bengal, Orissa, Bihar, M.P., etc. Farmers are generally growing roses, gerbera, carnation under flowers, and capsicum, tomato, cucumber under vegetables crops.

8.8.3 AUTOMATION OF MICRO IRRIGATION

Automation of micro irrigation system is a relatively new concept in India, though such types of systems are quite popular in developed countries where manual labor is costly and high irrigation efficiency is required to enhance the yield and quality of agricultural produce. Using automation one can control the irrigation valves, pumps, and fertilizer injectors automatically with minimum or even no manual interventions. Automation of irrigation is well justified where a large area to be irrigated is divided into small segments which are irrigated in sequence to match the flow of water available from its source. Its advantages over conventional micro irrigation system are saving in man power, energy and fertilizer, flexibility of operations, elimination of manual operations to open or close valves, possibility to change the frequency of irrigation and fertigation process, increase in water and fertilizer use efficiency, the system can be operated night and day, and the pump can be started/stopped exactly when required. A few studies indicate sensor-based irrigation-based scheduling in India. Luthra et al. (1996) designed and developed an automatic valve for auto irrigation system. In this system, soil water tension is sensed through a modified manometer-type tensiometer and their design provides control of irrigation at the pre-decided soil water tension and programmed timer. The developed auto irrigation system monitors soil water stress in the root zone continuously and controls irrigation as per the present value of soil water tension and duration of crop. Joshi et al. (1999) developed a soil moisture-based automated irrigation controller at Indian Institute of Technology (IIT), Kharagpur. The irrigation controller uses a tensiometer connected with a U-tube manometer to sense soil moisture potential. Automated irrigation controller has been designed and developed based on the principle of soil moisture potential.

A tensiometer in conjunction with mercury manometer and two sensors has been designed to detect the change in mercury level due to variation in soil moisture. The sensors give feedback to the electronic circuitry to actuate and de-actuate the irrigation pump. The developed irrigation controller was tested in laboratory and a field plot was irrigated through drip system operated with an electric motor-driven mono-block pump. The irrigation controller actuated and de-actuated the irrigation pump exactly at the present position of the sensors during the field test. The developed irrigation controller can be successfully used for automating the irrigation system.

Das et al. (2010) use a combination of wired and wireless sensors to collect sensory data such as soil pH, soil moisture, soil temperature, etc. Prathyusha et al. (2013) designed a microcontroller-based drip irrigation mechanism, which is a real-time feedback control system for monitoring and controlling all the activities of drip irrigation system more efficiently. The irrigation system controls the valves by using an automated controller to turn ON & OFF. This allows the farmer to apply the right amount of water at the right time, regardless of the availability of the labor to turn the valves or motor ON & OFF. Jitendra et al. (2013) developed a tensiometer-based soil moisture sensor at IARI, New Delhi and the integration of sensors network was done with telecommunication technologies having modified tensiometer, level sensor, controller, GSM receiver, transmitter, solenoid valve, water meter, and pump, to develop the automated irrigation system. The developed system was evaluated and tested successfully in a field planted with Okra (*Abelmoschus esculentus*) crop. This system has a vast potential for PF with sensor-based application in a fully automated system to interpolate over an area for spatial decision making needing to be tapped for making agriculture attractive in future.

8.8.4   SURFACE IRRIGATION AUTOMATION

There are various types of semi-automatic and automatic irrigation systems being developed and tested at different locations. Semi-automation of irrigated basins and borders is carried out using a water sensor feedback control system for surface irrigation. These studies mainly focus on the development and testing of different types of turnout gates and system operation and control. Irrigation automation is based on single-function drop-closed check gates in the supply ditch and weir-crest turnouts into the bay. Single-function gates are defined as those which either open or close only when tripped and manually reset. While dual-function gates, which have one ply that opens the gate and another that drops closed to terminate irrigation were used between the ditch checks. The use of mechanical and electrical timers and electric actuators to open and close these gates (trip cord system) were tested in the study. The system was found to be stable and requiring minimal maintenance.

The semi-automatic gates system was further extended with a water sensor feedback control system. The function of the sensor is to send a signal to terminate the irrigation. The signal was sent using wire or infrared telemetry to a station controller or receiver at the upper end of the bay. This signal actuates electric solenoids used to release gate latches to open or shut the control gates. The system was tested in a level basin irrigation system.

In a recent development, an automatic control system "Aquator" is developed by AWMA (2006). The Aquator system can be divided into two main parts, the "Base Station" and the "Field Units". The base station controller is generally located at the irrigator's home. It consists of the Aquator software program, a master transmitter, a roof mounted aerial, and a telephone dialer with battery back-up. A field unit is located at the bay to be remotely monitored or the irrigation structure (control gates) to be controlled. Each field unit (node) contains a radio receiver/transmitter (transceiver), relays, and controls electronics, aerial, solar panel, lead acid gel battery, battery voltage sensor, and weatherproof enclosure. When the field unit receives the signal from the base station it can open or close structures. Once the operation has been performed, a confirmation signal is sent back to the base station. The control system is independent of the type of gates which can be used for supply of water to the bays. Each node control system is costing about US$ 1100. The automation of surface irrigation by cut-off time or cut-off distance control was carried out by Niblack and Sanchez (2008).

The present operational irrigation controls systems are based on time-based turn on and turn off. There is no monitoring of soil moisture status, flow rates, and water storage and runoff. The logic to determine irrigation on-time is based on the irrigator's heuristic knowledge of the system. However, the system facilitates better operational conditions, reduces labor requirement, and improves the life style of the irrigator.

The limitations of the existing control systems are:

- The decision of "when to irrigate" is not informed but it is rather dependent on the irrigator's experience. The probability of over and under-irrigation could lead to loss of productivity especially in new fields.
- The application of irrigation water to each bay is time-based and based on the number of years of operation of the irrigation system by the irrigator. This could lead to lower application efficiencies.
- The control system also does not monitor or quantify the various water balance components of individual irrigation events such as inflow, soil moisture storage, infiltrated volume, and deep percolation (if any), or can evaluate seasonal performance or irrigation in the context of the whole enterprise.

Although limited work has been done in the area of PF, a lot of work has been done in the area of micro irrigation including sprinkler, drip and subsurface drip, sensor development, and protected cultivation of high-value vegetables and ornamental crops in the institute in the last 3 decades. However, there are still gaps and concerns that need to be addressed during the Twelfth Plan for fast expansion and sustainability of the PF technologies for small fields in the country.

8.9   EVALUATION OF PF

i. Economic analysis: Whether it is cost effective?
ii. Environmental assessment: Does it improve the quality of the environment or at least does not harm it?
iii. Rate of ToT (Transfer of Technology): Do farmers adopt it rapidly?
iv. Integration with industry and linkage with market through supply chain?

8.10   BENEFITS OF PF

- The concept of "doing the right thing in the right place at the right time" has a strong intuitive appeal which gives farmers the ability to use all operations and crop inputs more effectively.
- More effective use of inputs results in greater crop yield and/or quality, without polluting the environment.
- PA can address both economic and environmental issues that surround production agriculture today.

8.11   DRAWBACKS OF PF

- **High cost:** It has proven difficult to determine the cost benefits of PA management. At present, many of the technologies used are in their infancy, and pricing of equipment and services is hard to pin down.
- **Lack of technical expertise knowledge and technology:** The success of PA depends largely on how well and how quickly the knowledge needed

to guide the new technologies can be found. (India spends only 0.3% of its agricultural GDP in Research and Development.)

- Not applicable or difficult/costly for small land holdings.
- Heterogeneity of cropping systems and market imperfections.

8.12   OPPORTUNITIES AND CHALLENGES

PF can have a positive impact on the environmental quality. The opportunity exists to show producers how changing production practices will not place crops at risk and produce positive economic and environmental benefits. Conducting experiments on PA will require field or farm-scale studies and perhaps watershed-scale adoption of new management practices.

PF study will require attention to the following points at greater scale:

i. Appropriate questions that can be addressed at the field scale.
ii. Methods for measuring environmental end points that will demonstrate the efficacy of management practices.
iii. Commitment to multiple years of study to overcome meteorological variation.
iv. Adequate monitoring equipment for crop production, soil properties, and environmental quality in order to understand the changes occurring due to the management practices.
v. Use of comparison fields or farms in which no changes are made to provide a validation of the improved practices.
vi. Cooperation of producers to implement the practices with minor modifications across years so that variations can be isolated to the management practice and not producer influence.
vii. Database structure that includes geographic information layers and accurate GPS equipment to position any treatments in the same area across years.
viii. Funding sources that will allow for long-term studies across large areas.
ix. Evaluation of experimental design, implementation, and end results.
x. Commitment from the scientists, producers, and educators involved to maintain interest in the project over a sufficient period of time to allow the original objectives to be achieved.

8.13   FUTURE THRUST AREA FOR RESEARCH – TECHNOLOGY GAPS AND RESEARCH NEEDS

8.13.1   MICRO IRRIGATION

Design improvement is a continual process to suit the conditions of irrigation. A few of the challenges or future perspectives in development of micro irrigation system include:

i. Development of crop-specific package of practices for horticultural crops
Optimization of water and nutrient use in any crop requires the exact information on its periodical water and nutrient requirements because

micro irrigation systems are potent mechanisms to apply controlled quantities of water and nutrients at predetermined timings. This makes it a prerequisite that the production functions of horticultural crops are developed to determine the appropriate schedules for irrigation and fertigation. Therefore, crop-specific package of practices for horticultural crops grown under micro irrigation needs to be developed.

ii. Design of hybrid irrigation system

Huge schemes of the Government of India are being implemented in different states of the country for development of watersheds and development of water resources through community participation. But in all these schemes the emphasis remains only on the development of water resources and not on its efficient utilization. Similarly, in canal command areas farmers are increasingly adapting to construct water storage reservoirs to store canal water because no strict canal operation schedules are adhered to. Thus a number of water storage tanks are coming up through watershed development schemes and community water resource development schemes for storing canal water. All these tanks can be used as a catalyst for adopting micro irrigation systems.

iii. Design and development of sensors for automation

Micro irrigation enables a large degree of control of water application in a well-designed system. Water application is precise with a high degree of uniformity among drippers. It is possible, therefore, to apply water according to the exact demands of the plant as determined by changes in weather, crop growth stage, and soil water salinity. The amounts of fertilizer applied can also be adjusted to demand, according to the plant growth stage. Automation is employed to control the system's main valve in order to supply the desired amount of irrigation water. There is a need to develop low-cost sensors to achieve automation in micro irrigation. It is necessary to design and develop an indigenous sensor network-based irrigation system for increasing the agriculture water productivity.

iv. Low-head–low-energy drip irrigation system design

The low-head drip irrigation system (LHS) is a systematic development of a low-cost drip system which performs as drip, except that the water is applied on the soil surface or below the soil surface through discrete emitters with low discharge rates. The LHS provides an effective low energy and economical upgrade for flood/furrow irrigation. The LHS is specifically designed to enable farmers to grow vegetables in the courtyard minimizing front end investment, provide fast return on investment, reduce energy cost on pumping and pressurizing, move and reuse equipment easily, provide low system maintenance and management.

v. Designing emitters suitable for solar powered irrigation system

Uninterrupted supply of electricity is a major constraint to promote micro irrigation system. Solar-operated drip irrigation system can be a potential solution to enhance sustainability of energy usage. A specially designed emitter suitable for use with solar pumping system is needed to be developed.

vi. Designing emitter suitable for industrial/municipal waste water

With increasing industrialization and urbanization, a huge load of waste water is generated. This waste water can be treated and reused for irrigation purpose. Suitable emitter geometry needs to be designed for use of treated waste water.

vii. Drip design for soilless media

Introduction of new non-leakage feature in the emitter design has been useful in greenhouse using soilless substrate or highly drained media where irrigation is done frequently with a very short irrigation cycle. With the introduction of non-leakage function, the problem of low-head drainage is eliminated, achieving uniformity of water distribution in the short irrigation cycle.

viii. Micro irrigation design for intercropping

As regards the technology gaps in the micro irrigation system hardware, it may be pointed out that no appropriate micro irrigation system is available for intercropping and mixed cropping, particularly in orchards where good opportunity exists to increase the productivity of land by suitable intercropping. There is a need to develop information to recommend different micro irrigation systems under different soil, crop, and climatic conditions.

8.14 PROTECTED CULTIVATION

Protected structures have revolutionized the vegetable-production technology by way of quality vegetable seedling production using plastic trays and sterilized coco peat. Polyhouse production systems have extended the growing season of vegetables and made their availability year-round. It has opened the possibilities of growing exotic vegetables such as colored capsicums, cherry tomatoes, seedless watermelons, icebox melons, sweet cucumbers, colored cabbage, etc., besides regular crops like tomato, cucumber, eggplant, and climbing beans. Polyhouse technology has been proved to reduce the amount of irrigation water required for crop production and improve the nutrient use efficiency. To minimize the adverse effect of extreme climate conditions such as high temperature effects, dry winds, scorching sun, etc., protected structures are useful. Excess rainfall effects can be very well overcome by growing vegetables under rain-shelters. Polyhouse is also useful in reducing the incidence of pests and diseases, which are emerging as a bigger threat in changing climatic conditions.

Naturally ventilated greenhouses, which are less expensive compared to fan and pad greenhouses, are not suitable for hot climate or even for peak summer periods of semi-arid regions. Continuous cultivation of the same crop species season after season has resulted in the buildup of nematodes and soil borne pathogens which are difficult to manage. The availability of quality irrigation water for use is a problem in arid and semi-arid regions. Post harvest handling and transport facilities are insufficient for polyhouse-grown vegetable crops. Most of the polyhouse produce is for the riches/up-end market and effective large scale marketing channels and networks are not available.

Cost-effective structures have to standardize according to the prevalent local conditions. Research efforts are required to breed hybrids suitable for polyhouse cultivation. New crops, cropping systems based on market requirement, and GAP are to be standardized for the sustenance of polyhouse cultivation. Economy in critical inputs such as water fertilizers need to be addressed. Bio-intensive soil and crop management practices including organic production protocols need to be developed to overcome many biotic stresses. Post harvest handling and value addition are to be standardized to make this venture more profitable. Domestic and export market needs to be studied for the polyhouse grown produce.

R&D work on protected cultivation technology has been carried out in some parts of the country. The work has mainly focused on developing package of practices for different flower and vegetable crops. Better water and nutrient use efficiency in protected cultivation requires special attention through the incorporation of novel technologies such as super-absorbents. Considerable database is available on various parameters required for protected cultivation in different climatic conditions.

Despite successful development of the technology in R&D institutions, adoption of protected cultivation technology, especially at small growers' level, has not been satisfactory. There are sporadic instances of its success in Maharashtra and Karnataka but its benefits are yet to reach in the northern parts of the country. There is a huge domestic market in north India for high-value vegetables and flowers. Small farmers can greatly benefit from this market if they are provided with the right mix of protected cultivation technology to grow high-value crops. Apart from knowledge, skill, and financial constraints, there are discernible gaps across production to marketing, attributable to poor level of adoption of protected cultivation.

The protected cultivation technology aims to refine this technology at the farmers' fields in a holistic manner of production to consumption chain management – addressing crucial areas and technology gaps. Focused scientific studies would be taken up in areas such as production technology, indigenous varieties, pest mitigation and standardization of GAP protocols, utilization of novel technologies such as super-absorbents for better water and nutrient use efficiency, value addition, cool chain management, and establishment of marketing linkages including entrepreneurship skill development involving all the stakeholders. It aspires to standardize a perfect production technology and a model chain from production to consumption through skilled manpower among growers, traders, and stakeholders equipped with entrepreneurial acumen to optimize the production and marketing potentials.

High-quality vegetables and flowers have great potential both in domestic and overseas markets. The demand is ever-increasing. India has the advantage of producing these products at a relatively low cost if the production technology and location are carefully chosen. At present, adequate database on production technology is available which can be used as a take-off level for commercializing the technology on the farmers' fields. So far the technology has not been adopted at the small farmers' level due to mismatch between investment and returns. A successful commercial model can result by forming a cohesive group of farmers and it can produce enough volume for effective marketing. The project envisages developing such models in partnership mode. Availability of quality and true-to-type planting material of fruit crops has been a constraint in the area expansion of good quality fruit crops. It is proposed to develop scion banks and mother nurseries at cooperating centers. These will be linked to satellite nurseries which will grow planting material under protected conditions in all seasons on a commercial scale. Low productivity and poor quality of horticultural crops in our country is mainly due to poor water and nutrient use efficiency. Commercial use of super-absorbents can improve water and nutrient use efficiency, particularly in protected cultivation.

Some of the innovative researchable areas in the field of intensive/protected/hi-tech cultivation are:

- Extending the range of crops that can be grown under protective cultivation. At present it is restricted only to a few crops like capsicum, cucumber, tomato, gerbera, etc. There is great opportunity to standardize the cultivation practices under protective cultivation for crops like leafy vegetables, ladies finger, peas, beans, radish, carrot, melons, onions, ginger, etc.
- Developing mixed cropping practices to make best use of the space and other resources and also to bring a bio-park effect in order to minimize the plant-protection efforts.
- Use of natural enemies for pest management.
- Introducing soilless culture to overcome many hindrances cause by soil for nutrient availability.
- Standardizing practices like hydroponics, nutrient film technique, aeroponics, etc., simultaneously making use of the water for aqua culture also.
- Eco-friendly systems for green energy production like using bio-gas, bio-mass, solar, wind, or hybrid systems to meet the energy needed for running the systems in protected cultivation.

8.15   SURFACE IRRIGATION PROCESSES FOR AUTOMATION

The surface irrigation processes are governed by universal physical laws which, in turn, can be expressed as functions of a number of physical quantities. The physical quantities affecting the performance outcomes of an irrigation event are generally of two types: (i) System variables are those whose magnitude can be varied, within a wide band, by the irrigator; and (ii) System parameters include those physical or empirical quantities that measure the intrinsic physical characteristics of the system and very little modification is possible in those parameters. These parameters are variable with location, soil type, stage of crop, irrigation event, etc.

Generally, basin geometry, inflow rate, and cutoff time are considered as system variables while net irrigation requirement, surface roughness coefficient, slope, and infiltration parameters could be considered as system parameters. The performance parameters or objectives of an irrigation event could be described by three indices: (i) application efficiency; (ii) water requirement efficiency; and (iii) distribution uniformity.

8.16   THE POLICY APPROACH TO PROMOTE PF AT FARM LEVEL

- Identify the niche areas for the promotion of crop-specific PF.
- Creation of teams involving agricultural scientists, engineers, manufacturers, and economists to study the overall scope of PA.

- Promote the progressive farmers for PF technology who have sufficient risk bearing capacity.
- Encourage the farmers to study the spatial and temporal variability of the input parameters using primary data at field level.
- Provide complete technical backup support to the farmers to develop pilots or models, which can be replicated on a large scale.
- Pilot study should be conducted on the farmers' fields to show the results of PA implementation.
- Encourage the farmers to adopt water accounting protocols at farm level and to use micro level irrigation systems and water saving techniques.
- Government legislation restraining farmers from using indiscriminate farm inputs and thereby causing ecological/environmental imbalance would induce the farmers to go for alternative approaches.
- Creating awareness amongst farmers about the consequences of applying imbalanced doses of farm inputs like irrigation, fertilizers, insecticides, and pesticides.
- Policy support on procurement prices, efficient transfer of technology to the farmers, formulation of cooperative groups or self help groups since many of the PA tools are costly (GIS, GPS, RS, etc.).

8.17   CONCLUSIONS

PF gives farmers the ability to use crop inputs more effectively including fertilizers, pesticides, tillage, and irrigation water. More effective use of inputs means greater crop yield and/or quality without polluting the environment. However, it has proven difficult to determine the cost benefits of PA management. At present, many of the technologies used are in their infancy, and the pricing of equipment and services is hard to pin down. This can make our current economic statements about a particular technology dated. PA can address both the economic and environmental issues that surround production agriculture today. Questions remain about cost-effectiveness and the most effective ways to use the technological tools we now have, but the concept of "doing the right thing in the right place at the right time" has a strong intuitive appeal. Ultimately, the success of PA depends largely on how well and how quickly the knowledge needed to guide the new technologies can be found.

8.18   POLICY FOR PROMOTION OF PF IN INDIA

The approaches required to be adopted by the policy makers to promote PF at farm level are:

  i. Promote the PF technology for the specific progressive farmers who have sufficient risk bearing capacity as this technology may require capital investment.
 ii. Identification of niche areas for the promotion of crop-specific organic farming.
iii. Encourage the farmers to adopt water accounting protocols at farm level.

   iv. Promote use of micro level irrigation systems and water saving techniques.

   v. Encourage study of spatial and temporal variability of the input parameters using primary data at field level.

   vi. Evolve a policy for efficient transfer of technology to the farmers.

   vii. Provide complete technical backup support to the farmers to develop pilots or models, which can be replicated on a large scale.

   viii. Policy support on procurement prices, in formulation of cooperative groups or self help groups.

   ix. Designation of export promotion zones with necessary infrastructure such as cold storage, processing, and grading facilities.

8.19   FUTURE OF PF

Opportunities will continue for PF studies and technological intervention. Tools will become available to apply chemicals, fertilizers, tillage, and seed differentially to a field and collect the yield or plant biomass by position across the field. RS technology will allow us to observe variations within a field throughout the growing season, relative to the imposed management changes. Monitoring equipment exists for capturing the surface water and groundwater samples needed to quantify the environmental impact through surface runoff or leaching. The technology exists to capture the volatilization of nitrogen or pesticides from the field into the atmosphere from modified practices. The future direction of agriculture will depend upon the research community's ability to conduct this type of study, with confidence from the environmental and producer communities that changes will benefit the environment and increase the efficiency of agricultural production. There are depletions of ecological foundations of the agro-ecosystems, as reflected in terms of increasing land degradation, depletion of water resources and rising trends of floods, drought, and crop pests and diseases. There is an imperative socio-economic need to have enhanced productivity per units of land, water, and time.

   i. At present, 3 ha of rain-fed areas produce cereal grain equivalent to that produced in 1 ha of irrigated land. Out of 142 million ha of net sown areas, 92 million ha are under rain-fed agriculture in the country. PF technology can be applied in this area successfully.

   ii. From the equity point of view, even the record agricultural production of more than 200 Mt is unable to address the food security issue. Close to 60 Mt food grains in the storehouses of Food Corporation of India (FCI) is beyond the affordability and access to the poor and marginalized in many pockets of the country.

   iii. Globally, there are challenges arising from Globalization, especially the impact of World Trade Organization (WTO) regime on small and marginalized farmers.

   iv. Some other unforeseen challenges could be anticipated, e.g., the global warming scenario and its possible impact on diverse agro-ecosystems in terms of alterations in traditional crop belts, micro-level perturbations in hydrologic cycle, and more uncertain crop-weather interactions, etc.

BIBLIOGRAPHY

Abuj, M.D., Magar, A.P., Bombale, V.T., Suryawanshi, S.L. and Popale, P.G. 2010. Calibration of fertilizer tank. *International Journal of Agricultural Engineering*, 3(1), pp. 101–104.

Ajdary, K., Singh, D.K., Singh, A.K. and Khanna, M. 2007. Modelling of nitrogen leaching from experimental onion field under drip fertigation. *Agricultural water management*, 89(1–2), pp. 15–28.

Alila, P. and Srivastava, A.K. 2008. Slow release fertilizers and citrus: Emerging facts. *Agricultural Reviews*, 29(2), pp. 99–107.

Almas, L.K., Amosson, S.H., Marek, T. and Colette, W.A. 2003 *Economic Feasibility of Precision Irrigation in the Northern Texas High Plains*. Southern Agricultural Economics Association Annual Meeting, Mobile, AL.

Anjanappa, M., Venkatesh, J. and Kumara, B.S. 2013. Influence of organic, inorganic and bio fertilizers on flowering, yield and yield attributes of cucumber (cv. Hassan Local) in open field condition. *Karnataka Journal of Agricultural Sciences*, 25(4), pp. 493–497.

Antony, E. and Singandhupe, R.B. 2004. Impact of drip and surface irrigation on growth, yield and WUE of capsicum (Capsicum annum L.). *Agricultural water management*, 65(2), pp. 121–132.

AWMA. 2006. "Aquator" Automated irrigation, ltd. product brochure.

Bangar, A., and Chaudhari, B. 2001. Nutrient mobility in soil, uptake, quality andyield of suru sugarcane as influenced by fertigation through drip in medium vertisols. *Microirrigation*, (282), 480.

Bangar, A.R. and Chaudhari, B.C. 2004. Nutrient mobility in soil, uptake, quality and yield of Suru sugarcane as influenced by drip-fertigation in medium vertisols. *Journal of the Indian Society of Soil Science (India)*, 52(2), pp. 164–171.

Behera, S.K. and Panda, R.K. 2009. Effect of fertilization and irrigation schedule on water and fertilizer solute transport for wheat crop in a sub-humid sub-tropical region. *Agriculture, ecosystems & environment*, 130(3–4), pp. 141–155.

Bhat, R. and Sujatha, S. 2009. Soil fertility and nutrient uptake by arecanut (Areca catechu L.) as affected by level and frequency of fertigation in a laterite soil. *Agricultural Water Management*, 96(3), pp. 445–456.

Booker, J.D., Bronson, K.F., Trostle, C.L., Keeling, J.W. and Malapati, A. 2007. Nitrogen and phosphorus fertilizer and residual response in cotton-sorghum and cotton-cotton sequences. *Agronomy Journal*, 99, pp. 607–613.

Borah, S.C., Barbora, A.C. and Bhattacharyya, D. 2001. Effect of organic and inorganic manuring on growth, yield and quality of Khasi mandarin (Citrus reticulate Blanco.). *South Indian Horticulture*, 49, pp. 115–118.

Brahma, S., Barua, P., Saikia, L. and Hazarika, T. 2009. Studies on response of tomato to different levels of N and K fertigation inside naturally ventilated polyhouse. *Vegetable Science*, 36(3s), pp. 336–339.

Bronson, K.F., Silvertooth, J.C. and Malapati, A. 2007. Nitrogen fertilizer recovery efficiency of cotton for different irrigation systems. 2007 Proceedings Beltwide Cotton Conferences. [CD-ROM computer file]. National Cotton Council of America, Memphis, TN.

Chandrakumar, S.S., Thimmegowda, S., Srinivas, K., Reddy, B.M.C. and Devakumar, N. 2001. Performance of Robusta banana under nitrogen and potassium fertigation. *South Indian Horticulture*, 49, pp. 92–94.

Chandrashekara, C.P. 2009. Resource management in sugarcane (Saccharum officinarum L.) through drip irrigation, fertigation, planting pattern and LCC based N application, and area-production estimation through remote sensing. Doctoral dissertation, UAS Dharwad.

Chauvan, H.S. and Shukla, K.N. 1990. Proceedimgs of the XI International Congress the Use of Plastics in Agriculture, February 26–March 02. New Delhi, p. 57.

Chawla, J.K. and Narda, N.K. 2001. Economy in water and fertilizer use in trickle fertigated potato 1. *Irrigation and Drainage: The Journal of the International Commission on Irrigation and Drainage*, 50(2), pp. 129–137.

Clouse, R.W. 2006. Spatial application of a cotton growth model for analysis of sitespecific irrigation in the Texas High Plains. PhD dissertation, Texas A&M University.

Cornish, G.A. 2001. Pressurized irrigation technologies for small holders. In *Micro Irrigation*, H.P. Singh, S.P. Kaushik. A. Kumar, T.S. Murthy and J.C. Sameul (Eds.), pp. 102–115. New Delhi.

Das, I., Shah, N.G. and Merchant, S.N. 2010. *AgriSens: Wireless Sensor Network in Precision Farming: A Case study*. LAP Lambert Academic Publishing, Germany. ISBN: 978-3-8433-5525-4.

DeJonge, K.C. and Kaleita, A.L. 2006. Simulation of spatially variable precision irrigation and its effects on corn growth using CERES-Maize. ASABE Paper No 062119.

DeJonge, K.C., Kaleita, A.L. and Thorp, K.R. 2007. Simulating the effects of spatially variable irrigation on corn yields, costs and revenue in Iowa. *Agricultural Water Management*, 92, pp. 99–109.

Dingre, S.K., Pawar, D.D. and Lokre, V.A. 2012. Drip fertigation scheduling for enhancing productivity of onion seed in western Maharashtra. *Progressive Horticulture*, 44(2), pp. 271–275.

Fanish, S.A. and Muthukrishnan, P. 2011. Effect of drip fertigation and intercropping on growth, yield and water use efficiency of maize (Zea mays L.). *Madras Agricultural Journal*, 98(7&9), pp. 238–242.

Fanish, S.A.F.S.A., Muthukrishnan, P., Muthukrishnan, S.M. and Manoharan, S. 2011. Drip fertiga drip fertigation in maize (tion in maize (tion in maize (zea mays) based intercropping system. *Indian Journal of Agriculture Research*, 45(3), pp. 233–238.

Firake, N.N. and Kumbhar, D.B. 2002. Effect of different levels of N, P and K fertigation on yield and quality of pomegranate. *Journal of Maharashtra Agricultural Universities (India)*, 27(2), pp. 146–148.

Gnanamurthy, P., and Manickasundram, P. 2001. Advances in integrated nutrient management system for sustainable crop productivity and soil fertility. October 4.24,2001, Tamil Nadu Agricultural University, Coimbatore, pp. 110–116.

Gowda, N.C.B., Krishnappa, K.S. and Puttaraju, T.B. 2001. Dry matter production and fruit size in okra varieties as influenced by varying fertilizer levels. *Current Research University of Agricultural Sciences, Banglore*, 30(9), pp. 151–153.

Gupta, A. J., Ahmed, N., & Bhat, F. N. 2009. Enhancement of yield and its attributes of sprouting broccoli through drip irrigation and fertigation. *Vegetable Science*, 36(2), pp. 179–183.

Hasan, M., Sirohi, N.P.S., Kumar, V., Sharma, M.K. and Singh, A.K. 2003, October. Performance evaluation of different irrigation scheduling methods for peach through efficient fertigation system network. *VII International Symposium on Temperate Zone Fruits in the Tropics and Subtropics*, 662, pp. 193–197.

Hegde, D.M. and Srinivas, K. 1990. Plant water relations and nutrient uptake in French bean. *Irrigation Science*, 11(1), pp. 51–56.

Hemalatha, S., Maragatham, S., Radhika, K., and Kathrine, S. P. 2013. Fertigation for Crops and Nitrogen Fertigation for Sugarcane: A Review. *Research & Reviews: Journal of Agriculture and Allied Sciences*, 2(2), pp. 5–11, ISSN: 2319-9857.

Heermann, D.F., Hoeting, J., Thompson, S.E., Duke, H.R., Westfall, D.G., Buchleiter, G.W., Westra, P., Peairs, F.B. and Fleming, K. 2002 Interdisciplinary irrigated precision farming research. *Precision Agriculture*, 3, pp. 47–61.

INCID, 1994. Drip Irrigation in India, Indian National Committee on Irrigation and Drainage, New Delhi.

Ingle, H.V., Athawale, R.B., and Ingle, S.H. 2003. Effect of organic and inorganic fertilizer on yield and quality of Nagpur mandarin. *The Orissa Journal of Horticulture*, 31(1), pp. 10–13.

Ingle, H.V., Athawale, R.B., Ghawde, S.M. and Shivankar, S.K. 2001. Integrated nutrient management in acid lime. *South Indian Horticulture*, 49, pp. 126–129.

Irulappan, I., Abdul-Khader, J.B.M., and Muthukrishnana, S. 1984. Papaya research in Tamil Nadu. Proceedings national seminar on papaya and papain production. Faculty of Horticultuire, TNAU, Coimbatore, pp. 15–20.

Iyer, T.A., and Iyengar, T.R. 1956. A study of the decline of orange in Wynad. *South Indian Horticulture*, (4), pp. 70–81.

Jawaharlal, M., Swapna, C. and Ganga, M. 2012, February. Comparative analysis of conventional and precision farming systems for African marigold (Tagetes erecta L.). In International Conference on Quality Management in Supply Chains of Ornamentals QMSCO2012 970, pp. 311–317.

Jeyakumar, P., Amutha, R., Balamohan, T.N., Auxcilia, J. and Nalina, L. 2008, December. Fertigation improves fruit yield and quality of papaya. *II International Symposium on Papaya*, 851, pp. 369–376.

Jeyakumar, P., Kumar, N. and Soorianathasundaram, K. 2001. Fertigation studies in papaya (Carica papaya L.). *South Indian Horticulture*, 49, pp. 71–75.

Jeyabal, A., Palaniappan, S. and Chelliah, S. 2000. Effect of integrated nutrient management techniques on yield attributes and yield of sunflower (Helianthus annuus). *Indian Journal of Agronomy*, 45(2), pp. 384–388.

Jiong, L.G. 1990. Xiangtan Q-type automatic hydraulic flap gate. *Journal of irrigation and drainage engineering*, 116(2), pp. 211–218.

Jitendra, K., Patel, N., Rajput, T.B.S. and Sahoo, P.K. 2013. Development of soil moisture sensor for different methods of irrigation. Paper presented in 47th Annual Convention Indian Society of Agricultural Engineers and International Symposium on *"Bio Energy– Challenges and Opportunity"* Hyderabad. Jan 28–30, p. 150.

Joshi A., Tiwari, K.N. and Banerjee, S. 1999. Automated Irrigation controller. Patent Application No. 6/Cal/99 dated 04-1-1999, IIT Kharagpur.

Joshi A., Tiwari, K.N. and Banerjee, S. 2000. Granular Matrix soil moisture sensor. Patent Application No. 705/Cal/2000 dated 21-12-2000, IIT Kharagpur.

Joshi, A., Tiwari, K.N., and Banerjee, S. 2002. Low cost computer controlled automated irrigation system. Patent Application No. 307/Cal/2002 dated 15-5-2002, IIT Kharagpur.

Kadam, J.R., Dukre, M.V. and Firake, M.N. 1995. Nitrogen saving through Biwall subsurface irrigation in okra. *Journal of Mahrashtra Agricultural University*, 20(3), pp. 475–476.

Kadam, J.R., Dukre, M.V. and Firake, N.N. 1993. Effect of nitrogen application through drip irrigation on nitrogen saving and yield of okra. *Journal of Water Management*, 1(1), pp. 53–54.

Kadam, U.S., Deshmukh, M.R., Kadam, S.A. and Patil, H.M. 2008. Fertigation Management in Cauliflower (Brassica oleracea). *Environment, New Challenges*, p. 165.

Kapoor, V., Patil, S.K., Singh, U., Magre, H., Shrivastava, L.L., Mishra, V.N., Das, R.O., Samadhiya, V.K. and Diamond, R.B. 2008. Rice response to urea briquette containing diammonium phosphate and muriate of potash. *Agronomy Journal*, 100, pp. 526–536.

Keshavaiah, K.V. and Kumaraswamy, A.S. 1993. Fertigation and water use efficiency in potato under furrow and drip irrigation. *Journal of the Indian Potato Association*, 20(3), pp. 240–244.

Khan, S., & Abbas, A. 2007. Upscaling water savings from farm to irrigation system level using GIS-based agro-hydrological modelling. *Irrigation and Drainage*, 56(1), pp. 29–42.

Khan, S., Abbas, A., Gabriel, H.F., Rana, T. and Robinson, D. 2008. Hydrologic and economic evaluation of water-saving options in irrigation systems. *Irrigation and Drainage: The journal of the International Commission on Irrigation and Drainage*, 57(1), pp. 1–14.

Kumar, A. and Wani, S.H. 2001. Status and issues of fertigation in India. Micro-irrigation. Central Board of Irrigation and Power (CBIP), New Delhi, pp. 418–427.

Kumar, A., and Khanna, M. 1998. Low cost micro irrigation through bubbler irrigation system. *Journal of Indian Water Resources Society*. 18(4), pp. 25–28.

Kumar, A.R. 2004. Studies on the efficacy of sulphate of potash (SOP) on growth, yield and quality of banana. PhD Thesis submitted to Tamil Nadu Agricultural University, Coimbatore.

Kumar, A.R., and Kumar, N. 2007. Sulphate of potash foliar spray effects on yield, quality and better post harvest life of banana. *Better crops*, 91 (2), pp. 22–24.

Kumar, J. 2012. Development of sensor based scheduling in micro irrigation system. M.Tech. Thesis, Division of Agricultural Engineering, I.A.R.I., New Delhi.

Kumar, M. D., Turral, H., Sharma, B., Amarasinghe, U., and Singh, O. P. 2008. Water saving and yield enhancing micro irrigation technologies in India: when and where can they become best bet technologies? 7th Annual Partners' Meet of IWMI-Tata Water Policy Research program, ICRISAT, Hyderabad.

Kumar, N., Jeyakumar, P., and Manivannan, M. I. 2006. Balanced fertilization for sustainable yield and quality in tropical fruit crops. *Bangladesh Journal of Agricultural Research*, 4, pp. 69–79, Special Issue 2008.

Kumar, N., Meenakshi, N., Suresh, J. and Nosov, V. 2006. Effect of potassium nutrition on growth, yield and quality of papaya (Carica papaya L.). *Indian Journal of Fertilisers*, 2(4), p. 43.

Kumar, N., Suresh, J., and Selvi, R. 2007. Studies on efficacy of sulphate of potash on yield and quality of mango under tropical belt of India. Final report of TNAU-Kali & Sulz Gmbh, Kassel, Germany Collaborative Scheme. Horticultural College and Research Institute, Tamil Nadu Agricultural University, Periyakulam.

Kumar, S., Imtiyaz, M. and Kumar, A. 2009. Studying the feasibility of using microirrigation systems for vegetable production in a canal command area. *Irrigation and Drainage: The journal of the International Commission on Irrigation and Drainage*, 58(1), pp. 86–95.

Kumar, S., Imtiyaz, M., and Kumar, A. 2009. Studying the feasibility of using microirrigation systems for vegetable production in a canal command area. *Irrigation and Drainage*, 58(1), pp. 86–95.

Kumar, S., Kumar, A. and Singh, R. 2007. Microirrigation for onion cultivation in a canal command area. *Journal of Agricultural Engineering*, 44(1), pp. 33–37.

Kumar, S., Singh, R., Kaledhonkar, M.J., Nangare, D.D. and Kumar, A. 2013. Improving water productivity through micro-irrigation in arid Punjab regions. *Irrigation and Drainage*, 62(3), pp. 330–339.

Kumar, V., Gurusamy, A., Mahendean, P.P. and Mahendran, S. 2009. Optimization of water and nutrient requirement for yiel d maximization in hybrid rice under drip fertigation system. 8th International Micro Irrigation Congress.

Kumathe, S.S., Kumar, U.S., and Patil, P.B. 1998. Drip Irrigation for Kagzi Lime and Banana. In *AICRP on Agricultural Drainage*, A.K. Bhattacharya (Ed.), ICAR, New Delhi. p. 37.

Ladani, R.H., Khimani, R.A., Makawana, A.N. and Delavadia, D.V. 2004. Nutritional survey of mango cv. Kesar orchard in south saurashtra region of Gujarat. Improving Productivity, Quality, Post-harvest management and trade in horticultural crops. *First Indian Horticulture Congress*, 223, p. 222.

Luthra S.K., Kaledhonkar, M.J., and Gupta, S.K. 1996. Design and development of a valve with automatic controls for drip irrigation system. Seminar Proceedings IE (I) Agricultural Engineering Division, Karnataka State Centre. All India Seminar on Modern Irrigation Techniques, Bangalore, June 26–27, pp. 53–62.

Luthra, S.K., Kaledonkar, M.J., Singh, O.P. and Tyagi, N.K. 1997. Design and development of an auto irrigation system. *Agricultural Water Management*, 33(2–3), pp. 169–181.

Magar, S.S. 1995, April. The status of microirrigation in Maharashtra, India. *Proceedings of the Fifth International Microirrigation Congress*, 26, pp. 452–456).

Mahajan, G. and Singh, K.G. 2006. Response of greenhouse tomato to irrigation and fertigation. *Agricultural Water Management*, 84(1–2), pp. 202–206.

Mahalakshami, M. 2000a. Water and fertigation management studies in banana cv. Robusta (AAA) under normal planting and high density planting systems. Doctoral dissertation, Tamil Nadu Agricultural University; Coimbatore.

Mahalakshmi, M. 2000b. Water and Fertigation Management Studies in Banana cv. Robusta (AAA) under normal planting and high density planting system. Ph.D. Thesis, Tamil Nadu Agricultural University, Coimbatore.

Mahendran, P.P., Arulkumar, D., Gurusamy, A. and Kumar, V. 2011, October. Performance of nutrient sources and its levels on hybrid bhendi under drip fertigation system. 8th International Micro Irrigation Congress, p. 1.

Mahendran, P.P., Yuvaraj, M., Parameswari, C., Gurusamy, A. and Krishnasamy, S. 2013. Enhancing growth, yield and quality of banana through subsurface drip fertigation. *International Journal of Chemical Environmental Biological Sciences*, 1(2), pp. 391–394.

Mahmoud, H.H. 2013. Effect of different levels of planting distances, irrigation and fertigation on yield characters of main banana crop cv. Grand Naine. *Global Journal of Plant Ecophysiology*, 3(2), pp. 115–121.

Malash, N., Flowers, T.J., and Ragab, R. 2005. Effect of irrigation systems and water management practices using saline and non-saline water on tomato production. *Agricultural Water Management*, 78(1), pp. 25–38.

Malik, R.S., Kumar, K., and Bhandari, A.R. 1994. Effect of urea application through drip irrigation system on nitrate distribution in loamy sand soils and pea yield. *Journal of Indian Society of Soil Science*, 42(1), pp. 6–10.

Mandal, R.C. and Jana, P.K. 1998. *Water Resource Utilization & Micro-irrigation: Sprinkler & Drip System*. Kalyani Publishers, Ludhiana.

Mane, K.M. 1993. Economics of grapes cultivation under drip vs furrow method of irrigation in Maharashtra. M.Sc. (Agri) Thesis submitted to UAS, Dharwad.

Mane, M.S., Ayare, B.L., and Magar, S.S. 2008. Design of drip irrigation system. In *Principles of Drip Irrigation System*. Jain Brothers, New Delhi, p. 45. ISBN: 81-8360-077-8.

Manjunatha, M.V., Shukla, K.N., Chauhan, H.S., Singh, P.K. and Singh, R. 2001. Economic feasibility of micro irrigation system for various vegetables. In Proceeding of International Conference on Micro and Sprinkler Irrigation System held at Jalgaon, Maharashtra during, pp. 8–10.

Marais, A. 2001. Subsurface drip irrigation systems. In South African Irrigation Institute (SABI) congress.

Michael, A.M. 1978. *Irrigation: Theory and Practice*. Vikash Publishing House Pvt. Ltd., New Delhi, p. 801.

Mishra, J.N. and Paul, J.C. 2009. Impact of drip irrigation with plastic mulch on yield and return of brinjal crop. *Journal of Water Management*, 17(1), pp. 14–17.

Mishra, M.K., and Singh, M. 1991. *Potassium in Crop Production in India*. Potash Research Institute of India, Gurgaon, p. 32.

Mohan, N.K., and Ahmed, F. 1987. A decade of research on pineapple. *Research bulletin. Assam Agriculture University*, 9, pp. 2–12.

Muralidhara, H.R., Gundurao, D.S., Sarpashkar, A.M. and Ramaiah, R. 1994. Is drip irrigation viable for mulberry cultivation–An economic analysis? *Mysore Journal of Agriculture Science*, 28, pp. 256–266.

Murthy, V.V.N. 1998. *Land and Water Management Engineering*, Kalyani Publishers, New Delhi, pp. 586.

Muthukrishnan, P., and Fanish, S.A. 2011. Influence of drip fertigation on yield, water saving and water use efficiency in maize (Zea mays L.) based intercropping system. *Madras Agricultural Journal*, 98(7/9), pp. 243–247.

Naidu, C.N. 2004. Report of task force on microirrigation, Ministry of Agriculture, Government of India, January.

Naidu, Y., Meon, S., and Siddiqui, Y. 2013. Foliar application of microbial-enriched compost tea enhances growth, yield and quality of muskmelon (Cucumis melo L.) cultivated under fertigation system. *Scientia Horticulturae*, 159, pp. 33–40.

Naik, P.S., Singh, M. and Karmakar, P. 2013. *Adaptation options for sustainable production of cucurbitaceous vegetable under climate change situation. In Climate-Resilient Horticulture: Adaptation and Mitigation Strategies.* Springer, New Delhi, pp. 137–146.

Nakayama, and Bucks, 1986. *Management Principles and Fertilization in Trickle Irrigation.* Elsevier Applied Science Publishers, Amsterdam, p. 317.

Nalayini, P., Raj, S.P. and Sankaranarayanan, K. 2013. Water use efficiency, nutrient uptake and production potential of extra long staple Bt cotton-maize system with moisture conservation techniques and ET based irrigation. *Journal of Cotton Research and Development*, 27(1), pp. 45–49.

Nalina, L. 2002. Standardisation of Fertilizer Requirement for Tissue Cultured Banana Cv. Robusta (Aaa). Doctoral dissertation, Tamil Nadu Agricultural University, Coimbatore.

Narayanamoorthy, A. 1997a. Drip irrigation: A viable option for future irrigation development. *Productivity*, 38(3), pp. 504–511.

Narayanamoorthy, A. 1997b. Economic viability of drip irrigation: An empirical analysis from Maharashtra. *Indian Journal of Agricultural Economics*, 52(4), p. 728.

Narayanamoorthy, A. 2001. Impact of drip irrigation on sugarcane cultivation in Maharashtra. Agro-Economic Research Centre, Gokhale Institute of Politics and Economics, Pune, June.

Narayanamoorthy, A. 2004. Impact assessment of drip irrigation in India: The case of sugarcane. *Development Policy Review*, 22(4), pp. 443–462.

Narayanamoorthy, A. 2006. Potential for drip and sprinkler irrigation in India. Draft prepared for the IWMI-CPWF project on 'Strategic Analysis of National River Linking Project of India.

Narayan-Gowda, S.N., Siddalingaswamy, N., and Shivaprakash, R. 2007. Performance of drip ferti-irrigation or fertilization on mulberry leaf yield and quality. *Séricologia*, 47(2), pp. 215–224.

Narda, N.K. and Lubana, S.P. 1999. Growth dynamics studies of tomatoes under sub-surface drip irrigation. *Journal of Research Punjab Agricultural University*, 36(3–4), pp. 222–223.

Niblack, M. and Sanchez, C.A. 2008. Automation of surface irrigation by cut-off time or cut-off distance control. *Applied Engineering in Agriculture*, 24(5), pp. 611–614.

Nijbroek, R., Hoogenboom, G. and Jones, J.W. 2003 Optimizing irrigation management for a spatially variable soybean field. *Agricultural Systems*, 76, pp. 359–377.

Pandey, A.K., Singh, A.K., Kumar, A., and Singh, S.K. 2013. Effect of Drip Irrigation, Spacing and Nitrogen Fertigation on Productivity of Chilli (Capsicum annuum L.). *Environment & Ecology*, 31(1), pp. 139–142.

Pandey, V.K., Mahajan, V. 1999. Performance of drip irrigation on tomato. In Proceedings of National symposium on Progress in micro irrigation research in India. Water. Technology Centre for Eastern Region, Bhubaneswar July 1998, pp. 27–28.

Panigrahi, P., and Sahu, N.N. 2013. Evapotranspiration and yield of okra as affected by partial root-zone furrow irrigation. *International Journal of Plant Production*, 7(1), pp. 33–54.

Panigrahi, P., Sharma, R.K., Parihar, S.S., Hasan, M., and Rana, D.S. 2013. Economic Analysis of Drip-Irrigated Kinnow Mandarin Orchard under Deficit Irrigation and Partial Root Zone Drying. *Irrigation and Drainage*, 62(1), pp. 67–73.

Papadopoulos, I. 1987. Nitrogen fertigation of greenhouse-grown tomato. *Communications in Soil Science and Plant Analysis*, 18(8), pp. 897–907.

Papadopoulos, I. 1992. Phosphorous fertigation of trickle irrigated potato. *Fertilizer Research*, 31, pp. 9–13.

Parikh, M.M., Savani, N.G., Shrivastava, P.K., Avadaria, J.D., Thanki, J.D., Shah, G.B., Heller, G.H., Desai, S.G., Gohil, K.B., Patel, H.P., and Raman, S. 1996. In response of various crops to micro irrigation, mulching and fertigation. Proceedings of the All India Seminar on Modern Irrigation Technology, Bangalore, June 26–27, pp. 206–212.

Parikh, M.M., Shrivastava, P.J., Savani, N.G. and Raman, S. 1993. Feasibility study of drip in sugarcane. Proceedings of Workshop Sprinkler Drip Irrigation Systems, pp. 124–127.

Patel, J.C., and Patel, B.K. 2001. Journal of Indian Potato Association, 28, pp. 296–295.

Patel, N., and Rajput, T.B.S. 2000. Effect of fertigation on growth and yield of onion CBIP Publication No. 282: pp. 451–454.

Patel, N., and Rajput, T.B.S. 2001. Moisture front advance studies under twin emitter drip irrigation system. *Journal of Indian Water Resources Society*, 21(2), pp. 79–88.

Patel, N., and Rajput, T.B.S. 2001a. Fertigation of okra using commercially available granular fertilizers. Proceeding of international symposium on importance of potassium in nutrient management of sustainable crop production in India, held at New Delhi, pp. 3–5.

Patel, N., and Rajput, T.B.S. 2001b. Effects of fertigation on growth and yield of onion. In *Micro Irrigation*, H.P. Singh, S.P. Kaushik. A. Kumar, T.S. Murthy and J.C. Sameul (Eds.), pp. 451–454. New Delhi.

Patel, N., and Rajput, T.B.S. 2002a. Yield responses of some vegetable crops to different levels of fertigation. Paper presented in "Agriculture in Staying Global Scenario" bet Feb 21–23.

Patel, N., and Rajput, T.B.S. 2002b. Use of commercially available granular fertilizers for fertigation of Broccoli. Paper presented in 36th Annual Convocation ISAE held at IIT Kharagpur, Jan 28–30.

Patel, N., and Rajput, T.B.S. 2002c. Fertigation studies on pomegranate using commercially available granular fertilizer. Paper presented in 89th science congress held at Lucknow. Jan 3–7.

Patel, N., and Rajput, T.B.S. 2003. Yield response of some vegetable crops to different levels of fertigation. *Annals of Agriculture Research*, 24(3), pp. 542–545.

Patel, N., and Rajput, T.B.S. 2007. Effect of drip tape placement depth and irrigation level on yield of potato. *Agricultural Water Management*, 88(1), pp. 209–223.

Patel, N., and Rajput, T.B.S. 2008. Dynamics and modeling of soil water under subsurface drip irrigated onion. *Agricultural Water Management*, 95(12), pp. 1335–1349.

Patil, N.B. 1997. Weed Management in through drip irrigation. In *Drip Irrigation*, V.C. Patil, and S.S. Angadi, (Eds.), University of Agricultural Sciences, Bengaluru.

Patil, V.C., and Angadi, S.S. 1997. Drip irrigation, Pub. Agronomy Club, University of Agricultural Sciences, Dharwad, Karnataka, p. 140.

Paul, J.C., Mishra, J.N., Pradhan, P.L., and Panigrahi, B. 2013. Effect of drip and surface irrigation on yield, water-use-efficiency and economics of capsicum (capsicum annum l.) Grown under mulch and non mulch conditions in eastern coastal India. *European Journal of Sustainable Development*, 2(1), pp. 99–108.

Pawar, D.D., Dingre, S.K., and Surve, U.S. 2013. Growth, yield and water use in sugarcane (Saccharum officinarum) under drip fertigation. *Indian Journal of Agronomy*, 58(3), pp. 396–401.

Pawar, D.D., Dingre, S.K., Bhakre, B.D., and Surve, U.S. 2013. Nutrient and water use by Bt. cotton (Gossypium hirsutum) under drip fertigation. *Indian Journal of Agronomy*, 58(2), pp. 237–242.

Pawar, D.D., Raskar, B.S., Banger, A.R., Bhoi, P.G., and Shinde, S.M. 2001. Effects of water-soluble fertilizers through drip and planting techniques on growth, yield and quality of banana. In *Micro Irrigation*, H.P. Singh, S.P. Kaushik. A. Kumar, T.S. Murthy and J.C. Sameul (Eds.), pp. 515–519. New Delhi.

Prabhakar, M., Savanur, V., and Naik, C.L. 2001. Fertigation studies in hybrid tomato. *South Indian Horticulture*, 49, pp. 98–100.

Prasad, S. 1997. Development of criteria for optimum use of fertilizer through trickle/ drip irrigation system. Ph.D. thesis, I.A.R.I., New Delhi.

Prathyusha, K., Bala, G.S. and Ravi, K.S. 2013. A real time irrigation control system for precision agriculture using WSN in Indian agricultural sectors. *International Journal of Computer Science, Engineering and Applications (IJCSEA)*, 3(4), pp. 75–80.

Pugalendhi, L., Meenakshi, N. and Kavitha, M. 2007. Balanced fertilization for hybrid vegetable production. Role of balanced fertilization for horticultural crops, p. 72.

Punamhoro, P.B.N., Chowdhary, B.M., and Kandeyang, S. 2003a. Performance of different irrigation methods in okra (Abelmoschus esculentus (L.) Moench). *Journal of the Bangladesh Agricultural University*, 15(2), pp. 205–210.

Punamhoro, P.B.N., Chowdhary, B.M., and Kandeyang, S. 2003b. Performance of different irrigation methods in okra (Abelmoschus esculentus (L.) Moench). *Journal of the Bangladesh Agricultural University*, 15(2), pp. 205–210.

Raina, J.N., Suman, S., Kumar, P., and Spehia, R.S. 2013. Effect of Drip Fertigation With and Without Mulch on Soil Hydrothermal Regimes, Growth, Yield, and Quality of Apple (Malus domestica Borkh). *Communications in Soil Science and Plant Analysis*, 44(17), pp. 2560–2570.

Raina, J.N., Thakur, B.C., and Verma, M.L. 2013. Effect of drip irrigation and polyethylene mulch on yield, quality and water-use efficiency of tomato (Lycopersicon esculentum). *The Indian Journal of Agricultural Sciences*, 69(6).

Raj, A.F.S., Muthukrishnan, P. and Ayyadurai, P. 2013. Root characters of maize as influenced by drip fertigation levels. *American Journal of Plant Sciences*, 4(02), pp. 340–348.

Rajak, D., Manjunatha, M.V., Rajkumar, G.R., Hebbara, M. and Minhas, P.S. 2006. Comparative effects of drip and furrow irrigation on the yield and water productivity of cotton (Gossypium hirsutum L.) in a saline and waterlogged vertisol. *Agricultural Water Management*, 83(1–2), pp. 30–36.

Rajaraman, G. and Pugalendhi, L. 2013. Potential impact of spacing and fertilizer levels on the flowering, productivity and economic viability of hybrid Bhendi (Abelmoschus esculentus L. Moench) under drip fertigation system. *American Journal of Plant Sciences*, 4(09), pp. 1784–1789.

Rajaraman, G., Paramaguru, P., Aruna, P., Sudagar, I.P., and Ragothaman, G. 2011. Fertigation studies on leaf area and chlorophyll content in coriander (Coriandrum sativum L.). *Asian Journal of Horticulture*, 6(1), pp. 46–49.

Rajput, T.B.S. and Patel, N. 2002. Yield response of okra (Abelmoschus esculentus L.) to different levels of fertigation. *Annals of Agricultural Sciences*, 23(1), pp. 164–165.

Ram, T., Dalamu, R.S., Singh, K.N., Ram, D. and Jat, N.K. 2005. Precision Irrigation Systems in Agriculture: A Perspective. *Precision farming: A New Approach*, 8, p. 121.

Ramah, K., Santhi, P., and Ponnuswamy, K. 2010. Economic viability of drip fertigation in maize (Zea mays L.) based cropping system. *Madras Agricultural Journal*, 97(1–3), pp. 12–16.

Raman, S.S., Murthy, K.D., Ramesh, G., Palaniappan, S.P., and Chelliah, S. 2000. Effect of fertigation of growth and yield of gherkins. *Vegetable Science*, 27(1), pp. 64–66.

Ramniwas, R.A., Sarolia, D.K., Pareek, S., and Singh, V. 2012. Effect of irrigation and fertigation scheduling on growth and yield of guava (Psidium guajava L.) under meadow orcharding. *African Journal of Agricultural Research*, 7(47), pp. 6350–6356.

Ravi, N. and Jagadeesha, C.J. 2002. Precision Agriculture, Training course on Remote Sensing and GIS Applications in Agriculture, May 27th–7th June, 2002, RRSSC- Bangalore, pp. 225–228.

Ravichandrane, V., Kumar, N., Jeyakumar, P., Soorianathasundaram, K., and Vijayakumar, R.M. 2002. Influence of planting density and nutrient levels on growth and yield of papaya cv. CO_2. *South Indian Horticulture*, 55 (1), pp. 23–29.

Ray, S.S., Panigrahy, P. and Parihar, J.S. 2001. Role of Remote Sensing for precision farming with special Reference to Indian Situation Scientific Note SAC/RESA/ARG/AMD/SN/01/2001. Space Applications Center (ISRO), Ahmadabad, pp. 1–21.

Reddi, G., Sankara, H., and Yellamanda, R.T. 2002. *Efficient Use of Irrigation Water*. Kalyan Publishers, New Delhi, pp. 74–198.

Reddy, M., Ayyanagowdar, M. S., Nemichandrappa, M., Balakrishnan, P., Patil, M. G., Polisgowdar, B. S., and Satishkumar, U. 2013. Techno economic feasibility of drip irrigation for onion (Alluim cepa L). *Karnataka Journal of Agricultural Sciences*, 25(4), pp. 475–478.

Reddy, P. P. 2004. Vegetable research in India an IIHR perspective. Proceedings of Impact of Vegetable Research in India'by Sant Kumar, PK Joshi and Suresh Pal, pp. 45–58.

Rekha, K.B. and Mahavishnan, K. 2008. Drip fertigation in vegetable crops with emphasis on lady's finger (Abelmoschus esculentus (L.) Moench)-A review. *Agricultural Reviews*, 29(4), pp. 298–305.

Rekha, K.B., Reddy, M.G. and Mahavishnan, K. 2006. Nitrogen and water use efficiency of bhendi (Abelmoschus esculentus L. Moench) as influenced by drip fertigation. *Journal of Tropical Agriculture*, 43, pp. 43–46.

Rubeiz, I.G., 1990. Response of greenhouse cucumber to mineral fertilizers on a high phosphorus and potassium soil. *Journal of Plant Nutrition*, 13, pp. 269–273.

Sathya, S., Pitchai, G.J., Indirani, R. and Kannathasan, M. 2008. Effect of fertigation on availability of nutrients (N, P & K) in soil–A Review. *Agricultural Reviews*, 29(3), pp. 214–219.

Satyendra, K., Ashwani, K. and Rajbir, S. 2007. Micro irrigation for onion cultivation in a canal command area. *Journal of Agricultural Engineering*, 44(1). Print ISSN: 0256-6524.

Sen, S.K. 1985. Pineapple. In *Fruits of India. Tropical and Subtropical*, T.K. Bose (Ed.), Naya Prakasham, Culcutta, pp. 298–311.

Seshadri, K.V. and Madhavi, M. 2001. Effect of organic and inorganic manuring on twenty years old seedlings of sweet orange (Citrus sinensis (L.) Oosbeck.) cv. Sathugudi. *South Indian Horticulture*, 49, pp. 122–125.

Shantaraj, B.C. 1983. Comparison of drip, sprinkler and flood irrigation system for irrigating horticultural crops. Proceedings of second seminar on Drip irrigation held at TNAU, Coimbatore, pp. 112–117

Shanwad, U.K., Patil, V.C., Dasog, G.S., Mansur, C.P. and Shashidhar, K.C. 2002, October. Global positioning system (GPS) in precision agriculture. In The Asian GPS Conference, pp. 24–25.

Sharma, P., Kumar, S., Sharma, S.K., and Jhorar, R.K. 2013. Salt and water dynamics under daily drip irrigation with different saline water in cabbage [Brassica oleracea (L.) var. capitata]. *Annals of Biology*, 29(1), pp. 89–92.

Sharma, P., Kumar, S., Sharma, S.K., and Naresh, R. 2013. Response of different saline water on salt and water movement under drip irrigation in cabbage (Brassica oleracea L. var capitata). *Environment and Ecology*, 31(1), pp. 71–75.

Sharma, S., Patra, S.K., Roy, G.B. and Bera, S. 2013. Influence of drip irrigation and nitrogen fertigation on yield and water productivity of guava. *The Bioscan*, 8(3), pp. 783–786.

Shekinah, D.E., and Rakkiyappan, P. 2011. Conventional and Microirrigation Systems in Sugarcane Agriculture in India. *Sugar Tech*, 13(4), pp. 299–309.

Shinde, D.G., Patel, K.G., Solia, B.M., Patil, R.G., Lambade, B.M., and Kaswala, A.R. 2012. Clogging behaviour of drippers of different discharge rates as influenced by different fertigation and irrigation water salinity levels. *Journal of Environmental Research and Development*, 7(2A), pp. 917–922.

Shirgure, P., and Srivastava, A. 2013. Water management in citrus 1. *Evapotranspiration: Principles and Applications for Water Management*, 2013, p. 273.

Shirgure, P.S. and Srivastava, A.K. 2013. Nutrient-water interaction in citrus: Recent developments. *Agricultural Advances*, 2(8), pp. 224–236.

Shirgure, P.S. and Srivastava, A.K. 2013. Optimizing the potassium (K) dose of fertigation for the Nagpur mandarin (Citrus reticulate Blanco). *Agricultural Advances*, 2(8), pp. 243–249.

Shirgure, P.S., Srivastava, A.K. and Singh, S. 2001. Fertigation and drip irrigation in Nagpur mandarin (Citrus reticulata Blanco.). *South Indian Horticulture*, 49, pp. 95–97.

Shirgure, P.S., Srivastava, A.K., & Singh, S. 2000. Water Management in Citrus-A Review. *Agricultural Reviews- Agricultural Research Communications Centre India*, 21(4), pp. 223–230.

Shivashankar, K. and Khan, M.M. 1994. Fertigation studies with water soluble NPK fertilizer in crop production. Research highlights of fertigation projects, University of Agricultural Science, GKVK, Bangalore, pp. 1–6.

Shivshankar, K., Khan, M.M., and Farooqui, A.A. 1998. Fertigation studies with water soluble NPK fertilizers in crop production. Publication of University of Agricultural Sciences, Bangalore and Kemira Agro and Kemira OY Finland.

Shrestha, R.B., & Gopalakrishnan, C. 1993. Adoption and diffusion of drip irrigation technology: an econometric analysis. *Economic Development and Cultural Change*, 41(2), pp. 407–418.

Shukla, S.K., Yadav, R.L., Singh, P.N., and Singh, I. 2009. Potassium nutrition for improving stubble bud sprouting, dry matter partitioning, nutrient uptake and winter initiated sugarcane (Saccharum spp. hybrid complex) ratoon yield. *European Journal of Agronomy*, 30(1), pp. 27–33.

Singandhupe, R.B., Rao, G.G.S.N., Patil, N.G. and Brahmanand, P.S. 2003. Fertigation studies and irrigation scheduling in drip irrigation system in tomato crop (Lycopersicon esculentum L.). *European Journal of Agronomy*, 19(2), pp. 327–340.

Singh, A. K., Chakraborty, D., Mishra, P. and Singh, D. K. 2002. Nitrogen and potassium dynamics in fertigation systems. 17th World Congress of Soil Science, Bangkok, pp. 14–21.

Singh, A.K., Khanna, M., Chakraborty, D. and Kumar, A. 2001. Increasing water and nutrient use efficiency in broccoli through fertigation. In Proceeding of International Conference on Micro and Sprinkler Irrigation System held at Jalgaon, Maharashtra during (pp. 8–10).

Singh, D.K. 2004. Performance evaluation of subsurface drip irrigation systems. Unpublished Ph.D. thesis, P.G. School, IARI, New Delhi, p. 135.

Singh, D.K., Rajput, T.B.S., Sikarwar, H.S., Sahoo, R.N. and Ahmad, T. 2006. Simulation of soil wetting pattern with subsurface drip irrigation from line source. *Agricultural Water Management*, 83(1–2), pp. 130–134.

Singh, K. 1991. Nutrient requirement of vegetable crops. Khad Patrika (Hindi), pp. 23–24.

Singh, P., Singh, A.K. and Sahu, K. 2006. Irrigation and fertigation of pomegranate cv. Ganesh in Chhattisgarh. *Indian Journal of Horticulture*, 63(2), pp. 148–151.

Singh, S.D. 1978. Effects of Planting Configuration on Water Use and Economics of Drip Irrigation Systems 1. *Agronomy Journal*, 70(6), pp. 951–954.

Singh, S.R. 2000. Annual Report for 1999–2000 of the Directorate of Water Management Research, Patna.

Singh, U. 2006. Integrated nitrogen fertilization for intensive and sustainable agriculture. *Journal of Crop Improvement*, 15(2), pp. 259–288.

Singh, Y.V., and Saxena, A. 2001. Chilli yields as related to water and nitrogen management under drip irrigation system. In *Micro Irrigation*, H.P. Singh, S.P. Kaushik. A. Kumar, T.S. Murthy and J.C. Sameul (Eds.), pp. 468–471. New Delhi.

Sivakumar, V. 2007. Studies on influence of various nutrient levels applied through fertigation on growth, physiology, yield and quality of mango (Mangifera indica L) cv. Ratna under high density planting. Doctoral dissertation, Ph.D. (Hort.) Thesis, Tamil Nadu Agricultural University. Coimbatore.

Sivanappan, R.K. and Padmakumari, O. 1980. Drip irrigation. Tamil Nadu Agricultural University, Coimbatore, p. 70.

Sivanappan, R.K. 1979. Drip irrigation. *Irrigation Era*, 14(4), p. 13

Sivanappan, R.K. 1985. *Drip Irrigation in Action*, vol. II, ASAE, Michigan, pp. 736–740.

Sivanappan, R.K. 1990. Constraints and potential in popularizing drip irrigation. Nabard committee on the use of plastic in agriculture.

Sivanappan, R.K. 1993. Case studies with many farmers in Maharashtra state.

Sivanappan, R.K. 1994. Prospects of micro-irrigation in India. *Irrigation and Drainage Systems*, 8(1), pp. 49–58.

Sivanappan, R.K., Padmakumari, O. and Kumar, V. 1987. *Drip Irrigation*. Keerthi Publishing House Pvt. Ltd., Coimbatore, p. 412.

Sivanappan, R.K., Rajagopal, A. and Palaniswamy, D. 1978. Response of chillies to the drip irrigation [India]. *Madras Agricultural Journal (India)*, 65(9), pp. 576–579.

Smajstrla, A.G. and Locascio, S.J. 1996. Tensiometer-controlled, drip-irrigation scheduling of tomato. *Applied Engineering in Agriculture*, 12(3), pp. 315–319.

Solaimalai, A., Baskar, M., Sadasakthi, A., and Subburamu, K. 2005. Fertigation in high value crops-A Review. *Agricultural Reviews-Agricultural Research Communications Centre India*, 26(1), p. 1.

Spayd, S.E., Wample, R.L., Evans, R.G. 1994. Nitrogen fertilization of white Riesling grapes in Washington. Must and wine composition. *American Journal of Enology and Viticulture*, 45(1994), pp. 34–42.

Srinivas, K. 1998. Standardization of micro irrigation in fruit crops – An overview. In Proceedings Workshop on Micro irrigation and sprinkler irrigation systems 28–30 April, 1998, II-30, Central Board of Irrigation and Power, New Delhi.

Srivastava, A.K. 2013. Site specific nutrient management in citrus. *Agricultural Advances*, 2(2), pp. 53–67.

Srivastava, R.C., Verma, H.C., Mohanty, S., and Pattnaik, S.K. 2003. Investment decision model for drip irrigation system. *Irrigation Science*, 22(2), pp. 79–85.

Subbaiah, B.V. 1956. A rapid procedure for estimation of available nitrogen in soil. *Current Science*, 25, pp. 259–260.

Subramanian, J., Kulkarni, S., Benagi, V.I. and Kulkarni, S. 1997. Crop disease management with drip irrigation. In *Drip Irrigation*, V.C. Patil, and S.S. Angadi (Eds.), pp. 72–76. Bengaluru.

Sujatha, S. and Bhat, R. 2013. Impact of drip fertigation on arecanut–cocoa system in humid tropics of India. *Agroforestry systems*, 87(3), pp. 643–656.

Sukumar, C.R. 2003. Precision farming may turn a reality Financial Daily from THE HINDU group of publications.

Sundara, B. and Tripathi, B.K. 1989. Available N changes and N balance under multi ratooning of sugarcane varieties in tropical vertisol. *Proceedings of the International Society of Sugar Cane Technologists*, 23, pp. 80–88.

Swaminathan, M.S. 2002. Building a national nutrition security system. Paper presented at India–ASEAN Eminent Persons Lecture Series, 11 Jan. 2002.

Swietlik, D. 1992. Leaf water relation in a flood irrigated young grapefruit orchard: Implication for irrigation different soil volume. *Subtropical Plant Science*, 45, pp. 18–22.

Tiwari, K.N., Mal, P.K., Singh, R.M. and Chattopadhyay, A. 1998. Response of okra (Abelmoschus esculentus (L.) Moench.) to drip irrigation under mulch and non-mulch conditions. *Agricultural Water Management*, 38(2), pp. 91–102.

Tiwari, K.N., Singh, A. and Mal, P.K. 2003. Effect of drip irrigation on yield of cabbage (Brassica oleracea L. var. capitata) under mulch and non-mulch conditions. *Agricultural Water Management*, 58(1), pp. 19–28.

Tiwari, V.S., Raisadhan, S.R., Jagtap, D.D., and Pujari, C.V. 1999. Effect of organic and inorganic fertilizer on yield and quality of Sweet orange. International symposium on Citriculture, Nagpur, Nov. 23–27., p. 411.

Tripathi, R., Shahid, M., Nayak, A.K., Raja, R., Panda, B.B., Mohanty, S., Thilgham, K. and Kumar, A. 2013. *Precision Agriculture in India: Opportunities and Challenges*. Central Rice Research Institute, Cuttack.

Uriu, K. 1977. Congress on use of plastics in agriculture. Proceedings of 7th International, New Delhi, pp. 211–214.

Vasudeo, R., Naidu, R., Lakshmikantham, M. 1946. Ratooning sugarcane in Madras. *Madras Agricultural Journal*, (4), pp. 3–12.

Veeranna, H.K., Khalak, A., Farooqui, A.M. and Sujith, G.M. 2001. Effect of fertigation with normal and water soluble fertilizers compared to drip and furrow methods on yield, fertilizer and irrigation water use efficiency in chilli. *Micro Irrigation*, 2, pp. 461–466.

Veerannah, L. and Selvaraj, P. 1984. Studies on growth, dry matter partitioning and pattern of nutrient uptake in papaya. In Proceedings of National Seminar on Papaya and Papain Production, pp. 26–27.

Venugopal, R. and Rajkumar, M. 1998. Drip irrigation system–new concept in sugarcane culture at baramba of orissa state. VSI, pp. III.

Verma, R.S. 2002. *Sugarcane Ratoon Management, 202*. International Book Distributing Co. Pvt. Ltd., Lucknow.

Vijayakumar, G., Tamilmani, D. and Selvaraj, P.K. 2010. Irrigation and fertigation scheduling under drip irrigation in brinjal (Solanum melongena L.) crop. *International Journal of Bio-resource and Stress Management*, 1(2), pp. 72–76.

Vijayakumar, G., Tamilmani, D. and Selvaraj, P.K. 2010. Maximizing Water and Fertilizer Use Efficiencies under Drip Irrigation in Chili Crop. *Journal of Management & Public Policy*, 2(1).

Vijayakumar, K.R., Dey, S.K., Chandrasekhar, T.R., Devakumar, A.S., Mohankrishna, T., Rao, P.S. and Sethuraj, M.R. 1998. Irrigation requirement of rubber trees (Hevea brasiliensis) in the subhumid tropics. *Agricultural Water Management*, 35(3), pp. 245–259.

Vijayalakshmi, R., Veerabadran, V., Shanmugasundram, K. and Kumar, V. 2003. Micro-sprinkler irrigation and fustigation and land configuration as a best management technology package for groundnut. *8th International Micro Irrigation Congress*, p. 17.

Wood, M., Malano, H. and Turral, H. 1998. Real-time monitoring and control of on-farm surface irrigation systems. Department of Civil and Environmental Engineering, University of Melbourne.

Yule, I.J., Hedley, C.B. and Bradbury, S. 2008. Variable-rate irrigation. 12th Annual Symposium on Precision Agriculture Research & Application in Australasia, Sydney.

9 Low-Cost On-Farm Indigenous and Innovative Technologies of Rainwater Harvesting

P. K. Singh
Maharana Pratap University of Agriculture and Technology

Rohitashw Kumar
Sher-e-Kashmir University of Agricultural Sciences
and Technology of Kashmir (SKUAST-K)

CONTENTS

9.1   INTRODUCTION

India has been one of the few countries in the world which showed awareness of the need to conserve and care for the watershed resources of land, water, plants and animals in an integrated manner and the government has invested heavily on soil and water conservation (SWC) measures on watershed basis and many big projects are currently in operation. The results to-date of the government SWC programs have been disappointing (Vaidyanathan, 1991). SWC measures installed under special programs have rarely been maintained; on the contrary, there are many instances where farmers have destroyed these works soon after the departure of the implementing agency. Recent studies have shown, however, that in many regions the farmers' lack of interest in SWC programs has not been due to their lack of concern about erosion, but because the design of recommended technologies has not been suitable for their small farms (Kerr and Sanghi, 1992; Reij, 1991). It is now becoming clear that there are significant differences between farmers' and scientist's perceptions regarding soil erosion control (Chambers, 1991; Kerr and Sanghi, 1992). Many SWC interventions are not successful because they are not sufficiently rooted in the priorities and perceptions of the local farmers (Gupta, 1991; Reij, 1991; Fujisaka, 1989).

In this context, it is also not out of place to mention that program planners have time to time introduced a number of SWC measures/rainwater harvesting technologies which are not being tested in the specific areas under particular soil, slope, rainfall, socio-economic conditions and needs of the people. Similarly, some of the most adoptable and effective technologies are not being given due importance and left aside because they are slightly costlier, though for such effective technologies farmers could easily be motivated for their reasonable contribution. Studies have revealed that over the generations, farmers themselves have developed numerous indigenous SWC methods specific to particular soil, slope, rainfall, and socio-economic conditions (Kerr, 1991). It has also been observed in the area that farmers prefer to pay part of the cost of these indigenous practices even in villages where recommended SWC practices are offered with heavy subsidies.

Low cost indigenous technologies of rainwater harvesting have the potential to increase the productivity of arable lands by enhancing crop yields and by reducing the risk of crop failure in arid and semi-arid regions, where water shortage are common because of scanty rainfall and its uneven distribution. In arid and semi-arid regions, the occurrence and distribution of rainfall are not only uneven but also erratic, marked by prolonged rainless days. The rainfall fails especially at the time when it is required most for agriculture during the year. Under these circumstances, the concept of low cost community oriented indigenous rainwater harvesting technologies, both long term and short term, seem to be the only alternative by which the water scarcity problem can be mitigated and agricultural production can be increased substantially. The solution therefore, lies in harvesting rainwater through capturing, storing, and recycling it and later using it during prolonged parched periods.

9.2 INDIGENOUS TECHNOLOGIES OF SWC/RAINWATER HARVESTING

9.2.1 EARTHEN FIELD BUNDS

Very commonly found indigenous SWC technique where the farmers construct field bunds almost uniformly on field boundaries, which rarely correspond exactly to contour for minimizing soil erosion; demarcating field and ownership boundaries; producing fodder for animals and other items of economic importance (through suitable vegetative cover); protecting against trespassers and stray animals (through a combination of high bunds and thorny barriers); creating new fields or micro-environments (to reduce risk in rainfed agriculture); making field operations convenient, facilitating land partitioning for inheritance, etc.

In order to make them more effective and in achieving the desired benefits of SWC to the extent of the farmers' expectations, these bunds may be constructed by keeping the required top height same throughout the bund with a provision of waste weir at suitable site.

9.2.2 STONE BUNDS

Stone bunds are the most commonly used indigenous practice in highly sloping lands of limited depth of soil for the purpose of increasing crop productivity in rainfed areas. Simple stone bunds of varying sizes are constructed across the slope.

In such types of terraces bunds are formed gradually by allowing erosion on the upper parts of sloping fields and arresting the soil by creating a vegetative/stone barrier on the field boundary. By adopting this practice, land with limited depth of soil can safely be put under cultivation without further degradation in sloping areas. In this case the cost of construction is reduced and the decrease in yield in the regular bench terracing is minimized. Downward movement of soil is induced by up and down slope cultivation during the first 2–3 years. Presently, such terraces are known as Puerto Rican Terraces.

9.2.3 STONE WALL TERRACES (SWT)

In some of the highly sloping areas where soil depth is a limiting factor and also in the cultivable valleys, SWTs are very common particularly in those areas where stones are readily available in the area. Like stone bunds, the stone wall barriers are also put across the slope for developing terraces on downhill slopes and particularly in valleys. The downward movement of soil is induced in a similar fashion as stated above. A cross-section of SWT is decided by the farmers taking into account the slope of the land, rainfall, etc. This practice is also adopted in order to create additional cultivable lands by cutting the hill slopes and to concentrate the soil eroded from the adjoining lands at an appropriate site.

9.2.4 Rough Stone Slab Bunds

It is found to be a very effective, adoptable, and low cost indigenous technology in moderately sloping (0%–5%) arable lands where the small stone slabs are easily available at or near the site. In this system 30–45 cm high bunds of rough stone slabs, 5–10 cm thick and 45–60 cm long are put across the slope, uniformly all along the field boundaries. Stone slabs are thoroughly embedded in the soil one after the other in dug out furrows of 15–30 cm depth.

9.2.5 Rough Stone Bunds

In the absence of the slabs, simple stone pieces 10–20 cm thick, 45–60 cm long, and of varying widths are also used. In due course of time the small gaps in between two slabs/stones are covered by naturally occurring grasses, also acting as a filter strip. Some of the farmers prefer to have such bunds against smaller cross-sectional earthen bunds because in this system only a narrow strip of land goes out of cultivation and maintenance is almost nil.

9.2.6 Vegetative Peripheral Bunds/Barriers

Peripheral or boundary bunds/barriers of Agave sislana locally known as Ram bans/ Gul bans is a commonly used indigenous SWC technology in arid and semi-arid regions and the established bunds are found to be very effective. Barriers of Agave are also a very commonly used technique in many of the areas to stabilize the periphery of fields situated on the banks of big nallas or rivers.

9.2.7 Smaller Cross-Sectional Earthen Bunds Covered with Flat Stones or Pieces of Stone Slabs

In some of the hilly areas in moderately sloping lands; smaller cross-section earthen bunds of about 30–45 cm height are constructed across the slope almost on contours for enhancing in situ moisture conservation and also for checking soil erosion from arable lands. The top level is strictly maintained at a uniform level throughout the bund length and the top is covered with flat stones or pieces of rough stone slabs to keep the bunds safe from raindrops' impact and also from occasional damages caused by over-topping. Sometimes all the three sides of the bund are covered/ pitched with stones. As per the requirements of the area, a provision for safe disposal of excess runoff is also kept. The farmers used to maintain these bunds very carefully. In some of the areas these bunds are also established for controlling/stabilizing gullies.

9.2.8   Temporary Sediment Detention Dams (TSDD)

One of the ways adopted in the hilly areas of southern Rajasthan to concentrate eroded soil at appropriate locations is the construction of TSDD. In such areas most of the badly eroded lands are found in deep and narrow valleys, where due to high concentration of runoff the rate of soil erosion is very high. Under these situations construction of TSDD is adopted by the farmers. Suitable locations are those where the possibilities of sediment trapping is more. Initially a low height broad-based loose rock dam is constructed. The base width is decided keeping in view the rainfall pattern and expected runoff. Over the years the height of these barriers is increased and a new patch of cultivable land is created within the gullies/eroded valleys. The height of the dam is increased till the nallah/valley section reaches the extent where the gradient remains stable. In some of the areas such bunds have a height of 3–5 m or more. TSDD also acts as a temporary drop structure.

9.2.9   Diversion Ditches

Diversion ditches are small channels with banks on the downward slope side having desired gradient towards an outlet for the safe disposal of runoff from the upper reaches in the natural nallah, to prevent runoff from entering the lands of lower reaches which are already protected by some kind of soil conservation measures and to separate the arable lands from the non-arable lands. It is also one of the commonly used indigenous SWC technologies, in the hilly terrains of southern Rajasthan, and also in other parts of Rajasthan where a good amount of cultivated land exists in the lower reaches. To protect these lands from the damages caused by runoff water and channeling along the gradient towards the nallah, such diversion drains are being constructed. The cross-section and the type of ditches are mainly based on experiences. There are different types of drains considering the amount of runoff and other factors. These are as follows:

  i. Excavated ditches with required gradient in the base.
 ii. Excavated ditches supported by a suitable sized loose stone bund on the downward slope side. The required gradient is provided in the excavated ditch.
iii. Only loose stone bunds are installed and the desired gradient is provided by scraping the land after leaving 15–30 cm berm in the base on the upper side of slope.

9.2.10   Stone Wall for Nallah Bank Protection

This practice is adopted in those conditions where bank erosion is a problem, particularly in arable lands. This technique is primarily used only in those areas where stones are available at sites or very near to sites. Suitable cross-sectional wall of loose stone is constructed all along the bank or only at vulnerable sites. Erection of such protection wall is done starting from the bed of the nallah keeping appropriate foundation. The height of such walls depends on the depth of flow of water in the nallah. Sometimes these are also reinforced by planting suitable vegetative material such as Agave, Jatropha, and Mahadi, etc.

9.2.11 DHORA PALI

Field bunding is one of the common practices of SWC locally known as "Dhora pali". A bund of about 0.5 m² or even larger cross-section is constructed on the field boundaries in the arid zone. Sometimes waste weirs are also provided at suitable site. These areas are mainly put under kharif crops. In due course of time these bunds get stabilized by naturally occurring local grasses. Sometimes seeds of Dhaman grass are also sown during rainy season for stabilization. Venkateswarlu (1993) also reported that existing SWC practices in arid Rajasthan include large peripheral bunds about 1 m in height and 70–75 cm wide at the base. In some of the areas these bunds are strengthened with munj grass/agave.

9.2.12 KANA BANDI (MULCHING)

In desert areas, in order to keep the arable land productive, efforts are being made to protect the area from wind erosion. Kana bandi is done in the fields after kharif crops are harvested, particularly in those fields which are prone to erosion. The local material like sania, khinp, prunings of ker, ber, khejri, and phog and also local grasses such as sewan/munj are embedded in the soil, leaving about 30–40 cm length of the material vertically on the ground in lines 2–5 m apart. This practice checks the soil erosion to a great extent. Sometimes kana bandi is done in a square or rectangular manner (checker board fashion, 2–3 m²), particularly for stabilization of the sand dunes after rainy season, and the grass seeds are sown on the leeward side of the mulch. The grass grows and gradually replaces the mulch and controls the movement of sand. During kharif this organic material is incorporated in the soil, thereby also helping in increasing the organic matter content.

9.2.13 VILLAGE POND/TALAB

A common rural rainwater harvesting technology used throughout the semi-arid region of Rajasthan is the construction of ponds/nadis/tanks, etc. A pond is constructed at suitable sites mainly for domestic use and also for the recharging of groundwater. A suitable site from an economic viewpoint is selected by the villagers where the largest storage volume is obtained with the least amount of earth fill. Such conditions are generally found where the valley is narrow, the side slopes are relatively steep, and the slope of the valley floor will permit a large deep basin. Such sites tend to minimize the area of shallow waters. Surface runoff is the major source of feeding the ponds/talabs. Villagers also adopt some design criteria, viz., determination of the capacity, the size and shape of embankments, provision of emergency spillways, and provision for controlling seepage.

9.2.14   Talai – A Small Water Harvesting Structure

Talai is an indigenous water harvesting technique in the semi-arid regions of India, particularly for creating water points for cattle. In this system an earthen embankment of very low height (maybe of 1–2 m) is made at a suitable location in a nallah/natural drainage line, where a natural depression exists. The earth required in making the embankment is also taken out from the existing depression for increasing storage capacity. Presently this system is advocated and recommended in the name of SUNKEN PONDS, particularly in National Watershed Development Programme for Rain fed Areas (NWDPRA) projects.

9.2.15   Dry Stone Masonry Pond

Dry stone masonry ponds, between 1.5 and 2.5 m high, are constructed to collect and store water. In this type of structure the upstream and downstream walls are constructed 3–4 m apart by dry stone masonry after excavating a foundation of appropriate depth. The space in between these two walls is filled with locally available murrum or soil with proper compaction. The filling is done in layers of 20–30 cm height along with wetting and compaction. The earth fill is kept 10–20 cm above the top of the wall to provide an extra provision for natural settling over a period of time. Proper compaction is one of the important considerations to check seepage through the embankment and to ensure the stability of the structure. The length of the head wall extension depends on the specific site conditions. The height of such structures is restricted up to 2.5 m to avoid overturning due to water pressure. The width of the wall at the bottom is kept at 1.5 m and at the top it is only 0.5–0.6 m. The reduction in the width is maintained uniformly from bottom to top along the inner edge of the wall. The upper portion of the wall (0.30–0.5 m high) is constructed with cement mortar to avoid damage to the walls by stray cattle or human activities.

9.2.16   Ponds (Nada)

These large ponds are of two categories denoting both ownership and use. The nadas belonging to the Panchayat are for the specific purpose of providing drinking water for animals, while the private ones which have been constructed on kabile kasth lands are used for irrigation. These farm ponds are generally constructed by a group of farmers, whose lands remain temporarily submerged and after monsoon, i.e., in Rabi season, crops are sown as tank bed cultivation, when the water has evaporated or percolated. Stored water is sometimes drained through some indigenously developed surplussing arrangements for the sowing of Rabi crops.

9.2.17   Nadi (Semi-Arid/Aravali Region)

Nadi is a small traditional water harvesting structure constructed at an appropriate site to harvest the runoff water of relatively impervious non-arable uplands for the purpose of drinking water for animals, and for groundwater recharge of open dug wells situated in the lower reaches. These are also constructed to store water in the monsoonal nallahs in the upper reaches for various purposes and primarily for recharge of groundwater. The depth of such nadis generally does not exceed 3 m. These structures are constructed in two ways, depending upon the available funds. In the first system, both sides of the earthen embankment of appropriate width are supported by dry stone masonry walls. In the second system, the upside wall is pakka or masonry using lime or cement mortar. Masonry wall and earth fill is done in an arc shape having curvature raised by the locally available soil/murmur. Layer-wise wetting and compaction of soil is practised. The width of the earthen embankment and stone walls are decided by the villagers considering the size, topography, and other conditions of the catchment areas. A properly designed waste weir of surplussing arrangement is also provided at suitable site.

9.2.18   Nadi (Arid Regions)

In arid zone, the construction of Nadi is an age-old practice of water harvesting. These are small excavated or embankment village ponds, harvesting the meager precipitations to mitigate the scarcity of drinking water. These nadis hold water from 2 months to one full year after rains, depending on the catchment characteristics, the amount of rainfall received, its intensity, and distribution. Each village has one or more of such structures, depending on the demand of water and the availability of suitable sites. The capacity of such nadis is reduced in due course of time due to sediment deposition.

9.2.19   Tanka

Tanka, the most prevailing rainwater harvesting structure in the Indian desert, is a local term for the underground system. The traditional tanks are made by digging a hole of 3.0–4.25 m diameter in the ground and plastering it with lime mortar to a thickness of about 6 mm, followed by a cement plaster 3 mm thick. The top is covered with ber Thoms. The useful life of such a structure is about 3 years. The catchments are made in a variety of ways using locally available sealing materials like pond silt, murrum, coal ash, gravel, etc. Traditional tankas are temporary and are subject to leakage. Moreover, the catchment areas are not in accordance with the amount of rainfall received and runoff generated. The thorn cover does not prevent the water pollution and evaporation losses, and the capacity of such tankas is also not sufficient to fulfill the demands of a family for water throughout the year.

The CAZRI has designed an improved tanka, of 21,000 L capacity, which gets filled up with annual rainfall of 125 mm. The water is sufficient for a family of six persons throughout the year for drinking. It has a useful life of 25 years as it is constructed using cement masonry. The catchment area needed for this capacity is 778 m^2.

9.2.20 KHADIN

From a study of the farmers' water conservation practices, it is evident that they are acutely conscious of the value of rainwater and try to use it to grow at least one good crop during the year. Khadin is one such system, which is extensively used in the arid and semi-arid regions of Rajasthan. It is an indigenous water harvesting-cum-runoff farming structure. A khadin system is site-specific, needing a large natural, high runoff potential catchment in proximity of plain valley land with deep soils. The ratio of khadin catchment area, depending on the type of catchment is 1:12–1:15. These are constructed on low-lying lands where crops are raised by conserving rainwater from the rocky catchments. Cultivation in khadin is done by rationing runoff water over low-lying areas through the construction of a bund across the slope on the lower boundary line of the khadin land. The cross-section of the bund depends upon the soil type, the area of khadins, and the discharge from catchments. The water thus collected is allowed to percolate, after which an assured post-rainy season crop is grown. Sometimes crops are grown in kharif or Rabi depending upon the rainfall and runoff received in the khadins. For areas that will always be dependent on rainwater, this water harvesting practice has great relevance. Now the SWC scientists/engineers have also considered this indigenous technique as an important and useful water harvesting practice and have developed design criteria. Kolarkar et al. (1983) also reported that "khadins" or submergence tanks are an indigenous form of inundation farming in arid regions.

9.3 INNOVATIVE TECHNOLOGIES OF RAINWATER HARVESTING

9.3.1 ROOFTOP RAINWATER HARVESTING

Rooftop rainwater harvesting technique is applied mainly for domestic purposes or for groundwater recharging in the rural and urban areas. In this technique, the rainwater of the roof is either collected in the underground tanks or diverted to the wells/tube wells for groundwater recharging. Since the collected water is generally free from soil pollution, it can be used for drinking as well as domestic purposes. This technique is highly suitable for the low rainfall areas where the number of runoff-producing rainfall storms is limited and there is scarcity of drinking water.

9.3.2 SUBSOILING

Subsoiling is a system of deep tillage by which the subsoil is loosened and disturbed but is not inverted or brought to the surface. The term subsoiling has also been applied by some workers to any cultivation carried out in the soil below the normal ploughing depth. Subsoiling is possible with the help of deep soil loosening equipment, viz., chisel plough and subsoiler. Subsoiling is a totally mechanized operation. At present, subsoilers available in the market can be operated with any tractor equipped with a hydraulic lift. On suitable soils, chiseling is applicable if the restrictive soil layers are less than 45 cm deep, whereas subsoiling is applicable if the restrictive soil layers are more than 45 cm deep (Figure 6.6). Contour subsoiling is possible on slopes upto 30% but is most satisfactory on slopes below 22%–25% (Nag et. al., 1989, Singh and Mahnot, 1995, Singh and Mahnot, 2004).

9.3.3 CHAUKA SYSTEM

In Rajasthan, the Gram Vikas Navyuvak Mandal, Laporia (GVNML) has been very active in undertaking measures for improving the productivity of pasture and grazing lands significantly in their project area. All this is in keeping with the goal of GVNML, which is to "support integrated rural development on a sustainable basis". Since its inception, GVNML has been active in organizing and mobilizing rural communities to carry out activities such as repair of tanks, plantation programs, health, education, pastureland development, and SWC. Among others, GVNML has now been actively involved in developing village *gauchar* (common pasture lands), using ideas – technical and socially oriented – generated by the local people themselves. Importantly, this non-govermental organization has developed an innovative concept – the Chauka system – for reducing runoff and preventing soil erosion to augment in-situ moisture conservation, with gratifying success.

9.3.4 DOUBLE WALL CEMENT MASONRY STRUCTURE

This type of structure looks like an anicut. Both the upstream and downstream walls of the structure are constructed with cement masonry. The height of the structure and catchment area is usually restricted to 2.5–3.0 m and 100–150 ha, respectively. The base width of the upstream and downstream walls is generally taken as 1.0 and 0.8 m, respectively, whereas the top width of the upstream and downstream walls is restricted to 0.60 and 0.45 m, respectively. The width of the walls may be increased depending on the site conditions and the volume of water to be stored. For low-height structures (1.0–1.5 m) the base width of both the walls may be reduced by 20 cm. The width of the concrete bed is generally taken as 20 cm more than the base width of the masonry walls. The downstream wall or the falling side is tapered. The space in between these two walls is filled with locally available murrum or soil with proper compaction. The filling is done in layers of 20–30 cm height alongwith wetting and compaction. Proper compaction is an important consideration to ensure the stability of the structure.

9.3.5 PLASTIC-LINED FARM POND

Plastic lined farm ponds are particularly suitable for those areas where large quantity of water is lost through seepage, especially where the soil is gravelly and porous. In earthen dams there is also a common problem of seepage through the embankment. Under such circumstances, to check the seepage from all such types of farm ponds/earthen dams, plastic lining is a feasible solution. Polythene sheets of 200 μm may be used as lining material for seepage control in the ponds. The sheets are spread at the bottom and on the upstream side, up to the top width of the pond. An average

10 cm thick soil layer is also kept above the sheet to keep the sheet in proper place, to check external damage, and to protect it from exposure to the sun. A permanent and most effective lining material is brick and cement masonry, but it is costlier than other lining materials.

9.3.6 SUBSURFACE BARRIERS

Subsurface barriers are used to retain or arrest the seasonal subsurface flows and facilitate the abstraction of water through lined shallow wells, especially during periods of water scarcity. The objective is to place an impermeable barrier – either of clay or masonry across the river-bed, from the surface down to the bedrock or other solid impervious layer.

A trench of the required width is dug across the flow direction of the ground water. The earthwork involved may be carried out by manual labor since the excavation depths are generally not more than 3–6 m. Subsurface dams are generally constructed at the end of the dry season, when there is little water in the aquifer. There is usually some flow, however, and this must be pumped out during the construction work. After the construction of the dam, the trench is refilled with the excavated material. It is important that the refill is properly compacted by mechanical means and watering.

REFERENCES

Chambers, R. 1991. Farmer's Practices, Professionals, and Participation. In Kerr, J.M. (ed.) *Farmer's Practices and Soil Water Conservation Programmes. Summary Proceedings of a Workshop. 19–21 June.* ICRISAT, Patancheru.

Fujisaka, S. 1989. A method for farmer-participatory research and technology transfer: upland soil conservation in the Philippines. *Experimental Agriculture*, 25(4), pp. 423–433.

Gupta, A., 1991. Reconceptualising development and diffusion of technologies for dry regions. In Prasad C. and P. Das (ed.) *Extension Strategies for Rainfed Agriculture.* Indian Society of Extension Education, New Delhi.

Kerr, J.M. 1991. *Farmers Practices and Soil and Water Conservation Programmes: Summary Proceedings of Workshop. 19–21 June.* ICRISAT, Patancheru.

Kerr, J.M. and Sanghi, N.K. 1992. Indigenous Soil and Water Conservation in India's Semi-Arid Tropics. IIED International Institute for Environment and Development, Sustainable Agriculture Programme.

Kolarkar, A.S., Murthy, K.N.K., and Singh, N. 1983. Khadin A method for harvesting water. *Journal of Arid Environment*, 6, pp. 59–66.

Nag, K.N., Chandra, A. and Mahnot, S.C. 1989. Mechanization techniques for accelerating afforestation programmes on denuded hillocks. *AMA, Agricultural Mechanization in Asia, Africa and Latin America*, 20(3), pp. 78–80.

Reij, C., 1991. Indigenous soil and water conservation in Africa (No. 27). Sustainable Agriculture Programme of the International Institute for Environment and Development.

Singh and Mahnot. 2004. *Mechanical Soil Working Techniques for Soil and Water Conservation on Moderately Sloping Wasteland, Small Farm Mechanization.* ISAE, Rajasthan, pp. 98–101.

Singh, P.K. and Mahnot, S.C. 1995. Feasibility and cost effectiveness of mechanical soil working techniques for soil and water conservation measures on moderately sloping wasteland. *Indian Journal of Power and River Valley Development*, 45, pp. 106–109.

Vaidyanathan, A. 1991. *Integrated Watershed Development: Some Major Issues*. Founder's Day Lecture. Society for Promotion of Wasteland Development, New Delhi.

Venkateswarlu, J. 1993. *Crop Rotation in Relation to Maintenance of Soil Productivity in the Rainfed Semi- Arid Tropic of India*. FAO, Rome.

10 Impact of Climate Change on Food Safety

H. R. Naik

Sher-e-Kashmir University of Agricultural Sciences
and Technology of Kashmir (SKUAST-K)

CONTENTS

10.1   INTRODUCTION

Assuring food safety is a complex task. Food safety hazards can arise at any stage of the food chain, from primary production through to consumption. Foods are governed by food laws and regulations, which are collectively known as the food control system. The ultimate goal of this system is to ensure that food presented to consumers is safe and honestly presented. It is in the interest of all stakeholders to optimize the efficiency of the system in order to make the best possible public health impact with limited resources available. Major principles that underlie the strategies for improving the efficiency and effectiveness of food control are:

- That efforts are focused on issues that pose the greatest risk;
- That the responsibility for producing safe food rests unambiguously with the food businesses who are best placed to design and implement controls at the most appropriate point within the food production systems to *prevent* or minimize food safety risks;
- That the government establishes food safety requirements, facilitates industry's compliance with these and then ensures that the requirements are met through a range of regulatory and non regulatory measures.

The Fourth Assessment Report of the Intergovernmental Panel on Climate Change (IPCC, 2007) dispelled many uncertainties about climate change. Warming of the climate system is now unequivocal and according to IPCC the increase in global temperatures observed since the mid-20th century is predominantly due to human activities such as fuel burning and land use changes. Projections for the 21st century show that global warming will accelerate with predictions of the average increase in global temperature ranging from 1.8°C to 4°C. Other effects of climate change include trends towards stronger storm systems, increased frequency of heavy precipitation events, and extended dry periods. The contraction of the ice sheets will lead to rising sea-levels. These changes have implications on food production, food security, and food safety.

10.1.1   Effect on Food Crops and Animals

Crop production is extremely susceptible to climate change. It has been estimated that climate changes are likely to reduce yields and/or damage crops in the 21st century (IPCC, 2007). While the impact of biotic (microbial population of fungi, bacteria, viruses of the macro-environment, soil, air, and water) and abiotic factors (nutrient deficiencies, air pollutants, and temperature/moisture extremes) on crop production and food security are more obvious, it is important to note that these factors may also have a significant impact on the safety of the food crops. Another concern is the impact of climate change on the prevalence of environmental contaminants and chemical residues in the food chain. Climate change may affect zoonoses (diseases and infections which are naturally transmitted between vertebrate animals and man) in a number of ways. It may increase

    i. The transmission cycle of many vectors.
   ii. The range and prevalence of vectors and animal reservoirs.
  iii. In some regions it may result in the establishment of new diseases.

10.1.2 EFFECT ON FISHERIES

Climate change has implications for food safety. From a microbiological perspective, climate change exacerbates eutrophication (nutrient loading) causing phytoplankton growth, increased frequencies of harmful algal blooms (HABs), particularly of toxic species. Accumulation of these toxins by filter feeders (bivalve molluscs) and the subsequent consumption of these products have serious implications for humans. Furthermore, an increase in water temperature promotes the growth of organisms such as *Vibrio vulnificus* leading to an increased risk from handling or consuming fish grown in these waters (Paz et al., 2007). Climate change (in particular temperature increase) facilitates methylation of mercury and subsequent uptake by fish.

10.1.3 EFFECT ON FOOD HANDLING, PROCESSING, AND TRADING

Climate change affects not only primary production but also food manufacturing and trade. Emerging hazards in primary production could influence the design of the safety management systems required to effectively control those hazards and ensure the safety of the final product. Furthermore, increasing average temperatures could increase the hygiene risks associated with the storage and distribution of food commodities. It is important, therefore, that the food industry be vigilant to the need to modify hygiene programs. Reduced availability and quality of water in food handling and processing operations will also give rise to new challenges to hygiene management.

    Using the classic epidemiologic triad (host, agent, and environment), it is clear that climate, which impacts all three sectors of the triad, can have a dramatic effect on infectious disease. This is well documented and even predictable for some food and waterborne diseases of the developing world (e.g., bacillary dysentery and cholera) and perhaps less so for the developed world, where stringent public health measures (sewage disposal, clean water, and hygiene) moderate the risk of diarrhoeal disease. Evidence of the impact of climate change on the transmission of food and waterborne diseases comes from a number of sources, e.g., the seasonality of foodborne and diarrhoeal disease, changes in disease patterns that occur as a consequence of temperature, and associations between increased incidence of food and waterborne illness and severe weather events (Hall et al., 2002; Rose et al., 2001).

10.1.3.1 Sources and Modes of Transmission

The microflora of a food consists of the microorganisms associated with the raw material, those acquired during handling and processing and those surviving preservation techniques and storage. Bacteria, viruses, and parasitic protozoa (BVP) that are pathogenic to humans and frequently contaminate the food supply can be sub-categorized based on their ultimate source. They are predominantly associated with fecal matter (animals/human), on the skin, nose, and throat of healthy individuals and in nature. Based on the above categorization, the general scenarios by which foods become contaminated with pathogens include:

   i. Contact with human/animal sewage/feces
   ii. Contact with infected food handlers
   iii. Environmental contamination (from air, water, food contact materials, etc.)
   iv. Contact with raw foods, etc.

Such contamination can arise along any part of the farm-to-fork continuum and may arise from any number of sources. Climate constrains the range of infectious diseases, while weather, which is impacted by climate, affects the timing and intensity of outbreaks (Epstein, 2001). Therefore, the two early manifestations of climate change, particularly global warming, would be expansion in the geographic range and seasonality of disease, and the emergence of outbreaks occurring as a consequence of extreme weather events (Epstein, 2001).

10.1.3.2   Climatic Influences (e.g., Temperature, Humidity) on the Prevalence of Some Diseases

- Increases in disease notifications, particularly salmonellosis (D'Souza et al., 2004) and to a lesser extent campylobacteriosis (Kovats et al., 2005), are frequently preceded by weeks of elevated ambient temperature.
- Higher temperature and humidity in the week before infection has been correlated with decreased hospitalization rates for children diagnosed with rotavirus. This is particularly interesting because survival of the virus is favored at lower temperature and humidity (D'Souza et al., 2008). Rotavirus is considered a significant cause of foodborne illness (FAO, 2008).
- El Nino-associated rises in cholera have been documented for both Peru and Bangladesh, as have been increases in diarrhoeal disease in Peruvians (reviewed by Hall et al., 2002).
- Cholera is perhaps the best model for understanding the potential for climate-induced changes in the transmission of foodborne disease. *Vibrio cholerae* is the causative agent of this disease which produces substantial morbidity and mortality, particularly in the developing world.

Extreme weather conditions (e.g., flooding, drought, hurricanes, etc.) can have an impact on the transmission of disease. For example, periods of excessive precipitation and periods of drought influence both the availability and quality of water and have been linked to the transmission of water and food borne disease. Furthermore, extreme weather events can result in forced evacuation of refugees into close quarters. This frequently results in extreme stress, malnutrition, and limited access to medical care, all of which contribute to increased susceptibility and severity of disease.

TABLE 10.1
Examples of Some Zoonotic Agents That Are Expected to be Affected by Climate Change and Their Mode of Transmission

Virus	Host	Mode of Transmission to Humans
Rift Valley fever virus	Multiple species of livestock and wildlife	Blood or organs of infected animals (handling of animal tissue), unpasteurized or uncooked milk of infected animals, mosquitoes, hematophagous flies
Nipah virus	Bats and pigs	Directly from bats to humans through food in the consumption of date palm sap (Luby et al. 2006). Infected pigs present a serious risk to farmers and abattoir workers
Hendra virus	Bats and horses	Secretions from infected horses
Rotavirus	Humans	Fecal-oral route, spread through contaminated water and also by infected food-handlers who do not wash their hands properly
Hepatitis E virus	Wild and domestic animals	Fecal-oral. Pig manure is a possible source through contamination of irrigation water and shellfish in coastal waters

10.1.3.3   Effect of Climate Change on Zoonotic Disease

Zoonotic diseases are transmitted from animals to people in a number of ways. Some diseases are acquired by people through direct contact with infected animals or animal products and wastes. Other zoonoses are transmitted by vectors; while others are transmitted through the consumption of contaminated food or water (Table 10.1). The proliferation of zoonoses and other animal diseases may result in an increased use of veterinary drugs that could lead to increased and possibly unacceptable levels of veterinary drugs in foods (FAO, 2008).

10.1.3.4   Transmission of Bacteria

Bacterium	Host	Mode of Transmission
Salmonella	Poultry and pigs	Fecal/oral
Campylobacter	Poultry	Fecal/oral
E. coli O157	Cattle and other ruminants	Fecal/oral
Anaerobic sporeforming bacteria	Birds, mammals, and livestock	Ingestion of spores through environmental routes, water, soil, and feeds. This has been associated with outbreaks of anthrax in livestock and wild animals, blackleg (Clostridium chauvoei) in cattle, and botulism in wild birds after droughts. The meat and milk from cattle that have botulism should not be used for human consumption

(Continued)

Bacterium	Host	Mode of Transmission
Yersinia	Birds and rodents with regional differences in the species of animal, infected. Pigs are a major livestock reservoir	Handling pigs at slaughter is a risk to humans
Listeria monocytogenes	Livestock	In the northern hemisphere, listeriosis has a distinct seasonal occurrence in livestock probably associated with feeding of silage
Leptospirosis	All farm animal species	Leptospirae shed in urine to contaminate pasture, drinking water, and feed

Protozoan	Host	Mode of Transmission
Toxoplasma gondii	Cats, pigs, sheep	Cat feces are a major source of infection. Handling and consuming raw meat from infected sheep and pigs pose a zoonotic risk
Cyptosporidium and Giardia	Cattle, sheep	Fecal-oral transmission. (Oo)cysts are highly infectious and with high loadings, livestock feces pose a risk to animal handlers

10.1.3.5   Transmission of Protozoas

Parasites	Host	Mode of Transmission
Tapeworm (*Cysticercusbovis*)	Cattle	Fecal-oral
Liver fluke (*Fasciola hepatica*)	Sheep, cattle	Eggs are excreted in feces, and life cycle involves lymnaeid snail hosts. Human cases generally associated with the ingestion of marsh plants such as watercress

10.1.3.6   Transmission of Parasites
10.1.3.7   Climate Change Effects on BVP

1. Increase in the susceptibility of animals to disease
2. Increase in the range or abundance of vectors/animal reservoirs
3. Prolonging the transmission cycles of vectors

10.1.4   CLIMATE CHANGE AND ITS INFLUENCE ON MOULD AND MYCOTOXIN CONTAMINATION

Mycotoxins are a group of highly toxic chemical substances that are produced by toxigenic moulds that commonly grow on a number of crops. These toxins can be produced before harvest in the standing crop and many can increase, even dramatically, after harvest if the post-harvest conditions are favorable for further fungal growth.

TABLE 10.2

Moulds and Mycotoxins of Worldwide Importance

Mould Species	Mycotoxins Produced
Aspergillus parasiticus	Aflatoxins B1, B2, G1, G2
Aspergillus flavus	Aflatoxins B1, B2
Fusarium sporotrichioides	T-2 toxin
Fusarium graminearum	Deoxynivalenol (or nivalenol) Zearalenone
Fusarium moniliforme (F. verticillioides)	Fumonisin B1
Penicillium verrucosum	Ochratoxin A
Aspergillus ochraceus	Ochratoxin A

Human dietary exposure to mycotoxins can be directly through the consumption of contaminated crops. Mycotoxins can also reach the human food supply through livestock that have consumed contaminated feed. The problem of mycotoxin contamination of foods and the resulting public health impact is not new – it is likely that mycotoxins have plagued mankind since the beginning of organized crop production (FAO, 2001). At high doses mycotoxins produce acute symptoms and deaths but, arguably, lower doses that produce no clinical symptoms are more significant to public health due to the greater extent of this level of exposure. Particular mycotoxins may possess carcinogenic, immunosuppressive, neurotoxic, estrogenic, or teratogenic activity, some more than one of these. Table 10.2 lists mycotoxins that are of worldwide importance, meaning that they have been demonstrated to have significant impact on public health and animal productivity in a variety of countries. There are several other mycotoxins that are considered to be of regional significance (FAO, 2001).

Although the impact of climate change on fungal colonization has not been yet specifically and thoroughly addressed, temperature, humidity, and precipitation are known to have an effect on toxigenic moulds and on their interaction with the plant hosts. In general we know that fungi have temperature ranges within which they perform better and therefore increasing average temperatures could lead to changes in the range of latitudes at which certain fungi are able to compete. Since 2003, frequent hot and dry summers in Italy have resulted in increased occurrence of *A. flavus*, the most xerophilic of the *Aspergillus* genus, with consequent unexpected and serious outbreak of aflatoxin contamination, uncommon in Europe. Also, in United States serious outbreaks of *A. flavus* have been reported for similar reasons. Generally moist, humid conditions favor mould growth – moist conditions following periods of heavy precipitation or floods would be expected to favor mould growth. Generally speaking, conditions adverse to the plant (drought stress, stress induced by pest attack, poor nutrient status, etc.) encourages the fungal partner to develop more than under favorable plant conditions with the expectation of greater production of mycotoxins.

10.1.5 INFLUENCE OF CLIMATE CHANGE ON POST-HARVEST CONDITIONS

It is common for commodities to contain mycotoxigenic fungi at harvest. Up to the point of harvest, the status of the plant will play a major role in determining the degree of mycotoxin contamination. Thereafter, fungal development and mycotoxin production will be controlled by post-harvest handling techniques and practice. In the simplest terms, this will consist of some kind of cleaning, which may be conducted concomitantly with harvest, drying, and storage where stability is maintained by restricting the water availability to a level well below that required for fungal growth. Climate change could impinge on this part of the food chain, especially in regions where capital investment on such production infrastructure is lacking.

10.1.5.1 Effect on Post-Harvest Quality of Fruits and Vegetables

Climatic change results in greenhouse effect on fruits/vegetables. Higher temperature, CO_2 level, and Ozone depletion has a direct or indirect effect on the quality of horticultural produce at different stages from production to the consumer (Moretti et al., 2010).

1. **Photosynthesis: Due to global warming**, alterations in sugars, acids, flavonoids, and firmness in crops has been reported. For example:
   - Increased CO_2 level has caused tuber malformation, common scab, changed reducing sugar content in potatoes.
   - Increased ozone level resulted in
     1. Decreased photosynthesis process, decreased growth, and biomass accumulation.
     2. Increased vitamin C accumulation, decreased emission of volatile esters in strawberries.
     3. Increased β carotene, lutein, and lycopene in tomatoes.

10.1.5.2 Effect of Temperature on Fruits/Vegetables

- Photosynthesis, respiration, aquos relations in membrane stability affected.
- Normal physical process temperature (0°C–40°C). Best reported photosynthesisl respiration ratio > 10 at 15°C.
- Leaf to air vapor pressure difference (D) change affects photosynthesis.
- Enzyme catalyzed biochemical reactions affect photosynthesis.
- Harvest Index affected. Crops mature earlier.
- Rapid cooling to remove field heat to increase shelf life 2–3 fold for each 10°C increase in temperature. This reduces respiration rate and enzyme activity which slows ripening, senescence, maintains firmness, minimizes water loss, and pathogens but results in higher pulp temperature. The produce shall need more energy for cooling, thus raising product prices.

10.1.5.2.1 Effect of High Temperature

a. On Quality
- Flavor is affected, e.g., higher sugar content and lower organic acids of apple groups.
- Two times more firmness in crops grown under high temperature due to change in cell wall composition, cell number, and cell turgor properties.
- Increase in moisture content and certain fatty acids like palmitic, oleic, etc.

- Higher concentration of minerals like Ca, Mg, Cu, and K due to less water movement through the crop.
- Increase in antioxidant activity (flavonoids) in berries and but decrease in vitamin content.

b. **Physiological disorders**
  - Tomatoes grown at >30°C temperature affects their color development (yellowish white color rather than red), causes softening, increased respiration rate, and ethylene production.
  - Fruits exposed to 40°C have induced metabolic disorders, fungal and bacteria invasion.
  - Temperatures greater than 40°C result in sunburn of apples, water core and loss of texture, and decreased tolerance to low temperatures of the apples on storage.
  - Due to delay in winter, yields dropped significantly in crops like cauliflower and soybean.

10.1.5.2.2 Effect of CO_2 Exposure

Composition of air (78% N, 21% O, 0.93% Ar, and 0.03% CO_2) gets changed due to climate change resulting in Greenhouse combination effect of water vapor, CO_2, and minute amounts of other gases like (CH_4, Nitrous oxide, and Ozone) which absorb radiations leaving the earth's surface. CO_2 or other gases absorb the earth's infrared radiations, thus trapping heat and resulting in global warming. Their effect on crops is:

- Alter plant tissues in tenure of growth and physiological behavior.
- Photosynthesis, biomass product, sugars and organic acid content, stomatal – conductance, firmness, yield, light, water, nutrient use efficiency, and plant water potential.
- Decreased tuber formation in potatoes (63%) resulting in poor processing quality and lower tuber greening (12%). In total, 550 micro mol CO_2 mol concentration of CO_2 decreased glucose, fructose, and resistant starch concentrations reducing tuber quality due to decreased browning and acryl amide formation in French fries.
- 34% common scab reported.
- Increased CO_2 concentration, however, stimulates grapevine produce without affecting the quality of grapes.

10.1.5.3 Effect of High Ozone Formation

Ozone forms during periods of high temperature and solar irradiation, normally during the summer season. Concentration is maximum in the late afternoon and minimum in the early morning hours. Increased ozone may also be due to the movement of local winds or downdrafts from the stratosphere. The physiological effects are:

- The Stomatal conductance and ambient concentration is most important for O_3 uptake by plants. Ozone enters a plant through the stomata, causing direct cellular damage due to changes in membrane permeability.

- May or may not result in visible injury, reduced growth, and ultimately reduced yield.

10.1.5.3.1   Visible Injury

a. Change in pigmentation called bronzing.
b. Leaf chlorosis, e.g., yellowing green vegetable leaves affecting pH and qualities like appearance, color, flavor compounds.
c. Premature senesce.
d. Increase or decrease in yield of different crops due to impaired conductance.
e. Change in Carbon transport of root, tubers, bulbs results in less accumulation of starch and sugars.
f. Decreased biomass produce directly impacts size, appearance, etc.

10.1.5.3.2   Effect of Ozone on Crop Quality

- Crops like mushroom, seedless cucumber, and broccoli stored at 3°C–10°C and exposed to O_3 showed minimum response to storage temperatures.
- O_3 removes ethylene from the environment, and thus is useful in closed rooms, e.g., 0.4 HL/L O_3 removes 1.5–2.0 HL/L ethylene from an apple store room.
- O_3 increased vitamin C and decreased the emission of volatile esters in strawberries.
- O_3 has germicidal effect and overall no bad effect is observed.

10.1.6   Impacts of HABs

In recent decades, there has been an apparent increase in the occurrence of HABs in many marine and coastal regions (Figure 10.1, Hallegraeff, 1993). Toxin-producing HAB species are particularly dangerous to humans. A number of human illnesses are caused by ingesting seafood (primarily shellfish) contaminated with natural toxins produced by HAB organisms; these include amnesic shellfish poisoning (ASP), diarrheic shellfish poisoning (DSP), neurotoxic shellfish poisoning (NSP), azaspiracid shellfish poisoning (AZP), paralytic shellfish poisoning (PSP), and ciguatera fish poisoning. These toxins may cause respiratory and digestive problems, memory loss,

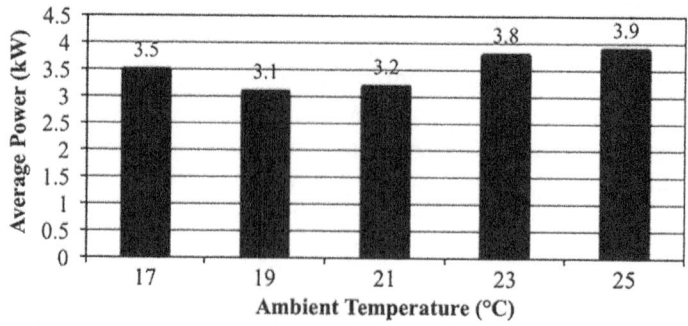

FIGURE 10.1   Ambient temperature and average power.

TABLE 10.3

Examples of HAB-Forming Algae and Their Effects on Food Safety (Faust & Gulledge 2002, Sellner et al., 2003)

Poisoning	Functional Group	Species
Diarrheic shellfish poisoning (DSP)	Dinoflagellates	*Prorocentrum* spp.
		Dinophysis spp.
		Protoperidinium spp.
Paralytic shellfish poisoning (PSP)	Dinoflagellate	*Alexandrium* spp.
Neurotoxic shellfish poisoning (NSP)	Dinoflagellate	*Gymnodinium* spp.
Amnesic shellfish poisoning (ASP)	Diatoms	*Pseudo-nitzschia* spp.
Ciguatera fish poisoning	Dinoflagellate	*Gambierdiscus* spp.

seizures, lesions and skin irritation, or even fatalities in fish, birds, and mammals (including humans) (Anderson et al., 2002; Sellner et al., 2003). Some of these toxins can be acutely lethal and are some of the most powerful natural substances known; additionally, no antidote exists to any HAB toxin (Glibert et al., 2005). Because these toxins are tasteless, odorless, and heat and acid stable, normal screening and food preparation procedures will not prevent intoxication if the fish or shellfish is contaminated (Baden et al., 1995; Fleming et al., 2006).

In addition to human health effects, HABs (Table 10.3) also have detrimental economic impacts due to closure of commercial fisheries, public health costs, and other related environmental and socio-cultural impacts (Trainer & Suddleson, 2005; NOAA-CSCOR, 2008).

10.1.6.1   HAB is Because of:

a. **Acidification of Waters**

It is possible that ocean acidification may cause changes in the HAB dynamics through changes in phytoplankton community composition, however, there are insufficient data to draw any conclusions about the impacts that increasing CO_2 might have on the growth and composition of HAB-causing marine phytoplankton.

b. **Impact of Sea-level Rise, Increased Precipitation, and Flash Floods on Harmful Algal Communities**

Sea-level rise, increased precipitation, and flash floods are most likely to affect harmful algal communities through increased nutrient release to coastal and marine waters. Two key nutrients required for phytoplankton growth, nitrogen (N) and phosphorus (P), are found in fertilizers and animal and human waste; however silicon (Si), which is only required for diatom growth, is not added to the environment through human activity. Increased concentrations of N and P without a corresponding increase in Si may cause changes in phytoplankton community composition, favoring dinoflagellates, which have no biological requirement for Si, at the expense of diatoms (Smayda, 1990). Such a shift in the phytoplankton community towards dinoflagellate dominance may result in increased numbers of HAB

species in regions prone to increased anthropogenic nutrients. Flash flooding and sudden storm events may release "pulses" of nutrients into coastal
waters. As the sea reclaims low-lying land, areas that are currently intensively farmed or urbanized may be drowned, causing the addition of nutrients, particularly N and P, to coastal systems. Additionally, as sea levels
rise, wetland habitats are lost. Wetlands act as natural filters for anthropogenic nutrients and are therefore important in regulating nutrient loads
to coastal waters. Wetlands and mangrove habitats also provide a natural
form of protection from storm surges and flooding (Nicholls et al., 2007).
Without these habitats, coastal waters may be more prone to increased levels of nutrients and nutrient imbalance.

10.1.7 ENVIRONMENTAL CONTAMINANTS AND CHEMICAL RESIDUES IN THE FOOD CHAIN

There are many pathways through which global climate change and variability may
impact environmental contamination and chemical hazards in foods. Contamination
of agricultural and pastureland soil with dioxins have been associated with climate
change–related extreme events, particularly with the increased frequency of inland
floods. Soil contamination can be attributed to remobilization of contaminated river
sediments, which are subsequently deposited on the flooded areas. In other cases,
contamination of the river water bodies, and subsequently of the flooded soils, may
have resulted from mobilization in upstream contaminated terrestrial areas such as
industrial sites, landfills, sewage treatment plants, etc. Results showed very high
levels of polychlorinated dibenzo-p-dioxins and dibenzofurans (PCDD/Fs) present
in soil in periodically flooded pastureland riverside of the dikes, and grazing on
the floodplains revealed a significant transfer of PCDD/Fs into milk (Umlauf et al.,
2005). While the uptake of contaminated soils during grazing is an important factor considering the transfer into the food chain, barn feeding of properly harvested
greens from the same floodplains is less critical (Umlauf et al., 2005). Sources of
chemical contamination of flood water included oil spills from refineries and storage
tanks, pesticides, metals, and hazardous waste. Several chemicals, such as hexavalent
chromium, manganese, p-cresol, toluene, phenol, 2, 4-D (an herbicide), nickel, aluminium, copper, vanadium, zinc, and benzidine were detected in flood water. Trace
levels of some organic acids, phenols, trace cresols, metals, sulfur chemicals, and
minerals associated with sea water were also detected (EPA, 2005). Concentrations
of most contaminants were within acceptable short-term levels, except for lead and
volatile organic compounds in some areas (Pardue et al., 2005).

10.1.8 CONTAMINATION OF WATERS

Higher water temperatures, increased precipitation intensity, and longer periods of
low flows exacerbate many forms of water pollution, including sediments, nutrients,
dissolved organic carbon, pathogens, pesticides, and salts (Kundzewicz et al., 2007).
In regions where intense rainfall is expected to increase, pollutants (pesticides, fertilisers, organic matter, heavy metals, etc.) will be increasingly washed from soils to

water bodies (Boorman, 2003). Higher runoff is expected to mobilize fertilisers and pesticides to water bodies in regions where their application time and low vegetation growth coincide with an increase in runoff (Soil and Water Conservation Society, 2003). Because of compaction, heavy rainfall after drought can result in more severe runoff and increased risk of certain types of contamination. Alternating periods of floods and drought can therefore aggravate the problem. Increasing ocean temperatures may indirectly influence human exposure to environmental contaminants in some foods (e.g., fish and mammal fats). Ocean warming facilitates methylation of mercury and subsequent uptake of methyl mercury in fish and mammals has been found to increase by 3%–5% for each 1°C rise in water temperature. Temperature increases in the North Atlantic are projected to increase the rates of mercury methylation in fish and marine mammals, thus increasing human dietary exposure (Booth and Zeller, 2005).

Sea-level rise related to climate change is expected to lead to saltwater intrusion into aquifers/water tables in coastal areas. This will extend areas of salinization of groundwater and estuaries, resulting in a decrease in freshwater availability for humans, agriculture, and ecosystems in coastal areas. One-quarter of the global population lives in coastal regions; these are water-scarce-less than 10% of the global renewable water supply (WHO, 2005) and are undergoing rapid population growth.

10.1.9 The Effect of Climate Change on the Cold-Chain

The food manufacturing industry utilizes chilling and freezing processes as a means of preserving foods. Refrigeration of these foods is continued during transportation, retail distribution, and home storage to maintain the foods at the desired temperatures. These are important steps in maintaining the safety, quality, and shelf life of foods for the consumer, and the processes from primary cooling through to domestic storage make up the "food cold-chain". If climatic change results in a substantial rise in average ambient temperatures, this will impose higher heat loads on all systems in the cold-chain. In systems that have the capacity to cope with these higher loads this will just require the refrigeration plants to run for longer periods and use more energy (James and James, 2010). In addition to the generation of CO_2 the refrigerants currently used in cold-chain have considerable global warming potential (GWP). Use of alternative refrigerants and alternative refrigeration cycles with a reduced GWP is the need of the hour.

About 20% of the global-warming impact of refrigeration plants is due to refrigerant leakage. The dominant types of refrigerant used in the food industry in the last 60 years have belonged to a group of chemicals known as halogenated hydrocarbons, e.g., chlorofluorocarbons (CFCs) and the hydrochlorofluorocarbons (HCFCs). Scientific evidence clearly shows that emissions of CFCs have been damaging the ozone layer and contributing significantly to global warming. The little data that are available suggest that currently the cold-chain accounts for approximately 1% of the CO_2 production in the world. However, this is likely to increase if global temperatures increase significantly. Until recently the major concern in the refrigeration industry regarding climate change has been the impact of refrigerants on the ozone layer and the replacement of current refrigerants with "greener" alternatives.

Energy efficiency is increasingly of concern to the food industry mainly due to substantially increased energy costs and pressure from retailers to operate zero carbon production systems. Reducing energy in the cold-chain has a big part to play since worldwide it is estimated that 40% of all food requires refrigeration and 15% of the electricity consumed worldwide is used for refrigeration. Simple solutions such as the maintenance of food refrigeration systems will reduce energy consumption. Repairing door seals and door curtains, ensuring that doors can be closed, and cleaning condensers produce significant reductions in energy consumption. In large cold storage sites, it has been shown that energy can be substantially reduced if door protection is improved, pedestrian doors fitted, liquid pressure amplification pumps fitted, defrosts optimized, suction liquid heat exchangers fitted, and other minor issues corrected.

New/alternative refrigeration systems/cycles, such as Trigeneration, Air Cycle, Sorption-Adsorption Systems, Thermoelectric, StirlingCycle, Thermoacoustic and Magnetic refrigeration, have the potential to save energy in the future if applied to food refrigeration.

10.2   ADDRESSING FOOD SAFETY IMPLICATIONS OF CLIMATE CHANGE

a. Understanding foodborne disease including zoonosis
b. Predictive models
c. Foodborne disease: surveillance/animal disease surveillance
d. Foodborne pathogens: monitoring and surveillance
e. Improved coordination among Public Health, Veterinary Health, Environmental Health, and Food Safety Services, i.e., one health concept
f. Prevention of mycotoxin contamination
g. Agricultural policy and public information review for mycotoxins
h. Good Horticultural, Agricultural, Animal husbandry, Aquaculture and Veterinary Practices
i. Replacement of hydroflourocarbon refrigerants in refrigeration.
j. Data exchange: good data exchange mechanisms are required at both national and international level. These should cover the distribution of animal and plant diseases, pests, ecological conditions including climate, and associated usage of pesticides, veterinary drugs, and chemotherapeutants will be needed to enable risk assessment, prevention, monitoring, and control.
k. Development of tools for rapid detection or removal of contaminants

10.3   SUMMARY AND CONCLUSIONS

Assuring food safety is a complex issue as it involves considerations from preproduction through to final home preparation of the food product. Recommendations on food safety management emphasize the need for broad input and coordination even though this remains a challenge in many countries. Recognizing, understanding, and preparing for the impacts of climate change further highlight the need to promote interdisciplinary approaches to address challenges affecting food safety, given the inter-relationships among environmental impacts, animal and plant health impacts, and food hygiene. These inter-relationships are further complicated by the broader public health implications of climate change as well as the food security implications. To address the challenges of climate change all the above-mentioned practices need to be followed in letter and spirit and also new crop varieties and technologies need to be generated through research.

REFERENCES

Anderson, D.M., Glibert, P.M. and Burkholder, J.M. 2002. Harmful algal blooms and eutrophication: nutrient sources, composition, and consequences. *Estuaries*, 25(4), pp. 704–726.

Baden, D., Fleming, L.E. and Bean, J.A. 1995. Handbook of Clinical Neurology: Intoxications of the Nervous System Part H. Natural Toxins and Drugs. Elsevier Press, Amsterdam, Netherlands, pp. 141–175.

Boorman, D.B. 2003. LOIS in-stream water quality modelling. Part 2. Results and scenarios. *Science of the Total Environment*, 314, pp. 397–409.

Booth, S. and Zeller, D. 2005.Mercury, food webs, and marine mammals: Implications of diet and climate change for human health. *Environmental Health Perspectives*, 113(5), pp. 521–526.

D'souza, R.M., Becker, N.G., Hall, G. and Moodie, K.B. 2004. Does ambient temperature affect foodborne disease? *Epidemiology*, pp. 86–92.

D'souza, R.M., Hall, G. and Becker, N.G. 2008. Climatic factors associated with hospitalizations for rotavirus diarrhoea in children under 5 years of age. *Epidemiology & Infection*, 136(1), pp. 56–64.

Environmental Protection Agency. 2005. Environmental Assessment Summary for Areas of Jefferson, Orleans, St. Bernard, and Plaquemines Parishes Flooded as a Result of Hurricane. Katrina. http://www.epa.gov/katrina/testresults/katrina_env_assessment_summary.htm.

Epstein, P.R. 2001. Climate change and emerging infectious diseases. *Microbes and Infection*, 3(9), pp. 747–754.

FAO. 2001. Manual on the Application of the HACCP System in mycotoxin Prevention and Control. FAO Food and Nutrition Paper.

FAO. 2008. Viruses in Food: Scientific Advice to support risk management activities. Microbiological Risk Assessment Series No.7.

FAO/IOC/WHO. 2005. Report of ad hoc Expert Consultation on Biotoxins in Bivalve Mollusco. September 2004, Rome.

Faust, M.A. and Gulledge, R.A. 2002. Identifying harmful marine dinoflagellates. *Contributions from the United States national herbarium*, 42, pp. 1–144. http://botany.si.edu/references/dinoflag/index.htm.

Fleming, L.E., Broad, K., Clement, A., Dewailly, E., Elmir, S., Knap, A., Pomponi, S.A., Smith, S., Gabriele, H.S. and Walsh, P. 2006. Oceans and human health: Emerging public health risks in the marine environment. *Marine pollution bulletin*, 53(10–12), pp. 545–560.

Glibert, P.M., Anderson, D.M., Gentien, P., Granéli, E. and Sellner, K.G. 2005. The global, complex phenomena of harmful algal blooms. *Oceanography*, 18, pp. 136–147.

Hall, G.V., D'Souza, R.M. and Kirk, M.D. 2002. Foodborne disease in the new millennium: out of the frying pan and into the fire? *Medical Journal of Australia*, 177(11/12), pp. 614–619.

Hallegraeff, G.M. 1993. Review of harmful algal blooms and their apparent global increase. *Phycologia*, 32, pp. 79–99.

IPCC (Intergovernmental Panel on Climate Change). 2007. Summary for policymakers. In *Climate Change 2007: Impacts, Adaptation and vulnerability. Contribution of Working Group II to the Fourth Assessment Report of the Intergovernmental Panel on Climate Change*, M.L. Parry, O.F. Canziani, J.P. Palutikot, P.J. van der Linden, and C.E. Hanson, eds. Cambridge University Press, Cambridge, UK.

James, S.J. and James, C. 2010. The food cold-chain and climate change. *Food Research International*, 43(7), pp. 1944–1956.

Kundzewicz, Z.W., Mata, L.J., Arnell, N.W., Doll, P., Kabat, P., Jimenez, B., Miller, K., Oki, T., Zekai, S. and Shiklomanov, I. 2007. Freshwater resources and their management. In *Climate Change 2007: Impacts, Adaptation and Vulnerability. Contribution of Working Group II to the Fourth Assessment Report of the Intergovernmental Panel on Climate Change*, M.L. Parry, O.F. Canziani, J.P. Palutikot, P.J. van der Linden, and C.E. Hanson, eds. Cambridge University Press, Cambridge, UK.

Kovats, R.S., Edwards, S.J., Charron, D., Cowden, J., D'Souza, R.M., EbiK, L., Gauci, C., Gerner-Smidt, P., Hajit, S., Hales, S., HernandezPezzi, G., Kriz, B., Kutsar, K., McKeown, P., Mellou, K., Meene, B., O'Brien, S., VanPelt, W. and Schmid, H. 2005. Climate variability and campylobacter infection: an international study. *International Journal of Biometeorology*, 49, pp. 207–214.

Luby, S.P., Rahman, M., Hossain, M.J., Blum, L.S., Husain, M.M., Gurley, E., Khan, R., Ahmed, B.N., Rahman, S., Nahar, N. and Kenah, E. 2006. Foodborne transmission of Nipah virus, Bangladesh. *Emerging Infectious Diseases*, 12(12), pp. 1888–1894.

Moretti, C.L., Mattos, L.M., Calbo, A.G. and Sargent, S.A. 2010. Climate changes and potential impacts on postharvest quality of fruit and vegetable crops: a review. *Food Research International*, 43(7), pp. 1824–1832.

Nicholls, R.J., Wong, P.P., Burkett, V.R., Codignotto, J.O., Hay, J.E., McLean, R.F., Ragoonaden, S. and Woodroffe, C.D. 2007. Coastal systems and low-lying areas. In *Climate Change 2007: Impacts, Adaptation and Vulnerability Contribution of Working Group II to the Fourth Assessment Report of the Intergovernmental Panel on Climate Change*, M.L. Parry, O.F. Canziani, J.P. Palutikot, P.J. van der Linden, and C.E. Hanson, eds. Cambridge University Press, Cambridge, UK, pp. 315–356.

NOAA-CSCOR. 2008. Economic Impacts of Harmful Algal Blooms (HABs) fact sheet. http://www.cop.noaa.gov/stressors/extremeevents/hab/current/HAB_Econ.html

Pardue, J.H., Moe, W.M., McInnis, D., Thibodeaux, L.J., Valsaraj, K.T., Maciasz, E., Van Heerden, I., Korevec, N. and Yuan, Q.Z. 2005. Chemical and microbiological parameters in New Orleans floodwater following Hurricane Katrina. *Environmental Science & Technology*, 39(22), pp. 8591–8599.

Paz, S., Bisharat, N., Paz, E., Kidar, O. and Cohen, D. 2007. Climate change and the emergence of Vibrio vulnificus disease in Israel. *Environmental Research*, 103(3), pp. 390–396.

Rose, J.B., Epstein, P.R., Lipp, E.K., Sherman, B.H., Bernard, S.M. and Patz, J.A. 2001. Climate variability and change in the United States: potential impacts on water-and foodborne diseases caused by microbiologic agents. *Environmental health perspectives*, 109(Suppl 2), pp. 211–221.

Sellner, K.G., Doucette, G.J. and Kirkpatrick, G.J. 2003. Harmful algal blooms: causes, impacts and detection. *Journal of Industrial Microbiology and Biotechnology*, 30(7), pp. 383–406.

Smayda, T.J. 1990. Novel and nuisance phytoplankton blooms in the sea: evidence for a global epidemic. In: *Toxic Marine Phytoplankton*, E. Granéli, B. Sundström, L. Edler, and D.M. Anderson eds. Elsevier, New York, pp. 29–40.

Soil and Water Conservation Society. 2003. Soil erosion and runoff from cropland. *Report from the USA, Soil and Water Conservation Society*, 63, p. 49.

Trainer, V.L., and Suddleson, M. 2005. Monitoring approaches for early warning of domoic acid events in Washington State. *Oceanography*, 18, pp. 228–237.

Umlauf, G., Bidoglio, G., Christoph, E.H., Kampheus, J., Krüger, F., Landmann, D., Schulz, A.J., Schwartz, R., Severin, K., Stachel, B. and Stehr, D. 2005. The situation of PCDD/Fs and dioxin-like PCBs after the flooding of river Elbe and Mulde in 2002. *Acta Hydrochimica et Hydrobiologica*, 33(5), pp. 543–554.

WHO. 2005. *Ecosystems and Human Well-Being: Health Synthesis. A Report of the Millennium Ecosystem Assessment.* World Health Organization, Geneva, p. 54.

11 Microbial Assisted Soil Reclamation for Sustainable Agriculture in Climate Change

Jyotsna Kiran Peter, Uday Shankar Pandey, Arjun Karmakar, Anjulata Suman Patre, and Gaussuddin
Sam Higginbottom University of Agriculture, Technology and Sciences

CONTENTS

11.1 INTRODUCTION

Soil is a dynamic entity (Delgado and Gomez, 2016) that encompasses physical (solids, air, and water), chemical (organic and inorganic forms), and biological (Macro: micro – flora and fauna) components. The components of solid phase of soil include organic and inorganic matter. All living and dead cell biomass contribute to the organic matter in the soil, while a range of inorganic substances are present in the soil such as rock, silicates, minerals, salts, ions, etc. The plant health is reflective of good soil condition. Gaseous phase is present in discrete pores which influences the functional and geochemical cycling of nutrients in the lithosphere, hydrosphere, and atmosphere. Soil microorganisms constitute less than 0.5% (w/w) of the soil mass, but they play a key role in soil properties and processes (Yan et al., 2015; Kang et al., 2014). In addition, the emission of Carbon dioxide (CO_2) from soils, which includes respiration from soil organisms and roots, contributes approximately 10% to the atmospheric CO_2.

Rhizosphere is the narrow sphere of the plant root highly influenced by the roots of plant. This zone is nutrient rich in comparison to bulk soil and exhibits intense biological and chemical activities (Parashar et al., 2014). The sphere of the root of plant is always under dynamic conditions that allow numerous interactions between the microbial cells harbored in the region plant roots, soil abiotic factors. The soil supports the growth of microorganisms in numerous ways and in return these microbial flora enrich the soil and promote plant growth. They are also indicators of soil as a living entity. Soil respiration is one such process that acts as a dual indicator. The estimation of soil respiration relates to the microbial load associated with plant activities and the emission of greenhouse gases (GHGs) reflected from global or specific area, and indicates alarming climate change.

Microorganisms also serve the purpose of maintaining soil health and plant growth. Many of the rhizospheric microorganisms are beneficial to plants and are regarded as biofertilizers or plant growth promoting rhizobacteria (PGPRs) due to their various attributes like production of plant growth hormones, releasing antimicrobial agents to control various phytopathogens, such as siderophores by *Pseudomonas aeruginosa*, BT toxin by *Bacillus thuringiensis*, etc. The chapter aims to explore the vivid roles of bacterial exopolysaccharides (EPS) and biosurfactants, how these molecules interact with plant roots, their physiology and metabolism, and their interaction with phytopathogens.

11.2   SOIL RESPIRATION: AN INDICATOR OF SOIL HEALTH AND CLIMATE CHANGE

The Global geochemical carbon cycling is the basis of the GHGs emission of which 20% emission is from agricultural fields (Figure 11.1). Soil microflora contributes to 99% of the CO_2 arising as a result of the decomposition of organic matter, while the contribution of soil fauna is much less. Root respiration, however, contributes to 50% of the total soil respiration. Soil respiration is an important component of the global carbon cycle; even a small variation of soil respiration prominently influences the atmosphere's CO_2 concentration and the soil organic carbon storage (Zhang et al., 2015). Jiang et al. (2015) revealed that soil respiration under cold resistant crops was more sensitive to temperature changes, compared to thermophilic crops in agro ecosystems. A total of 55% of CO_2 is released from fossil fuels (Figure 11.2). Soil basal respiration (SBR) of microbial biomass is a major attribute related to soil fertility (Romero Friere et al., 2016; Niemeyer et al., 2012) and is a common indicator of soil quality (Romero Friere et al., 2016; ISO, 2002). Hence the respiratory measurements are significant indicators of soil quality and stress to soil microorganisms (Azarbad et al., 2013; Dai et al., 2004).

A comparative evaluation of soil respiration in rhizosphere and bulk soil of *Psidium guajava* (Guava) duly treated with test antagonist *Aspergillus niger* from two different depths (0–15 cm and 15–30 cm) had showcased a strong correlation between the type of substrate (substrate considered: peptone as protein supplement and glucose as sugar supplement along with control soil that has no added substrate), and soil respiration. Soil respiration was measured through the Alkali Trap method. In case of soil sampled at 0–15 cm, irrespective of soil treatment with antagonist and days of incubation, a significant effect of the substrate was observed on total soil microbial respiration over the control; the highest value recorded was in the

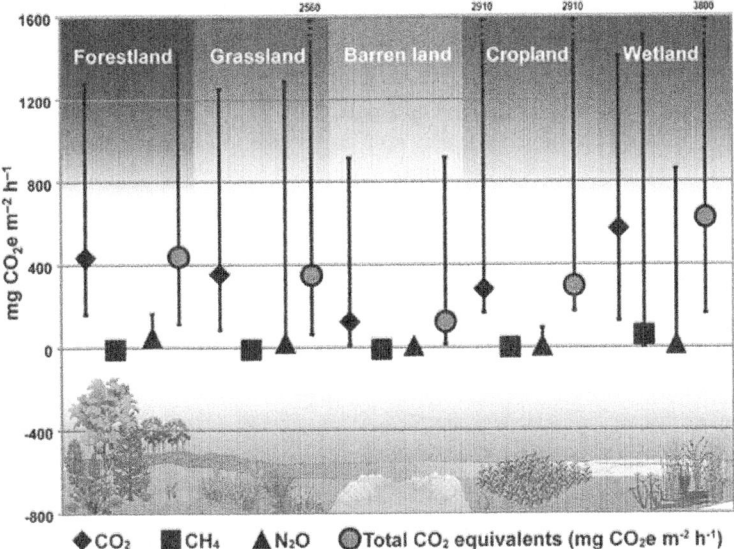

FIGURE 11.1 GHG emissions (CO_2-eq) of CO_2, N_2O, and CH_4 from soils with different land cover: grassland ($n = 47$), forestland ($n = 22$), barren land ($n = 17$), cropland ($n = 41$), and wetland ($n = 67$). Median values for the sub-collectives are shown with the symbols; the range is indicated with solid lines. Only consistent data discussed in this paper have been integrated. Some maximum values exceed the upper concentration limits. (*Source*: Oertel et al., 2016.)

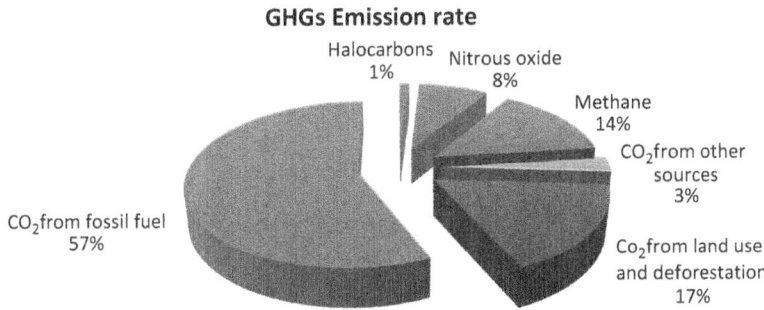

FIGURE 11.2 Types of GHG emissions in environment. (*Source*: Jat et al., 2015; IPCC, 2007a, 2007b.)

case of protein-supplemented soil sample ($227.82\,mg\ CO_2$ from 0 to 15 cm depth and $225.25\,mg\ CO_2$ from 15 to 30 cm soil depth). The highest value of soil organic content was read in case of guava field which was left untreated with fungal antagonist (0.50% and 0.41% from 0 to 15 cm and 15 to 30 cm depths, respectively). A significant positive correlation existed between the CO_2 evolved at 6 days and from 0 to 15 cm soil depth, and the amount of CO_2 evolved in 4 and 5 days from 15 to

30 cm soil depth. The increase in the amount of CO_2 exhibited a linear trend up to 7 days in case of 0–15 cm and 15–30 cm depth soil samples. Soil application of fungal antagonist did not result in a significant change in soil respiration irrespective of the substrates and days of incubation (Figures 11.3–11.5).

CO_2 is produced in soil by roots, soil organisms, and to a small extent by the chemical oxidation of carbon containing naturals. Lamberty et al. (2016) capitalized a long-term reciprocal soil transplant experiment to examine the response of dry

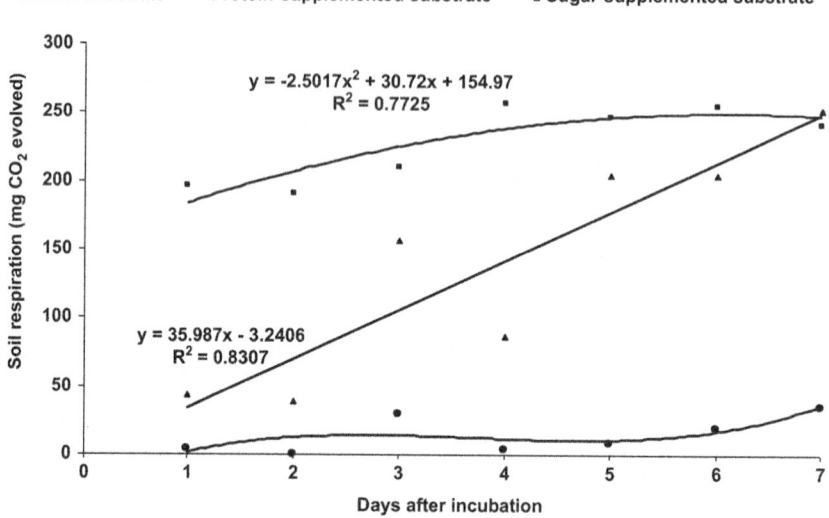

FIGURE 11.3 Pattern of respiration of soil (amount of CO_2 evolved) from 1 to 15 cm depth under different substrates.

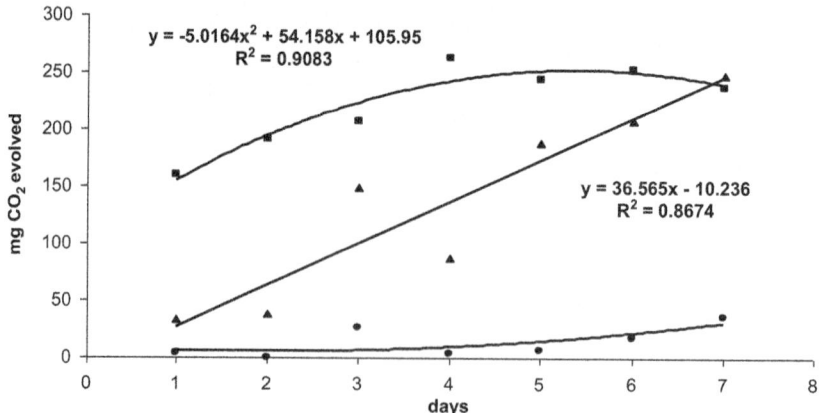

FIGURE 11.4 Pattern of respiration of soil (amount of CO_2 evolved) from 15 to 30 cm depth under different substrates.

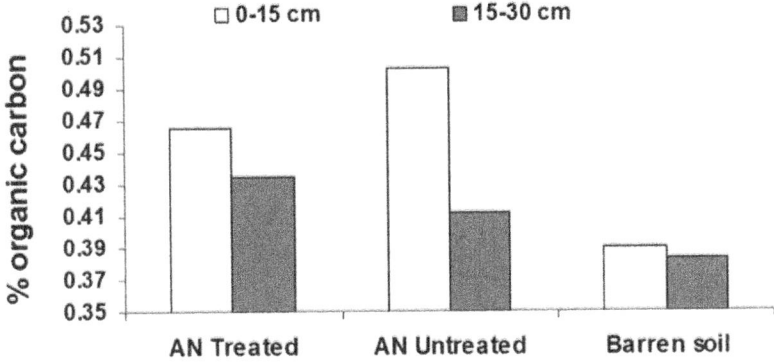

FIGURE 11.5 Percentage of organic carbon in different soils.

land soils to climate change. Over a laboratory incubation of 100 days, reciprocal transplanted soils respired roughly with equal cumulative amounts of carbon as non-transplanted controls from the same site. Soils transplanted from the hot, dry, lower site to the cooler and wetter upper site exhibited almost no respiratory response to temperature but soils originally from the upper, cooler sites had higher respiration rates. This prompted significant differences in microbial activity and supported the idea that environmental shifts can influence soil carbon through metabolic changes and suggest that the microbial populations responsible for soil heterotrophic respiration may be constrained in different ways as shorter or longer terminal microbial dynamics under changing climate. Carey et al. (2016) showed the respiratory release of CO_2 from soil has revealed no significant differences in the temperature sensitivity of soil respiration between control and warm plants in all biomes with the exception of deserts and boreal forests. The analysis added a unique cross biome perspective on the temperature response of soil respiration, showing how soil C dynamics change with climate warming. Lai et al. (2012) investigated an arid region with widespread saline/alkaline soils and evaluated soil respiration of different agricultural and natural ecosystems. The research was based on soil respiration of five ecosystems, namely, soil temperature, soil moisture, soil pH, soil electrical conductivity, and soil organic content in field. The compared values of natural ecosystem revealed that the mean seasonal soil respiration rates of agriculture ecosystem were 96%–386% higher, and agricultural ecosystem exhibited lower CO_2 absorption by saline/alkaline soil. Soil temperature and moisture revealed 48%, 86%, 84%, 54%, and 54% seasonal variation of soil respiration in the five ecosystems, respectively. The relationship between soil respiration and electrical conductivity was observed to be negative, rather a weak correlation existed between soil respiration, pH and soil organic carbon contents. Guntiñas et al. (2013) had examined the sensitivity of soil respiration in moisture and temperature in three soils from Atlantic temperature arid humid zone (Galicia, NW Spain) used for different purposes (forest grassland and cropland). The research revealed least sensitivity of grassland soil to changes in temperature and moisture *via* constant emission model whilst the Q_{10} model showed that the sites under three types of use responded similarly, with the cropland soil being the least sensitive. The investigation provided an optimal useful tool for planning the best type of soil use to minimize the effect of climate change and emission of CO_2 from soil.

11.3 EXPLOITING MICROBIAL EPS FOR ESTIMATING THEIR ROLE IN PLANT GROWTH MANAGEMENT AND COMBATING PLANT PATHOGENS

Soil harbors ample microorganisms that stimulate plant growth and help them survive under varied ecological and environmental conditions. Microorganisms are either directly or indirectly involved in influencing the plant and soil health. Microorganisms are the simple option to enhance the productivity fertility of soil with sophisticated cellular machines and mechanisms to respond to their environment. Microbial cells are a great source of microbial metabolites and cell components that trigger a cascade of positive and negative interactions with their suitable candidates in soil. Among the various metabolites, microbial EPS are of immense importance to the own bacterial cell as well as they are the interacting molecules with the exterior of the microbial cell. Microbial EPS are part of the microbial membranes or storage materials. EPS of microbial origin are contributors of soil organic matter and even these metabolites serve as the nutrient molecules for a few microorganisms, otherwise they elicit a toxic effect on the plant pathogens residing in the soil. EPS renders the inhibitory action of microorganisms *via* different mechanisms and eventually prevents the biofilm formation of few microorganisms. Microbial EPS are soluble or insoluble polymers secreted by microorganisms (Kumar et al., 2007).

An investigation was carried out to evaluate the effects of partially purified exopolysaccharide (PPEPS) of soil bacteria namely, *Bacillus subtilis*, *Pseudomonas aeruginosa*, and *Serratia marscecens* and their consortia at the research farm of Sam Higginbottom University of Agriculture Technology and Sciences during the summer season leading to the suggestive roles of PPEPS on *Spinacia oleracia* sown during March and harvested during April–May. The soil was sterilized with formalin and left for a month to remove unwanted life forms. The plot was treated with seven different treatments of PPEPS from sole and consortia at the rate of 2 g/mL used for root dipping of seedlings at 10 days after sowing (DAS) in nursery. The seedlings were thereafter transplanted to field with the rhizosphere inoculated with 10 mL (2% Aq) PPEPS again. Plant growth was monitored till the harvest stage (Figure 11.6) and to investigate the *in vitro* effect of PPEPS on plant pathogenic fungi (*Fusarium oxysporum*, *Rhizoctonia solani*, *Alternaria solani*, and *Helminthosporium* sp) agar well-diffusion assay was performed using 50 µL/mL PPEPS. The results revealed the promotion of plant growth at varied sources of EPS (Table 11.1) and showed antifungal activity towards the selected ones.

Solvent purified EPS from individual bacteria, namely *Bacillus subtilis*, *Pseudomonas aeruginosa*, and *Serratia marcescens* showed the inhibitory action on test plant pathogenic fungi viz. *Fusarium oxysporum* (highest zone of inhibition through *Pseudomonas aeruginosa*: 6.5 mm followed by consortium 1 than

Plant growth of spinach at 30 days

FIGURE 11.6   Growth response of *Spinacia oleracia* treated as seedling dip method with PPEPS from individual and consortia of selected native soil bacteria.

other treatments), *Rhizoctonia solani* was highly inhibited by consortium 4 followed by consortium 1, revealing that the combination of PPEPS of *Pseudomonas aeruginasa* and *Serratia marcescens* is effective against it while it was weakly inhibited by other treatments, *Alternaria solani* was again most effectively inhibited by PPEPS of *Pseudomonas aeruginosa* followed by consortium 1 and the rest of the treatments were unable to inhibit its growth at 50 µL/mL dimethyl sulfoxide (DMSO) concentration, and *Helminthosporium* sp was highly inhibited by consortium 1 showing 3.33 mm of zone of inhibition comparable to sole PPEPS from the three bacteria, among which the most effective was PPEPS of *Pseudomonas aeruginosa*, the rest of the consortium could not inhibit its growth (Table 11.2).

The EPS or biosurfactants of the microbial origin develop a signaling system among rhizobial microflora and plant roots. A microbe–microbe or microbe–plant interaction is based on these molecules present in the rhizosphere of the plant roots. These microbial molecules exhibit a wide range of interacting attributes that contribute to improve soil health restitution and favors plant growth. Naseem and Bano (2014) characterized PGPR explicating their EPS role in the drought tolerance of maize. The soil sample was taken from 6 in. depth of an arid area (Rawalpindi Distt.) with temperature range (40°C–52°C in summer) and (below 0°C in winters) falling under severe drought with high winds and scanty rainfall (100–250 mm during July–September). EPS producing strains stimulated protein and sugar contents in leaves under stress conditions. The plants inoculated with EPS producing bacteria showed decreased activity of Ascorbate peroxidase (APX), CAT, and Glutathione peroxidase (GTX) enzymes under drought stress. Soil moisture content was 10%–12% in untreated control plants that declined to 5%–7% under drought stress. The highest moisture content was recorded for rhizosphere soil inoculated with PGPR strains in combination with respective EPS both under drought stress and unstressed conditions. *Pseudomonas peneri* and *Pseudomonas aeruginosa* with their respective EPS showed the highest increase in soil moisture content by 68% and 67%, respectively. Skorupska et al. (2006) presented a review highlighting specific complex interactions existing between soil bacteria belonging to *Rhizobium, Sinorhizobium,*

TABLE 11.1

Growth Response of *Spinacia Oleracia* Treated as Seedling Dip Method with PPEPS from Individual and Consortia of Selected Native Soil Bacteria

Growth Response of *Spinacia Oleracia* Treated With Variable PPEPS From Sole Bacteria and Consortia

Abbr	Treatments	Average Seed Germination (%)	Plant Height (cm) at 35 DAS	No of Leaves at 35 DAS	Leaf Length (cm) at 35 DAS	Leaf Width (cm) at 35 DAS	Root Length (cm) at 35 DAS	Chlorophyll Contents at 30 DAS of Leaves	Fresh Weight (g) of Leaves at 30 (Harvest)
T0	PPEPS – *Bacillus subtilis*	79.0	16.933 ± 1.419[c]	9 ± 1.0[e]	6.6 ± 0.6[d]	4.433 ± 0.35[c]	7.733 ± 0.37[b]	1.217 ± 0.18[d]	6.626 ± 0.75[e]
T1	PPEPS of *Pseudomonas aeruginosa*	95.8	19.266 ± 2.64[de]	13.333 ± 1.52[cd]	10.366 ± 0.66[c]	5.233 ± 0.28[c]	10.833 ± 0.51[a]	1.246 ± 0.14[d]	10.746 ± 1.21[c]
T2	PPEPS of *Serratia marcescens*	99.4	29.7 ± 3.05[a]	25.333 ± 1.52[a]	14.7 ± 1.05[a]	9.1 ± 0.62[a]	8.2 ± 0.75[b]	2.033 ± 0.31[a]	21.282 ± 1.26[a]
T3	Consortium C1-(PS+SM)	93.8	22.9 ± 2.28[c]	18 ± 2.0[b]	12.366 ± 1.95[abc]	4.8 ± 0.81[c]	8 ± 0.36[b]	1.466 ± 0.15[bcd]	8.806 ± 0.47[d]
T4	Consortium (C2- PS+BS)	95.8	22.033 ± 2.05[cd]	12 ± 1.0[d]	12.066 ± 1.42[bc]	6.866 ± 0.55[b]	6.9 ± 1.55[b]	1.603 ± 0.08[bc]	9.826 ± 0.49[cd]
T5	Consortium (C3- BS+SM)	98.8	24.8 ± 1.15[bc]	12.333 ± 2.08[d]	12.766 ± 1.80[ab]	6.833 ± 0.51[b]	8.2 ± 0.74[b]	1.544 ± 0.104[bc]	8.583 ± 0.91[b]
T6	Consortium (C4- BS+ PS+SM)	98.2	24.7 ± 0.69[bc]	15.333 ± 1.52[bc]	14.166 ± 1.56[ab]	7.066 ± 0.35[b]	7.6 ± 0.62[b]	1.703 ± 0.17[b]	13.336 ± 0.67[b]
T7	DMSO	98.2	27.8 ± 1.6[ab]	24.333 ± 1.52	13.966 ± 1.32[ab]	7.733 ± 0.49[b]	8.133 ± 0.49[b]	1.376 ± 0.05[cd]	14.186 ± 0.57[b]
SD		3.342	4.202417	5.930574	2.62618	1.586224	1.14924	0.265804	4.612473

(Continued)

TABLE 11.1 (Continued)
Growth Response of *Spinacia Oleracia* Treated as Seedling Dip Method with PPEPS from Individual and Consortia of Selected Native Soil Bacteria

Growth Response of *Spinacia Oleracia* Treated With Variable PPEPS From Sole Bacteria and Consortia

Abbr Treatments	Average Seed Germination (%)	Plant Height (cm) at 35 DAS	No of Leaves at 35 DAS	Leaf Length (cm) at 35 DAS	Leaf Width (cm) at 35 DAS	Root Length (cm) at 35 DAS	Chlorophyll Contents at 30 DAS of Leaves	Fresh Weight (g) of Leaves at 30 (Harvest)
F_{cal}	0.009	12.787	42.925	10.872	27.444	6.717	7.468	89.256
F_{tab}	0.99	1.737	3.383	4.832	9.072	0.008	0.004	1.26
F_{test}	NS	S	S	S	S	S	S	S
$CD_{(5\%)}$	141.7	3.521	2.710	2.385	0.907	0.295	0.295	1.468

Ps: *Pseudomonas aeruginosa*; Bs: *Bacillus subtilis*; Sm: *Serratia marsœcens*; CD: critical difference; NS: non significant; S: Significant.
Letters in superscript tells most significant value as 'a' followed by lesser significant value 'b' and as follows for 'c', 'd' and 'e' upon Tukey's pairwise comparision of the data.

TABLE 11.2

In Vitro Antifungal Act of PPEPS from Individual and Consortia on Selected Plant Pathogenic Fungi

		Zone of Inhibition (mm) PPEPS at Concentration of 50 µl/ml DMSO			
Abbre	Treatments	*Fusarium oxisporium*	*Rhizoctonia solani*	*Alternaria solani*	*Helminthosporium SP*
T0	PPEPS – *Bacillus subtilis*	2.666 ± 1.154701^c	2.666 ± 0.57735^{bc}	0 ± 0^c	1.333 ± 0.57735^c
T1	PPEPS of *Pseudomonas aeruginosa*	**6.5 ± 0.5^a**	0 ± 0^d	**4.333 ± 0.57735^a**	2.333 ± 0.57735^b
T2	PPEPS of *Serratiamarscecens*	2 ± 1.0^c	1.666 ± 1.154701^{bc}	0 ± 0^c	1.333 ± 0.57735^c
T3	Consortium C1-(PS+SM)	4.5 ± 0.5^b	2.833 ± 0.763763^b	1.666 ± 0.57735^b	**3.333 ± 0.57735^a**
T4	Consortium (C2- PS+BS)	3 ± 1.0^c	2.333 ± 0.57735^{bc}	0 ± 0^c	0 ± 0^d
T5	Consortium (C3- BS+SM)	2.166 ± 0.763763^c	1.5 ± 0.5^c	0 ± 0^c	0 ± 0^d
T6	Consortium (C4- BS+ PS+SM)	2.666 ± 0.57735^c	**6.666 ± 0.57735^a**	0 ± 0^c	0 ± 0^d
T7	DMSO	0 ± 0^d	0 ± 0^d	0 ± 0^c	0 ± 0^d
SD		1.608687	2.060081	1.653748	1.303079
F_{cal}		11.435	28.141	86.166	26.75
F_{tab}		0.001	4.822	3.047	6.656
F_{test}		**S**	**S**	**S**	**S**
$CD_{(5\%)}$		1.447	1.179	0.544	0.763

Ps: *Pseudomonas aeruginosa*; Bs: *Bacillus subtilis*; Sm: *Serratia marscecens*; CD: critical difference; NS: non significant; S: Significant.

The significance values (a–d) are based on the statistic data using One Way ANOVA followed by Tukey's post hoc (HSD) pairwise comparison among variables. The highest significant value is assigned the letter 'a' followed by 'b' and so on. The least significant is given 'd'.

Mesorhizobium, Phylorhizobium, Bradyrhizobium, and *Azorhizobium* with their host leguminous crop resulting in the development of root nodules. The review is focused on EPS crucial for invasion leading to formation of undetermined type nodules of legumes (Clover, Vetch, Peas, etc.). Rhizobial EPS are species-specific heteropolysaccharides polymers consisting of sugars substituted with non-carbohydrate moieties. One of the known rhizobial EPS is Succinoglycan (EPS I) produced by *S. meleloti* made of octasaccharide units with one Galactose and seven glucose residues (molar ratio: 1:7) joined by β-1, 3, β-1, 4, β-1, 6 Glycosidic linkages. Another category of EPS is Galactoglucan (EPS II); this is produced only under phosphate starvation or mutation of gene *muc*R and *exp*R. *S. meleloti* mutants in EPS I production elicited non-infected pseudonodules even induced plant defense response on *Madicago sativa*. Phenolic compounds were more in pseudonodules rather than in wild type nodules, while the addition of low molecular weight EPS I enabled *S. meliloti* mutant to infect the host plant and acted as suppression of defense system. In *R. leguminoserum* bv *trifolii*, *Trifolium* symbiosis, rhizobia completely deficit in EPS biosynthesis induced empty nodules with deposition of polyphenol materials, necrosis of plant cells and thick outermost cell layers indicating plant defense reactions. Bacteroids were not fully developed with inability to fix N_2. In another case, *Bradyrhizobium japonicum*, as per Glycine max symbiosis, *exo*B mutant produced structurally transformed induced nodules with significant quantity of phytoalexinsglyceolin hoarded. This kind of accumulation is evident in infection by *Phytophthora megasperme* as antimicrobial agent against it. Qurashi and Sabri (2012) ascertained the stimulatory influence of bacterial EPS on the growth of chick pea and soil aggregation under salt stress and influence of EPS on biofilm formation by *Halomonas variabilis* and *Planococcus rifietoensis*. *P. rifietoensis* exhibited higher EPS secretion at increasing salt concentration (1.5 M–2 M) in comparison to *H. variabilis*. Bacterial incubation included 85% seed germination, 50% seedling length, 52% fresh weight, and 58% dry weight. Soil aggregation increased at high salt concentration (100 mM) and was more pronounced at rhizosphere. The binding property of EPS causes soil particles to cement and strengthens aggregate formation that supports plant growth. Kyungseok et al. (2008) elicited rhizobacterial EPS-induced resistance in Cucumber against *Colletotrichum orbiculare*. Vardharajula et al. (2011) confirmed drought tolerant plant growth promoting *Bacillus* spp influences growth, osmolytes antioxidant status of Maize under drought stress. The drought tolerant *Bacillus* spp identified were *B. amyloliquifaciens*, *B. licheniformis*, *B. thuringiensis*, *Paenibacillus favisporus*, and *Bacillus subtilis* based on 16 S r DNA gene sequences. These *Bacillus* spp. enhanced osmoregulated proline, sugars, free amino acids, and declined electrolyte leakage. Strains also bring to light reduced activity of antioxidant enzymes, APX, Catalase, and GTX. Jones (2012) had observed that increased production of succinoglycan is enough to increase the productivity of the symbiosis between *S. meliloti* 1021 and the host plant *M. truncatula* A17. This suggests that the native level of succinoglycan produced by *S. meliloti* 1021 is a limiting factor for invasion of this plant host. The study recommends to modify rhizobial strains that are used in the inoculation of legume host crop plants so that they produce higher levels of symbiotic EPS and to determine whether this can increase crop yields.

Rhizosphere harbors a distinct density and diversity of microbial flora and even the plant associated bacteria and yeasts constitutively produce biosurfactant. These biomolecules play a vivacious function in motility, signaling, and biofilm formation, signifying that biosurfactant governs plant–microbe interaction. In agriculture, biosurfactants can be used for plant pathogen elimination and for increasing the bioavailability of nutrients for beneficial plant-associated microbes. Biosurfactants can widely be applied for improving the agricultural soil quality by soil remediation (Sachdev and Cameotra, 2013).

A plethora of research have emerged with plenty of characteristic features of microbial surfactants that are sole EPS or a part of it involved in dynamic interaction between the microbe and plant, or say microbe–microbe interaction.

Pseudomonas aeruginosa is a Gram-negative bacterium highly distributed in soil and associated with plants growth promotion directly or indirectly in numerous ways which exudes rhamnolipids (RLs) belonging to class Glycolipid of biosurfactants composed of a hydrophilic head formed by one or two rhamnose molecules i.e. monorhamnolipid (monoRL) and dirhamnolipid (diRL), and a hydrophobic tail containing one or two fatty acids. (Deziel et al., 2001; Monteiro et al., 2007). RLs are EPS which are surface-active glycolipids (Maier and Soberón-Chávez 2000; Nitschke et al., 2005; Soberón-Chávez, 2004; Soberón-Chávez et al., 2014) composed of L-rhamnose and 3-hydroxyalkanoic acid. RLs are synthesized when one or two rhamnose sugar molecules fuse with one or two β-hydroxy 3-hydroxy fatty acids (Banat et al., 2000). Two RL biosurfactants, L-rhamnopyranosyl-Lrhamnopyranosyl-β-hydroxydecanoyl-β-hydroxydecanoate or Rha-Rha C_{10}-C_{10} and L-rhamnopyranosyl-L-rhamnopyranosyl-β-hydroxydecanoyl-β-hydroxydodecanoate or Rha–Rha C_{10}-C_{12} have been reported to be produced by *P. aeruginosa* B189 strain (Thanomsub et al., 2007).

There are four types of RLs:

1. Mono-rhamnolipids (Rh_1): one rhamnose sugar attached to two molecules of β-hydroxydecanoic acid
2. Dirhamnolipids (Rh_2): two rhamnose sugars attached to two molecules of β-hydroxydecanoic acid
3. Tri-rhamnolipids (Rh_3): one rhamnose sugar attached to one molecule of β-hydroxydecanoic acid
4. Tetra-rhamnolipids (Rh_4): two rhamnose sugars attached to one molecule of β-hydroxydecanoic acid

Various attributes of RLs:

1. **Surface tension reduction:** RLs from *Pseudomonas aeruginosa* decrease the surface tension of water from roughly 72 millinewtons (mN)/m to 25–30 mN/m, and the interfacial tension of water/hexadecane to <1 mN/m (Singh and Cameotra, 2004).
2. **Biodegradable and environment friendly:** RLs are easily degradable and particularly suited for environmental applications such as bioremediation and dispersion of oil spills. (Mulligan, 2005; Chen et al., 2013).
3. **Biodegradability, non-toxicity, and mutagenicity:** RLs biosurfactants have been shown to be nontoxic and non-mutagenic, compared to the toxicity and mutagenicity associated with chemically derived surfactants.
4. **Excellent emulsifiers:** RLs also exhibit excellent emulsification properties and have the highest emulsification index against toluene (86.4%) (Monteiro et al., 2007). RLs also have excellent emulsifying power with a variety of hydrocarbons and vegetable oils (Abalos et al., 2001)
5. **Alternative of oil recovery:** Biosurfactants as RLs are mainly used in the petrochemical industry to enhance oil recovery and for hydrocarbon.

6. **Inhibitory action of RLs zoospores:** RLs have been determined for the inhibition of zoospore forming plant pathogens that have acquired resistance to commercial chemical pesticides (Hultberg et al. 2008a), and another investigation has shown that RLs stimulated plant immunity which is considered as an alternative strategy to reduce the infection by plant pathogens (Vatsa et al. 2010). Plant growth-promoting *Pseudomonas putida* produces biosurfactants that caused lysis of zoospores of the oomycete pathogen *Phytophthora capsici*; causative agent of damping-off of cucumber (Kruijt et al. 2009). RLs exhibit high zoosporicidal activity, most likely through zoospore lysis, against various zoosporic phytopathogens, including species from the *Pythium*, *Phytophthora*, and *Plasmopara* genera.

7. **Insecticidal action of RLs:** Recent investigation has also established RLs as an insecticidal compound (An et al., 2011; Jean and Nestor, 2015).

8. **Antifungal activity:** Antifungal properties of biosurfactant produced by strains of *Pseudomonas fluorescens* is well documented in literature (Nielsen and Sørensen 2003). Hultberg et al. (2008b) have reported that fluorescent pseudomonads with the biosurfactant producing ability can inhibit the growth of fungal pathogens such as *Pythium ultimum* (causative agent of damping off and root rot of plants), *Fusarium oxysporum* (causes wilting in crop plants), and *Phytophthora cryptogea* (causes rotting of fruits and flowers). *Pseudomonas* sp. are reported as biocontrol agents against *Verticillium microsclerotia*; a causative agent of Verticillium wilt mainly in potatoes. The biosurfactant produced by this *Pseudomonas* sp. is considered to play a major role in the inhibition of *in vitro* viability of *Verticillium* sp. (Debode et al. 2007). Strains of *Pseudomonas* sp. terminate the growth of pathogenic fungi *Rhizoctonia solani* (causes several plant diseases) and *Phythium ultimum* (causes damping-off and root rot of plants) by the production of dual functioning compounds tensin, viscosin, and viscosinamid. The dual function includes biosurfactant and antifungal activity (Andersen et al. 2003).

9. **Induction of non-specific immunity to plants:** They are involved in non-specific immunity in plants, induce resistance in plants, and are also active in other plant species. RLs are capable of stimulating defense genes in tobacco, also potent protectors in monocotyledonous plants against biotrophic fungi (Mulligan, 2005).

10. **RLs assisted composting:** Apart from these antiphytopathogenic properties, addition of biosurfactant is also known to accelerate the compositing process by providing favorable conditions for microbial growth (Zhang et al. 2011a, 2011b).

11. **Biofilm forming ability:** Microbial factors such as motility, ability to form biofilm on root surface, and release of quorum sensing molecules are required to establish association with the plant. It is reviewed that quorum sensing molecules such as acyl homoserine lactone (AHL) are required for synthesis of antifungal compounds by the rhizobacteria. Studies also indicate that the concentration of these molecules is high in rhizosphere as compared to that in bulk soil (soil away from plant roots) suggesting the role of AHL and AHL-like molecules in rhizosphere competence (ability of beneficial

microorganism to colonize the root surface). These AHL are also reported to contribute in the regulation of EPS essential for biofilm formation (Newton and Fray 2004; Loh et al. 2002). Dusane et al. (2010) have recently reported that the biosurfactant (RL) produced by *Pseudomonas* spp. regulates the process of quorum sensing (cell to cell communication). It is also reported that biosurfactants affect the motility of microorganisms, participate in signaling and differentiation as well as in biofilm formation (Ron and Rosenberg 2011; Berti et al. 2007; Van Hamme et al. 2006; Kearns and Losick 2003).

12. **Microbicidal action:** Furthermore, RLs harbor antiviral, algicidal, mycoplasmicidal, and antiamoebal properties (Cosson et al. 2002; Itoh et al. 1971; Wang et al. 2005). Krzyzanowska et al. (2012) mentioned the biocontrol efficacy of biosurfactant from *Pseudomonas* and *Bacillus* spp isolated from rhizosphere soil that exhibited the inhibition of soft rot causing *Pectobacterium* and *Dickeya* spp.

13. **Swarming motility:** In *P. aeruginosa*, swarming motility was relatively recently reported (Déziel et al., 2001; Köhler et al., 2000; Rashid and Arthur, 2000) and for which it requires flagella and the production of a wetting agent, at least hydroxyalkanoyloxy acid (HAA's), and ideally also RLs (Caiazza et al. 2005; Köhler et al. 2000). A notable exception was noted by Shrout et al. (2006) who were able to see swarming of an rhlAB mutant. Recently, Tremblay et al. (2007) reported that HAAs and di-RLs actually modulate the development of the fractal-like patterns formed by migrating *P. aeruginosa* swarming colonies. They found that di-RLs promote tendril formation and migration, acting like self-produced chemotactic attractants, while HAAs play the opposite role, repelling swarming tendrils. Mono-RLs seem to act solely as wetting agent. These findings extended the previous work by Caiazza et al. (2005) who identified a role for RLs in maintenance of the swarming.

The genus *Bacillus* spp. are rod shaped endospore forming Gram-positive bacteria belonging to Low % moles G + C bacteria (Kingdom: Bacteria; Phylum: Firmicutes; Class: Bacilli; Order: Bacillales; Family: Bacillaceae) (Maughan and Auwera, 2011). On artificial solid medium like nutrient agar (simple medium for culture of bacteria) *Bacillus* spp produces mucoid glossy colonies that indicate *Bacillus* cells are rich in EPS. The genera *Bacillus* secrete biosurfactant of belonging to the class Lipopeptides (LPs) which are lipids attached to a short polypeptide chain. The types of LPs produced by bacteria are Surfactin, Iturin (iturins, mycosubtilins and bacillomycins), Fengycin, Lychensin, etc. (Joshi et al., 2008). Its structure is characterized by a heptapeptidic moiety linked to a β-hydroxyl fatty acids. A lactone bridge between the β-hydroxyl function of the acid and the carboxy-terminal the peptide confers a cyclic structure to this molecule (Dufour et al., 2005; Abdel-Mawgoud et al., 2008). Many of the species of *Bacillus* produce swarming growth on solid medium. *Bacillus* species excrete a broad spectrum of bioactive peptides with great potential for biotechnological and biopharmaceutical applications. *Bacillus* sp. and strains produce a variety of EPS such as levan, β-1, 3-Glucan (Gummadi and Kumar, 2005), and heteropolymers mainly composed of neutral sugar (Larpin et al., 2002), uronic acid, uncommon sugar (Kodali et al., 2009), or sugar–protein conjugate

(Zheng et al., 2008). Some EPS from *Bacillus* have shown excellent emulsifying, flocculating, heavy metal removal capacity, or pharmaceutical activity (Arena et al., 2006; Kodali et al., 2009; Salehizadeh and Shojaosadati, 2003; Zheng et al., 2008).

Surfactin: Surfactin is a peptide lipid with a molecular weight of about 1,050 and is composed of L-aspartic acid, L-Glutamic acid, L-Valine, L-Leucine (1:1:1:2:2) and fatty acids. Surfactin is produced by *Bacillus subtilis* which is a mixture of several hydroxy-fatty acids with chain length of 13–15 carbon atoms. The main component is 3-hydroxy-13-methyltetradecanoic acid. Surfactin crystals appeared rectangular with blunt corners and were arranged perpendicular to each other making a plus sign (Abdel-Mawgoud et al., 2008).

Iturin: The LPs belonging to the iturin family are potent antifungal agents which can be used as biopesticides for plant protection. The surfactin and iturin compounds are cyclic lipoheptapeptides which contain a β-hydroxy fatty acid and a β-amino fatty acid, respectively, as lipophilic components (Vater et al., 2002; Joshi et al., 2008). Among the LPs belonging to the iturin family, iturin A is the most studied compound. It is a heptapeptide interlinked with α-amino-acid fatty acid with carbon chain length from C_{14} to C_{17} produced by *Bacillus subtilis* strains reported to have antifungal activities.

Fengycin: It belongs to LP class of biosurfactants and possesses antifungal activity. Fengycin, a cyclic lipodecapeptide containing a β-hydroxy fatty acid with a side-chain length of 16–19 carbon atoms offers an efficient antifungal activity. In a recent research conducted by Roy et al. (2013) isolated fengycin from *Bacillus thuringiensis* SM1 using 16 S r DNA analysis and explored its self assembled fengycin which offered antifungal activity against *Candida albicans* and *Aspergillus niger*, it even inhibited bacterial strains namely, *Staphylococcus epidermidis* and *Escherichia coli* at a concentration range of 1 mg/mL to 1,95 g/mL.

Lychensin: A surfactin-related compound is lichenysin, a LP discovered in the supernatant of *Bacillus licheniformis* culture. Lichenysin has Glutamine amino-acid in position 1 while surfactin has Glutamic acid.

Various attributes of LPs from *Bacillus* spp

1. Antifungal activity towards phytopathogenic fungi: LP biosurfactant produced by the strains of *Bacillus* exhibits growth inhibition of phytopathogenic fungi therefore, these biosurfactants can be used as biocontrol agent. *Brevibacillus brevis* strain HOB1 produces surfactin isoform and this LP biosurfactant has demonstrated strong antibacterial and antifungal property which can be exploited for control of phytopathogens (Haddad et al., 2008). *Colletotrichum gleosporiodes*, causative agent for anthracnose on papaya leaves is reported to be controlled by biosurfactant producing *Bacillus subtilis* isolated from soil (Kim et al., 2010). *B. subtilis* is considered the major producer of those antibiotic peptides and a *B. subtilis* strain producing iturin A and surfactin was shown to be effective for the control of damping-off caused by *R. solani* in tomato plants. As *B. subtilis* 20B showed (Joshi et al., 2008) no chitinase production but have antifungal activity against mycelia growth of *C. indicum, A. burnsii, F. oxysporium, F. udum, T. herzanium* and *R. bataticola* it can be hypothesized that it

produces some antifungal compound. The co-production of surfactin and iturins has been reported in *B. subtilis* strains (Ahimou et al., 2000). Thus, produced antifungal compound can be further characterized and can be applied as a biocontrol agent.

2. Reduction of surface or interface tension: Kumar et al. (2017) produced and optimized iturin A through response surface methodology and examined the surface tension reduction by iturin A of *Bacillus amyloliquifasciens* RHNK22 that was able to show highest reduction in surface tension from 60.50 to 26.12 mN/m in 24 h and 29.04 mN/m in 48 h in comparison with other isolates for various hydrocarbons.

3. Antibacterial activity of LPs: El-Sheshtawy et al. (2015) discussed the role of biosurfactant obtained from *Bacillus licheniformis* simultaneously was capable of microbial enhanced oil recovery and also reduced growth of sulfate reducing bacteria using 1% crude biosurfactants.

4. Antifungal activity: In a very recent study by Jha et al. (2016), lipopeptide extracted from *Bacillus* was characterized as Fengycin through TLC, MALDI-TOFF, and FTIR analysis and the compound revealed its antifungal nature towards the tested plant pathogenic fungi namely, *Rhizoctonia solani*, *Chrysosporium indicum*, *Alternaria burnsii*, *Fusarium oxysporium*, and *Rhizoctonia bataticola* while the inhibition of *Aspergillus niger* and *Penicillium chrysogenum* was not attained. Kim et al. (2010) reported coproduction of biosurfactant LPs – Iturin A, Fengycin, and Surfactin A, from *B. subtilis* CMB32, and its inhibitory effect on fungal pathogen *Colletotrichum gloeosporioides*. The co-production of surfactants and fungicide is an interesting characteristic with potential practical applications in petroleum, environmental, and agricultural industries. A recent research by Kumar et al. (2017) emphasized the use of biosurfactants in agriculture as who approached to characterize LP class of biosurfactant as Iturin A from *Bacillus amyloliquifasciens* RHNK22 to showcase highly potent antifungal activity against *Sclerotium rolfsii* and *Macrophomina phaseolina*. Zhaolin et al. (2015) investigated biocontrol effects of *Bacillus licheniformis* W10 bacterial suspension and its antifungal protein on peach brown rot caused by *Monilinia fructicola* in storage peach fruits and the effects on fruit quality. Korean ginseng (*Panax ginseng*) is one of the most important perennial herb plants grown and used in Asia. Ginseng anthracnose, caused by the fungus *Colletotrichum panacicola*, has a severe impact on yields, necessitating the development of an organic control method. Inoculation of *Panax ginseng* plants with *B. subtilis* significantly suppressed the number of disease lesions of *C. panacicola* and was as effective as the chemical fungicide iminoctadine tris (albesilate). The antifungal activity of *B. subtilis* against *C. panacicola* was observed on a co-culture medium (Ryu et al., 2014).

5. Antiviral activity: Lipopeptides from *Bacillus* spp show antiviral property (Banat et al., 2010). Antimicrobial LPs produced by *B. subtilis* fmbj inactivated cell-free virus of porcine parvovirus, pseudorabies virus, new castle disease virus, and bursal disease virus, while it effectively inhibited replication and infectivity of the new castle disease virus and bursal disease virus

but had no effect on pseudorabies virus and porcine parvovirus (Huang et al. 2006). The more effective inactivation of enveloped viruses, such as retroviruses and herpes viruses, compared to non-enveloped viruses, suggests that this inhibitory action may be mainly due to the physico–chemical interactions between the virus envelope and the surfactant.

6. Hydrocarbon biodegradation: A research was conducted by Yadav et al. (2016) that dealt with the isolation and characterization of biosurfactant producing *Bacillus* sp. from diesel fuel contaminated site. The biosurfactant belonged to class LP and was found potent for degradation of diesel.

7. Heavy metal removal from contaminated sites: Mulligan and Gibbs (2004) evaluated the feasibility of using surfactin from *Bacillus subtilis* for the removal of heavy metals from contaminated soil and sediments. Batch soil washing experiment was performed containing high levels of heavy metals and hydrocarbons (890 mg/kg of zinc, 420 mg/kg of copper, and 12.6% oil and grease). Water removed minimal amounts of copper and zinc (less than 1%) and 0.25% surfactin removed 70% copper and 22% zinc.

8. Reduction of surface tension and emulsification: Surfactin produced by *Bacillus subtilis* was capable of surface tension reduction [critical micelle concentration (CMC) = 70 M in 0.1 M Tris-HCl buffer pH 8] and emulsified a wide variety of hydrocarbons *viz* Tridecane, Pristane, Diesel, Kerosene, Hexadecane, Toluene, Kerosene, Heptane, Benzene, Crude oil as 56.8 ± 0.6, 55.6 ± 1.9, 55 ± 0.3, 59.01 ± 0.4, 55 ± 0.8, 52.8 ± 1.6, 53.2 ± 2.2 and 67.6 ± 1.0, respectively (De-Faria et al., 2011).

9. Antiadhesive activity: Gomes and Nitschke (2012) evaluated RL and surfactin that potentially reduced adhesion and removed biofilms in sole and mixed cultures of food pathogenic bacteria. Different concentrations of the biosurfactants, surfactin from *Bacillus subtilis* and RLs from *Pseudomonas aeruginosa*, were evaluated to reduce the adhesion and to disrupt biofilms of food-borne pathogenic bacteria. Individual cultures and mixed cultures of *Staphylococcus aureus*, *Listeria monocytogenes*, and *Salmonella enteritidis* were studied using polystyrene as the model surface. The pre-conditioning with surfactin 0.25% reduced by 42.0% the adhesion of *L. monocytogenes* and *S. enteritidis*, whereas the treatment using RLs 1.0% reduced by 57.8% the adhesion of *L. monocytogenes*, and by 67.8% the adhesion of *S. aureus* to polystyrene.

10. Antibiofilm forming action: Surfactin is able to inhibit biofilm formation of *Salmonella typhimurium*, *S. enterica*, *E. coli*, and *Proteus mirabilis* in polyvinyl chloride wells, as well as vinyl urethral catheters (Mireles et al., 2001).

Taxonomic hierarchy of *Serratia* states that it is a Gram-negative bacteria belonging to kingdom: Bacteria; Phylum: Proteobacteria; Class: Gammaproteobacteria; Order: Enterobacteriales; Family: Enterobacteriaceae; Genus: *Serratia*. The main habitat of *Serratia marscecens* is water but they are widely distributed in terrestrial environments and associated with plants. Few strains of *Serratia* are able to produce a non diffusible red pigment known as Prodigiosin. Both pigmented and non-pigmented strains of *Serratia* are biosurfactant producing. The biosurfactant produced by them are termed as serrawettin characterized as wetting agent. Serrawettin are aminolipids (Falkow et al., 2006). The lipopolysaccharide (LPS) of *Serratia* acts as endotoxin present in the outer membrane of the cell and it possesses bactericidal action due to the presence of O-side chain of LPS. The LPS of any bacteria has three components: (i) Lipid A, (ii) Core polysaccharide, and (iii) O- side chain or O-antigen. In presence of Fe^{2+}, some strains of *Serratia marcescens* produce a water-soluble pink pigment pyrimine, L-2-(2-pyridyl-D/-pyrroline-5-carboxylic acid), which diffuses into the agar surrounding the colonies (Khanafari et al., 2006).

Serrawettin: Serrawettin belongs to LP class of biosurfactants. Serrawettin is a wetting agent that helps the bacteria to adhere to surfaces. These biosurfactant are also related to the swarming growth of the genus peculiarly visible in solid agar medium. The serrawettin also provides hydrophobicity to the cell. This property of *Serratia marscecens* make it a suitable candidate for the bioremediation of oil and hydrocarbon contamination as it readily adheres at air: water and oil: water interfaces as well as adhesion to solid substances rendering to hydrophobic chemical modifications. Serrawettins are non-ionic biosurfactants with three molecular species *viz.* Serrawettin, W_1: Cyclo (D-3-hydroxydecanoyl-L-seryl)$_2$; W_2: d-3-hydroxydecanoyl-D-leucyl-L-theonyl-D- phenylalanyl -L-isoleucyl lactone; and W_3: cyclopsipeptide composed of five amino acids and one dodecanoic acid (Matsuyama et al., 2011).

Various attributes of Serrawettins from *Serratia* spp

1. **Emulsification:** Anyanwu et al. (2011) in his studies on LP biosurfactant from *Serratia marscecens* NSK-1 recovered from petroleum contaminated sites revealed the emulsifying nature of the LP biosurfactant capable of forming stable emulsions with several vegetable oils. Emulsification activity of the biosurfactant was highest with a vegetable oil, soybean oil with emulsification index of 98%, and least with kerosene, a hydrocarbon showing emulsification index of 50%. Peter et al. (2014) examined the emulsification activity of a consortium of *Pseudomonas aeruginosa*, *Serratia marscecens*, and *Bacillus subtilis* on vegetable oils (Mustard oil, Soybean oil, Coconut, and Jasmine oil) and hydrocarbons (petrol, diesel, and kerosene). The biosurfactants from individual and consortium based were potent to emulsify the wide variety of substrates.

2. **Antinematode action:** Pradel et al. (2007) described the inhibitory action of serrawettin extracted from *Serratia marscecens* against a nematode namely, *Caenorhabdis elegans*. The study revealed the lawn avoidance behavior of the compound serrawettin towards *C. elegans*. Lawn avoidance deprives the nematode of bacterial food, which is otherwise consumed continuously. Serrawettin W_2 repelled *C. elegans* due to the biosurfactant effect of surface tension reduction.

3. **Wetting agent:** The secretion of exolipid produced by *Serratia marscecens* has wetting property from non-pigmented *Serratia* cells. Three lipids obtained owing to the lipid activity were W_1, W_2, and W_3 which were extracted through chromatography.

4. **Antibacterial activity:** *Serratia* species that are good source of serrawetins also acts as antibacterial and antifungal agents showing growth inhibition in artificial medium for few test bacterial pathogens. It disturbs the cell wall composition that lead to death of cell.

5. **Chitinolytic activity:** Extra cellular enzymes of the genus possess chitinolytic activity that is employed as a natural antifungal agent.

6. **Hydrocarbon degradation:** Wongsa et al. (2004) isolated *Serratia marscecens* showing which capacity spectrum of hydrocarbon degradation (gasoline, kerosene, diesel, and lubricating oil). It was able to degrade 90%–95% diesel and kerosene supplemented to Mineral Salt media as a sole source of carbon in 2 to 3 weeks duration. The degradation attained was estimated to be an effort of WatG while the lubricating oil was only 60% degraded in two weeks of time.

Hence, these green surfactants are important parameters for microbes to achieve a beneficial association with the plant roots and improve the growth of the plant. Furthermore, these biosurfactants produced by rhizobacteria increase the bioavailability of hydrophobic molecules which may serve as nutrients. Biosurfactants produced by soil microbes provide wettability to soil and support proper distribution of chemical fertilizers in soil, thus assisting plant growth promotion. Reviewing the functions of biosurfactant indicates the essential role of these green compounds for sustainable agriculture. As per Sachdev and Cameotra (2013) the chemical compositions of biosurfactants reported that potent biocontrol agents can be altered by changing the production scheme. This approach may lead to biosynthesis of highly target-specific green surfactant/s. The high prevalence of biosurfactants and biosurfactant producing bacteria in rhizosphere is a positive indication for its potent role in sustainable agriculture.

11.4   CONCLUSION

In conclusion, the microflora is the representation of soil health and the growth of a plant. In the soil, rhizosphere represents vivacious functions and interaction of soil microflora and in soil and with plant parts such that they are a great source of communication among them. They enhance plant growth with various influencing attributes, many of which are unfolded over time and many are yet to be unfolded; biosurfactants and microbial EPS are one of them. Soil respiration is high throughput indicator of soil dynamic conditions especially due to microbial and plants. It is greatly beneficial in indication of GHGs emission (CO_2) from soil and can be related to climate change.

ACKNOWLEDGMENTS

The authors are thankful to ICAR, HARP division, Jharkhand and Department of Industrial Microbiology, Jacob Institute of Biotechnology and Bioengineering, Sam Higginbottom University of Agriculture Technology and Sciences for soliciting the requirements and facilitation of the research.

REFERENCES

Abalos, A., Pinazo, A., Infante, M.R., Casals, M., Garcia, F. and Manresa, A., 2001. Physicochemical and antimicrobial properties of new rhamnolipids produced by Pseudomonas an eruginosa AT10 from soybean oil refinery wastes. *Langmuir*, 17(5), pp. 1367–1371.

Abdel-Mawgoud, A.M., Aboulwafa, M.M. and Hassouna, N.A.H., 2008. Characterization of surfactin produced by Bacillus subtilis isolate BS5. *Applied biochemistry and biotechnology*, 150(3), pp. 289–303.

Ahimou, F., Jacques, P. and Deleu, M., 2000. Surfactin and iturin A effects on Bacillus subtilis surface hydrophobicity. *Enzyme and Microbial Technology*, 27(10), pp. 749–754.

An, C.J., Huang, G.H., Wei, J. and Yu, H., 2011. Effect of short-chain organic acids on the enhanced desorption of phenanthrene by rhamnolipid biosurfactant in soil–water environment. *Water research*, 45(17), pp. 5501–5510.

Andersen, J.B., Koch, B., Nielsen, T.H., Sørensen, D., Hansen, M., Nybroe, O., Christophersen, C., Sørensen, J., Molin, S. and Givskov, M., 2003. Surface motility in Pseudomonas sp. DSS73 is required for efficient biological containment of the root-pathogenic microfungi Rhizoctonia solani and Pythium ultimum. *Microbiology*, 149(1), pp. 37–46.

Anyanwu, C.U., Obi, S.K.C., and Okolo, B.N., 2011. Lipopeptide biosurfactant production by Serratia marcescens NSK-1 strain isolated from petroleum-contaminated soil. *Journal of Applied Sciences Research*, 7(1), pp. 79–87.

Arena, A., Maugeri, T. L., Pavone, B., Iannello, D., Gugliandolo, C. and Bisignano, G., 2006. Antiviral and immunoregulatory effect of a novel exopolysaccharide from a marine thermotolerant Bacillus licheniformis. *International Immunopharmacology*, 6(1), pp. 8–13.

Azarbad, H., Niklińska, M., van Gestel, C.A., van Straalen, N.M., Röling, W.F. and Laskowski, R., 2013. Microbial community structure and functioning along metal pollution gradients. *Environmental Toxicology and Chemistry*, 32(9), pp. 1992–2002.

Banat, I.M., Franzetti, A., Gandolfi, I., Bestetti, G., Martinotti, M.G., Fracchia, L., Smyth, T.J. and Marchant, R., 2010. Microbial biosurfactants production, applications and future potential. *Applied Microbiology and Biotechnology*, 87(2), pp. 427–444.

Banat, I.M., Makkar, R.S. and Cameotra, S.S., 2000. Potential commercial applications of microbial surfactants. *Applied Microbiology and Biotechnology*, 53(5), pp. 495–508.

Berti, A.D., Greve, N.J., Christensen, Q.H. and Thomas, M.G., 2007. Identification of a biosynthetic gene cluster and the six associated lipopeptides involved in swarming motility of Pseudomonas syringae pv. tomato DC3000. *Journal of Bacteriology*, 189(17), pp. 6312–6323.

Caiazza, N.C., Shanks, R.M. and O'toole, G.A., 2005. Rhamnolipids modulate swarming motility patterns of Pseudomonas aeruginosa. *Journal of Bacteriology*, 187(21), pp. 7351–7361.

Carey, J.C., Tang, J., Templer, P.H., Kroeger, K.D., Crowther, T.W., Burton, A.J., Dukes, J.S., Emmett, B., Frey, S.D., Heskel, M.A. and Jiang, L., 2016. Temperature response of soil respiration largely unaltered with experimental warming. *Proceedings of the National Academy of Sciences*, 113(48), pp. 13797–13802.

Chen, Q., Bao, M., Fan, X., Liang, S. and Sun, P., 2013. Rhamnolipids enhance marine oil spill bioremediation in laboratory system. *Marine Pollution Bulletin*, 71(1–2), pp. 269–275.

Cosson, P., Zulianello, L., Join-Lambert, O., Faurisson, F., Gebbie, L., Benghezal, M., Van Delden, C., Curty, L.K. and Köhler, T., 2002. Pseudomonas aeruginosa virulence analyzed in a Dictyostelium discoideum host system. *Journal of Bacteriology*, 184(11), pp. 3027–3033.

Dai, J., Becquer, T., Rouiller, J.H., Reversat, G., Bernhard-Reversat, F. and Lavelle, P., 2004. Influence of heavy metals on C and N mineralisation and microbial biomass in Zn-, Pb-, Cu-, and Cd-contaminated soils. *Applied Soil Ecology*, 25(2), pp. 99–109.

Debode, J., Maeyer, K.D., Perneel, M., Pannecoucque, J., Backer, G.D. and Höfte, M., 2007. Biosurfactants are involved in the biological control of Verticillium microsclerotia by Pseudomonas spp. *Journal of Applied Microbiology*, 103(4), pp. 1184–1196.

De-Faria, A.F., Teodoro-Martinez, D.S., de Oliveira Barbosa, G.N., Vaz, B.G., Silva, Í.S., Garcia, J.S., Tótola, M.R., Eberlin, M.N., Grossman, M., Alves, O.L. and Durrant, L.R. 2011. Production and structural characterization of surfactin (C14/Leu7) produced by Bacillus subtilis isolate LSFM-05 grown on raw glycerol from the biodiesel industry. *Process Biochemistry*, 46(10), pp. 1951–1957.

Delgado, A. and Gómez, J.A. 2016. The soil. Physical, chemical and biological properties. In Villalobos, F., Fereres, E. (eds.) *Principles of Agronomy for Sustainable Agriculture*. Springer, Cham, pp. 15–26.

Déziel, E., Comeau, Y. and Villemur, R. 2001. Initiation of biofilm formation byPseudomonas aeruginosa 57RP correlates with emergence of hyperpiliated and highly adherent phenotypic variants deficient in swimming, swarming, and twitching motilities. *Journal of Bacteriology*, 183(4), pp. 1195–1204.

Dufour, S., Deleu, M., Nott, K., Wathelet, B., Thonart, P. and Paquot, M. 2005. Hemolytic activity of new linear surfactin analogs in relation to their physico-chemical properties. *Biochimica et Biophysica Acta (BBA)-General Subjects*, 1726(1), pp. 87–95.

Dusane, D.H., Zinjarde, S.S., Venugopalan, V.P., Mclean, R.J., Weber, M.M. and Rahman, P.K. 2010. Quorum sensing: implications on rhamnolipid biosurfactant production. *Biotechnology and Genetic Engineering Reviews*, 27(1), pp. 159–184.

El-Sheshtawy, H.S., Aiad, I., Osman, M.E., Abo-Elnasr, A.A. and Kobisy, A.S. 2015. Production of biosurfactant from Bacillus licheniformis for microbial enhanced oil recovery and inhibition the growth of sulfate reducing bacteria. *Egyptian Journal of Petroleum*, 24(2), pp. 155–162.

Falkow, S., Rosenberg, E., Schleifer, K.H. and Stackebrandt, E. (Eds.) 2006. *The Prokaryotes. Proteobacteria: Gamma Subclass*. The genus Serratia. Vol 6, chapter: 3.3.11, Springer, pp. 219–244.

Gomes, M.Z.d.v and Nitschke, M. 2012. Evaluation of rhamnolipid and surfactin to reduce the adhesion and remove biofilms of individual and mixed cultures of food pathogenic bacteria. *Food Control*, 25(2), pp. 441–447.

Gummadi, S.N. and Kumar, K. 2005. Production of extracellular water insoluble β-1, 3-glucan (curdlan) fromBacillus sp. SNC07. *Biotechnology and Bioprocess Engineering*, 10(6), pp. 546–551.

Guntiñas, M.E., Gil-Sotres, F., Leirós, M.C. and Trasar-Cepeda, C. 2013. Sensitivity of soil respiration to moisture and temperature. *Journal of Soil Science and Plant Nutrition*, 13(2), pp. 445–461.

Haddad, N.I., Wang, J. and Mu, B., 2008. Isolation and characterization of a biosurfactant producing strain, Brevibacilis brevis HOB1. *Journal of Industrial Microbiology & Biotechnology*, 35(12), pp. 1597–1604.

Huang, Xianqing, Zhaoxin Lu, Haizhen Zhao, Xiaomei Bie, FengXia Lü, and Shujing Yang. Antiviral activity of antimicrobial lipopeptide from Bacillus subtilis fmbj against pseudorabies virus, porcine parvovirus, newcastle disease virus and infectious bursal disease virus in vitro. *International Journal of Peptide Research and Therapeutics*, 12, no. 4 (2006): 373–377.

Hultberg, M., Bergstrand, K.J., Khalil, S. and Alsanius, B. 2008a. Characterization of biosurfactant producing strain of fluorescen pseudomonads in soilless cultivation system. *Antonie Van Leewenhoek*, 94(2), pp. 329–334.

Hultberg, M., Bergstrand, K.J., Khalil, S. and Alsanius, B., 2008b. Production of biosurfactants and antibiotics by fluorescent pseudomonads isolated from a closed hydroponic system equipped with a slow filter. *Antonie Van Leeuwenhoek*, 93, pp. 373–380.

IPCC. 2007a. Intergovernmental Panel on Climate Change (IPCC). Fourth Assessment Report: Climate Change, Geneva.

IPCC. 2007b. Summary for Policymakers. Climate Change: The Physical Science Basis. Contribution of Working Group I to the Fourth Assessment Report of the Intergovernmental Panel on Climate Change, Cambridge University Press, Cambridge, United Kingdom.

ISO. 2002. Soil quality – determination of abundance and activity of soil microflora using respiration curves. International Organization for Standardization, International Standard ISO No.17155.

Itoh, S., Honda, H., Tomita, F. and SUZUKI, T., 1971. Rhamnolipids produced by Pseudomonas aeruginosa grown on n-paraffin (mixture of C12, C13 and C14 fractions). *The Journal of Antibiotics*, 24(12), pp. 855–859.

Jat, M.L., Dagar, J.C., Sapkota, T.B., Singer, Y., Govaerts, B., Riduara, S.L., Saharawat, R.K., Tetarwal, J.P., Jat, R.K., Hobbs, H. and Sterlomg, C., 2015. Chapter 3: Climate Change and Agriculture: Adaptation Strategies and Mitigation Opportunities for Food Security in South Asia and Latin America. *Advances in Agronomy*, 137, pp. 127–235.

Jean, P.S. and Nestor, I. 2015. Agriculture and food applications of rhamnolipids and its production by Pseudomonas aeruginosa. *Journal of Chemical Engineering and Process Technology*, 6(2), pp. 1–8.

Jha, S.S., Joshi, S.J., and Geetha, S.J. 2016. Lipopeptide production by Bacillus subtilis R1 and its possible applications. *Brazilian Journal of Microbiology*, 47(4), pp. 955–964.

Jiang, J., Guo, S., Zhang, Y., Liu, Y., Liu, Q., Jiang, J., Wang, R., Wang, Z., Li, N., and Li, R., 2015. Changes in temperature sensitivity of soil respiration I phases of a three crop rotation system. *Soil and Tillage Research*, 150: pp. 139–146.

Jones, K.M., 2012. Increased Production of the exopolysaccharide succinoglycan enhances Sinorhizobium meliloti 1021 symbiosis with the host plant Medicago truncatula. *Journal of Bacteriology*, 194(16), pp. 4322–4331.

Joshi, S., Bharucha, C. and Desai, A.J., 2008. Production of biosurfactant and antifungal compound by fermented food isolate Bacillus subtilis 20B. *Bioresource Technology*, 99(11), pp. 4603–4608.

Khanafari, A., Assadi, M.M. and Fakhr, F.A., 2006. Review of prodigiosin, pigmentation in Serratia marcescens. *Online Journal of Biological Sciences*, 6(1), pp. 1–13.

Kang, S.M., Khan, A.L., Waqas, M., You, Y.H., Kim, J.H., Kim, J.G., Hamayun, M. and Lee, I.J., 2014. Plant growth-promoting rhizobacteria reduce adverse effects of salinity and osmotic stress by regulating phytohormones and antioxidants in Cucumis sativus. *Journal of Plant Interactions*, 9(1), pp. 673–682.

Kearns, D.B. and Losick, R., 2003. Swarming motility in undomesticated Bacillus subtilis. *Molecular Microbiology*, 49(3), pp. 581–590.

Kim, P.I., Ryu, J., Kim, Y.H. and Chi, Y.T., 2010. Production of biosurfactant lipopeptides Iturin A, fengycin and surfactin A from Bacillus subtilis CMB32 for control of Colletotrichum gloeosporioides. *Journal of Microbiology and Biotechnology*, 20(1), pp. 138–145.

Kodali, V.P., Das, S. and Sen, R., 2009. An exopolysaccharide from a probiotic: biosynthesis dynamics, composition and emulsifying activity. *Food Research International*, 42(5–6), pp. 695–699.

Köhler, T., Curty, L.K., Barja, F., Van Delden, C. and Pechère, J.C., 2000. Swarming of Pseudomonas aeruginosa is dependent on cell-to-cell signaling and requires flagella and pili. *Journal of Bacteriology*, 182(21), pp. 5990–5996.

Kruijt, M., Tran, H. and Raaijmakers, J.M., 2009. Functional, genetic and chemical characterization of biosurfactants produced by plant growth-promoting Pseudomonas putida 267. *Journal of Applied Microbiology*, 107(2), pp. 546–556.

Krzyzanowska, D.M., Potrykus, M., Golanowska, M., Polonis, K., Gwizdek-Wisniewska, A., Lojkowska, E. and Jafra, S., 2012. Rhizosphere bacteria as potential biocontrol agents against soft rot caused by various Pectobacterium and Dickeya spp. strains. *Journal of Plant Pathology*, 94(2), pp. 367–378.

Kumar, A.S., Mody, K. and Jha, B., 2007. Bacterial exopolysaccharides–a perception. *Journal of Basic Microbiology*, 47(2), pp. 103–117.

Kumar, P.N., Swapna, T.H., Khan, M.Y., Reddy, G. and Hameeda, B., 2017. Statistical optimization of antifungal iturin A production from Bacillus amyloliquefaciens RHNK22 using agro-industrial wastes. *Saudi Journal of Biological Sciences*, 24(7), pp. 1722–1740.

Kyungseok, P., Kloepper, J. and Ryu, C. 2008. Rhizobacterial exopolysaccharides elicit induced resistance on cucumber. *Journal of Microbiology and Biochemistry*, 18(6), pp. 1095–1100.

Lai, L., Zhao, X., Jiang, L., Wang, Y., Luo, L., Zheng, Y., Chen, X. and Rimmington, G.M., 2012. Soil respiration in different agricultural and natural ecosystems in an arid region. *PloS one*, 7(10), p. e48011.

Lamberty, B.B., Bolton, H., Fansler, S., Heredia-Langner, A., Liu, C., McCue, L.A., Smith, J. and Bailey, V., 2016. Soil respiration and bacterial structure and function after 17 years of a reciprocal soil transplant experiment. *PLoS One*, 11(3), p. e0150599.

Larpin, S., Sauvageot, N., Pichereau, V., Laplace, J.M. and Auffray, Y., 2002. Biosynthesis of exopolysaccharide by a Bacillus licheniformis strain isolated from ropy cider. *International journal of food microbiology*, 77(1–2), pp. 1–9.

Loh, J., Pierson, E.A., Pierson III, L.S., Stacey, G. and Chatterjee, A., 2002. Quorum sensing in plant-associated bacteria. *Current Opinion in Plant Biology*, 5(4), pp. 285–290.

Maier, R.M. and Soberon-Chavez, G., 2000. Pseudomonas aeruginosa rhamnolipids: biosynthesis and potential applications. *Applied Microbiology and Biotechnology*, 54(5), pp. 625–633.

Matsuyama, T., Tanikawa, T. and Nakagawa, Y., 2011. Serrawettins and other surfactants produced by Serratia. In Biosurfactants (pp. 93–120). Springer, Berlin, Heidelberg.

Maughan, H. and Van-der Auwera, G., 2011. Bacillus taxonomy in the genomic era finds phenotypes to be essential though often misleading. *Infection, Genetics and Evolution*, 11(5), pp. 789–797.

Mireles, J.R., Toguchi, A. and Harshey, R.M., 2001. Salmonella enterica serovar Typhimurium swarming mutants with altered biofilm-forming abilities: surfactin inhibits biofilm formation. *Journal of Bacteriology*, 183(20), pp. 5848–5854.

Monteiro, S.A., Sassaki, G.L., de Souza, L.M., Meira, J.A., de Araújo, J.M., Mitchell, D.A., Ramos, L.P. and Krieger, N., 2007. Molecular and structural characterization of the biosurfactant produced by Pseudomonas aeruginosa DAUPE 614. *Chemistry and Physics of Lipids*, 147(1), pp. 1–13.

Mulligan, C.N., 2005. Environmental applications for biosurfactants. *Environmental Pollution*, 133: pp. 183–198.

Mulligan, C.N. and Gibbs, B.F., 2004. Types, production and applications of biosurfactants. *Proceedings-Indian National Science Academy Part B*, 70(1), pp. 31–56.

Naseem, H. and Bano, A., 2014. Role of plant growth-promoting rhizobacteria and their exopolysaccharide in drought tolerance of maize. *Journal of plant interactions*, 9(1), pp. 689–701.

Niemeyer, J.C., Lolata, G.B., de Carvalho, G.M., Da Silva, E.M., Sousa, J.P. and Nogueira, M.A., 2012. Microbial indicators of soil health as tools for ecological risk assessment of a metal contaminated site in Brazil. *Applied Soil Ecology*, 59, pp. 96–105.

Newton, J.A. and Fray, R.G., 2004. Integration of environmental and host-derived signals with quorum sensing during plant–microbe interactions. *Cellular Microbiology*, 6(3), pp. 213–224.

Nielsen, T.H. and Sørensen, J., 2003. Production of cyclic lipopeptides by Pseudomonas fluorescens strains in bulk soil and in the sugar beet rhizosphere. *Applied and Environmental Microbiology*, 69(2), pp. 861–868.

Nitschke, M., Costa, S.G. and Contiero, J., 2005. Rhamnolipid surfactants: an update on the general aspects of these remarkable biomolecules. *Biotechnology Progress*, 21(6), pp. 1593–1600.

Oertel, C., Matschullat, J., Zurba, K., Zimmermann, F., and Erasmi, S., 2016. Greenhouse Gas Emission from soils- A review. *Chemier der Erde- Geochemistry*, 76(3), pp. 327–352.

Parashar, P., Kapoor, N., and Sachdeva, S., 2014. Rhizosphere: its structure, bacterial diversity and significance. *Reviews in Environmental Science and Bio/Technology*. 13(1) pp. 63–77.

Peter, J.K., Rao, A.K. and Kumari, R., 2014. Consortium based Biosurfactant development for degradation and emulsification of oils and Petroleum Hydrocarbons. *International Journal of Engineering and Computer Science*, 3(6): pp. 6476–6490.

Pradel, E., Zhang, Y., Pujol, N., Matsuyama, T., Bargmann, C.I. and Ewbank, J.J., 2007. Detection and avoidance of a natural product from the pathogenic bacterium Serratia marcescens by Caenorhabditis elegans. *Proceedings of the National Academy of Sciences*, 104(7), pp. 2295–2300.

Qurashi, A.W. and Sabri, A.N., 2012. Bacterial exopolysaccharide and biofilm formation stimulate chickpea growth and soil aggregation under salt stress. *Brazilian Journal of Microbiology*, 43(3), pp. 1183–1191.

Rashid, M.H., and Arthur, A., 2000. Inorganc polyphosphate is needed for swimming, swarming, and twitching motilies of *Pseusomonas aeruginosa*. *Proceedings of the NATIONAL Academy of Sciences of the USA*, 97(9), pp. 4885–4890.

Romero-Freire, A., Aragón, M.S., Garzón, F.M. and Peinado, F.M., 2016. Is soil basal respiration a good indicator of soil pollution? *Geoderma*, 263, pp. 132–139.

Ron, E.Z. and Rosenberg, E., 2011. Natural roles in biosurfactants. *Environmental Microbiology*, 3, pp. 229–236.

Roy, A., Mahata, D., Paul, D., Korpole, S., Franco, O.L., and Mandal, S.M., 2013. Purification, biochemical characterization and self-assembled structure of a fengicin-like antifungal peptide from Bacillus thuringiensis starin SM1. *Frontiers in Microbiology*, 4, pp. 1–6.

Ryu, H., Park, H., Suh, D.S., Jung, G.H., Park, K. and Lee, B.D., 2014. Biological control of Colletotrichum panacicola on *Panax ginseng* by Bacillus subtilis HK-CSM-1. *Journal of Ginseng Research*, 38(3), pp. 215–219.

Sachdev, D. and Cameotra, S.S., 2013. Biosurfactants in agriculture. *Applied Microbiology and Biotechnology*, 97, pp. 1005–1016.

Salehizadeh, H. and Shojaosadati, S.A., 2003. Removal of metal ions from aqueous solution by polysaccharide produced from Bacillus firmus. *Water Research*, 37(17), pp. 4231–4235.

Shrout, J.D., Chopp, D.L., Just, C.L., Hentzer, M., Givskov, M. and Parsek, M.R., 2006. The impact of quorum sensing and swarming motility on Pseudomonas aeruginosa biofilm formation is nutritionally conditional. *Molecular Microbiology*, 62(5), pp. 1264–1277.

Singh, P. and Cameotra, S.S., 2004. Potential applications of microbial surfactants in biomedical sciences. *TRENDS in Biotechnology*, 22(3), pp. 142–146.

Skorupska, A., Janczarek, M., Marczak, M. Mazur, A. and Krol, J. 2006. Rhizobial exopolysaccharide: genetic control and symbiotic function. *Microbial Cell Factories*, 5(7), pp. 1–19.

Soberón-Chávez, G. 2004. Biosynthesis of rhamnolipids. In Ramos, J.L. (ed.), Pseudomonas. Kluwer Academic/Plenum Publishers, New York, pp. 173–189.

Soberón-Chávez, G., Lépine, F. and Déziel, E. 2014. Production of rhamnolipids by Pseudomonas aeruginosa. *Applied Microbial Biotechnology*, 68, pp. 718–725.

Thanomsub, B., Pumeechockchai, W., Limtrakul, A., Arunrattiyakorn, P., Petchleelaha, W., Nitoda, T. and Kanzaki, H., 2007. Withdrawn: Chemical structures and biological activities of rhamnolipids produced by Pseudomonas aeruginosa B189 isolated from milk factory waste. *Bioresource Technology*, 98(5), pp. 1149–1153.

Tremblay, J., Richardson, A.P., Lépine, F. and Déziel, E., 2007. Self-produced extracellular stimuli modulate the Pseudomonas aeruginosa swarming motility behaviour. *Environmental microbiology*, 9(10), pp. 2622–2630.

Van Hamme, J.D., Singh, A. and Ward, O.P., 2006. Physiological aspects: Part 1 in a series of papers devoted to surfactants in microbiology and biotechnology. *Biotechnology Advances*, 24(6), pp. 604–620.

Vardharajula, S., Zulfikar Ali, S., Grover, M., Reddy, G. and Bandi, V., 2011. Drought-tolerant plant growth promoting Bacillus spp.: effect on growth, osmolytes, and antioxidant status of maize under drought stress. *Journal of Plant Interactions*, 6(1), pp. 1–14.

Vater, J., Kablitz, B., Wilde, C., Franke, P., Mehta, N. and Cameotra, S.S., 2002. Matrix-assisted laser desorption ionization-time of flight mass spectrometry of lipopeptide biosurfactants in whole cells and culture filtrates of Bacillus subtilis C-1 isolated from petroleum sludge. *Applied and Environmental Microbiology*, 68(12), pp. 6210–6219.

Vatsa, P., Sanchez, L., Clement, C., Baillieul, F. and Dorey, S., 2010. Rhamnolipid biosurfactants as new players in animal and plant defense against microbes. *International Journal of Molecular Sciences*, 11(12), pp. 5095–5108.

Wang, X., Gong, L., Liang, S., Han, X., Zhu, C. and Li, Y., 2005. Algicidal activity of rhamnolipid biosurfactants produced by Pseudomonas aeruginosa. *Harmful Algae*, 4(2), pp. 433–443.

Wongsa, P., Tanaka, M., Ueno, A., Hasanuzzaman, M., Yumoto, I. Okuyama, H., 2004. Isolation and characterization of novel strains of Pseudomonas aeruginosa and Serratia marcescens possessing high efficiency to degrade gasoline, kerosene, diesel oil, and lubricating oil. *Current Microbiology*. 49(6), pp. 415–422.

Yadav, A.K., Manna, S., Pandiyan, K., Singh, A., Kumar, M., Chakdar, H., Kashyap, P.L. and Srivastava, A.K., 2016. Isolation and characterization of biosurfactant producing Bacillus sp. from diesel fuel-contaminated site. *Microbiology*, 85(1), pp. 56–62.

Yan, N., Marshner, P., Wenhong, C., Zuo, C. and Wei, Q. 2015. Influence of salinity and water content on soil microorganisms. *International Soil and Water Conservation Research*, 3(4), pp. 316–323.

Zhang, C., Wang, S. and Yan, Y., 2011a. Isomerization and biodegradation of beta-cypermethrin by Pseudomonas aeruginosa CH7 with biosurfactant production. *Bioresource Technology*, 102(14), pp. 7139–7146.

Zhang, F., Gu, W., Xu, P., Tang, S., Xie, K., Huang, X. and Huang, Q., 2011b. Effects of alkyl polyglycoside (APG) on composting of agricultural wastes. *Waste Management*, 31(6), pp. 1333–1338.

Zhang, Y., Guo, S., Liu, Q., Jiang, J., Wang, R. and Li, N., 2015. Responses of soil respiration to land use conversions in degraded ecosystem of the semi-arid Loess Plateau. *Ecological Engineering*, 74, pp. 196–205.

Zhaolin, J., Huiwen, H., Huijuan, Z., Feng, H., Yunhui, T., Zhengwen, Y. and Jingyou, X., 2015. The Biocontrol Effects of the Bacillus licheniformis W10 Strain and Its Antifungal Protein against Brown Rot in Peach. *Horticultural Plant Journal*, 1(3), pp. 131–138.

Zheng, Y., Ye, Z. L., Fang, X. L., Li, Y. H., Cai, W. M., 2008. Production and characteristics of a bioflocculant produced by Bacillus sp. F19. *Bioresource Technology*, 99(16), pp. 7686–7691.

12 Production of Temperate Fruits in Jammu & Kashmir under Climate Change Scenario

M. K. Sharma, F. A. Banday, and Amit Kumar
Sher-e-Kashmir University of Agricultural Sciences &
Technology of Kashmir (SKUAST-K)

CONTENTS

12.1   INTRODUCTION

Because of a wide range of agro-climatic conditions in Jammu and Kashmir (J&K), a large number of fruit crops are being grown. Fruit production is the backbone of the state's economy, providing livelihood to more than 35 lakh people out of a total population of 1.25 crore (2011 census). Over 7 lakh families are involved in this sector. The state generates about Rs. 6,000 crore income from fresh fruits, and over Rs. 200 crore from dry fruits. A variety of fruit crops ranging from sub-tropical to temperate fruits are grown in the state, which include apple, pear, peach, plum, apricot, cherry, grapes, citrus, olive, mango, ber, strawberry among fresh fruits, and walnut and almond among dry nuts (Table 12.1). The statistics related to the area, production, and productivity of previous years in the state is given in Tables 12.1 to 12.13.

The temperate fruit production in the Jammu and Kashmir is confronted with the major problem of low productivity mainly due to the following reasons which will further aggravate under climate change scenario. These problems should be managed in a scientific and systematic manner in order to sustain fruit production in the state.

12.2 CULTIVATION OF OLD DEGENERATED VARIETIES

In the temperate fruit culture of the state, very old varieties still occupy a major share in terms of the area; however, their yield and quality have degenerated over the years. In apples, Red Delicious, Royal Delicious, Maharaji, Golden Delicious, Cox's Orange Pippin, in pears, Kashmiri Nakh and Fertility, in peaches, July Elberta and Quetta, in plums Santa Rosa, in apricots, New Castle, in cherries, Double and Misri, and in strawberries, Senga Sengana and Confitura have the major share. However, among the nut fruits grown in the state, seedling selections form the major share.

TABLE 12.1
Kind-Wise Area and Production of Fruits in J&K (2015–2016)

Kind	Area (ha)	Production (MT)	Productivity (MT/ha)
Apple	161,773	1,966,417	12.16
Pear	14,475	105,935	7.32
Apricot	6,097	14,142	3.32
Peach	2,615	5,953	2.28
Plum	4,279	11,658	2.72
Cherry	2,816	10,244	3.64
Grapes	315	1,299	4.12
Litchi	872	2,264	2.60
Citrus	14,392	33,961	2.36
Olive	707	25	0.04
Mango	12,660	23,856	1.88
Ber	5,390	10,752	1.99
Anola	1,967	3,269	1.66
Guava	2,451	8,530	3.48
Kiwi	7	-	-
Other	10,376	408	0.04
Total fresh	241,182	2,217,584	9.19
Walnut	88,900	266,133	2.99
Pecan	415	81	0.20
Almond	7,132	7,060	0.99
Total dry	96,495	276,415	2.86
Total	337,677	2,493,999	7.39

Source: Anonymous (2016).

TABLE 12.2
Area & Production Trends in Fruits in J&K (Last 6 Decades)

Year	Area (ha)	Prod. (MT)
1953–1954	12,400	16,000
1980–1981	131,008	563,028
1990–1991	176,297	769,949
2000–2001	219,039	931,800
2004–2005	258,311	1,331,861
2005–2006	265,229	1,373,574
2006–2007	283,085	1,492,638
2007–2008	295,141	1,636,203
2008–2009	305,562	1,690,181.5
2009–2010	315,205	1,712.409
2010–2011	325,133	2,221,990
2011–2012	342,791	2,161,169
2012–2013	347,223	1,742,142
2013–2014	355,094	2,073,948
2014–2015	359,088	1,542,676
2015–2016	337,677	2,493,999

TABLE 12.3
Area Expansion Trends in Fruits in J&K (Last 13 Years)

Year	Area (lakh ha)	% Increase
2001–2002	2.21	-
2002–2003	2.32	4.97
2003–2004	2.42	4.31
2004–2005	2.58	6.61
2005–2006	2.65	2.68
2009–2010	3.15	18.87
2011–2012	3.43	8.89
2012–2013	3047	1.30
2013–2014	3.55	2.30
2014–2015	3.59	1.12
2015–2016	3.38	−5.96

These varieties are not high yielders and are highly susceptible to insect-pests, diseases, and inclement weather conditions. In apples, the Delicious varieties have a strong tendency toward alternate bearing after a few years of commercial fruit production. In apples the varieties Starkrimson, Vance Delicious, Oregon Spur, Silver Spur, Super Chief, Fuji, Gala RedLum, and Red Velox are quite promising. In almonds, Merced, Non Pareil, Ne-Plus-Ultra, IXL, Jordanolo are good varieties. In strawberries, Chandler, Camarosa, and Jutogh special are good varieties. The varieties released by SKUAST-Kashmir i.e. Lal Ambri, Sunhari, Akbar, Firdous, Shireen,

TABLE 12.4
Fruit Production Trends in J&K (Last 13 Years)

Year	Production (MT)	% Increase
2001–2002	10.97	-
2002–2003	11.46	4.46
2003–2004	12.73	11.08
2004–2005	13.31	4.56
2005–2006	13.73	3.16
2009–2010	14.93	8.69
2011–2012	21.61	44.74
2012–2013	17042	−19.40
2013–2014	20.74	19.00
2014–2015	15.43	−25.62
2015–2016	24.94	61.66

TABLE 12.5
Fruit Productivity Trends in J&K (Last 15 Years)

Year	Productivity (MT/ha)	% Increase
2001–2002	4.96	16.70
2002–2003	4.94	−0.40
2003–2004	5.26	6.48
2004–2005	5.16	−1.90
2005–2006	5.18	0.39
2009–2010	5.43	4.83
2011–2012	6.30	16.02
2012–2013	5.02	−20.32
2013–2014	5.84	16.33
2014–2015	4.30	−26.37
2015–2016	7.39	71.86

Source: Anonymous (2016).

Shalimar Apple-1, and Shalimar Apple-2 in apples, Waris, Parbat, Makhdoom, and Shalimar in almonds, and Hamdan and Sulaiman in walnuts are high yielders with good quality characters.

12.3　POOR-QUALITY PLANTING MATERIAL

Nursery men are still producing planting material of old degenerated varieties. However, the plant material of new superior varieties is not available in adequate quantity. The bud wood banks of these new, outstanding varieties need to be established to meet the growing demand. The propagation methods of different temperate fruit crops with assured success also need to be standardized, particularly in nut crops.

TABLE 12.6
Area, Production, and Productivity of Fruits in Different Districts of Kashmir Division (2015–2016)

Districts	Area (ha)	Production (MT)	Productivity (MT/ha)
Srinagar	3,293	24,079	7.31
Ganderbal	14,465	80,753	5.58
Budgam	20,408	197,586	9.68
Baramulla	29,964	560,610	18.71
Bandipora	7,690	59,613	7.75
Kupwara	28,460	322,583	11.33
Anantnag	33,627	269,357	8.01
Kulgam	25,568	242,711	9.49
Shopian	26,166	268,050	10.24
Pulwama	24,925	176,835	7.09
Total	214,566	2,223,607	10.36

Source: Anonymous (2016).

TABLE 12.7
Area, Production, and Productivity of Fruits in Different Districts of Jammu Division (2015–2016)

Districts	Area (ha)	Production (MT)	Productivity (MT/ha)
Jammu	12,163	26,013	2.14
Samba	5,869	11,409	1.94
Kathua	14,680	38,562	2.63
Udhampur	11,505	29,430	2.56
Reasi	6,951	13,167	1.89
Doda	9,157	49,248	5.38
Kishtwar	5,953	23,834	4.00
Ramban	7,690	19,743	2.58
Rajauri	14,472	42,867	2.96
Poonch	14,200	26,834	1.89
Total	91,150	281,107	3.08

Source: Anonymous (2016).

12.4 LOW-DENSITY PLANTINGS

Old apple orchards are planted at a distance of 6–7 × 6–7 m or more, accommodating less number of plants per unit of area. In other temperate fruits also, the planting distance is quite higher than required. With the introduction of the concept of high density planting, more plants can be accommodated per unit of area, thus obtaining

TABLE 12.8

Area, Production, and Productivity of Fruits in Different Districts of Ladakh Division (2015–2016)

Districts	Area (ha)	Production (MT)	Productivity (MT/ha)
Kargil	2,067	3,061	1.48
Leh	1,650	7,654	4.64
Total	3,717	10,715	2.88

Source: Anonymous (2016).

TABLE 12.9

Area and Production of Fruits in Different Divisions of J&K (2015–2016)

Divisions	Area (%)	Production (%)
Jammu	35.35	11.27
Kashmir	63.54	88.30
Ladakh	1.11	0.43

Source: Anonymous (2016).

TABLE 12.10

Percentage of Total Area and Production Under Different Fruits in J&K (2015–2016)

Types of Fruits	Area (%)	Production (%)
Sub-tropical	11.53	3.04
Temperate	88.47	96.96

higher yields of better quality fruits. Trees on seedling rootstocks give heterogeneous performance with very less yields (Table 12.14).

Different clonal rootstocks like M-7, M-9 and its clones, MM-106, MM-111, in pear OHxF-34, 51, 69, 87, 230, 233, Oregon 211, 249, in peach Siberian C and Rubira, in plum St. Julien, Pixy, in cherry Gisela 5, 6 and Charger, in apricot Hybd. P 2038 and St. Julein P 6703, can be used in the raising of high density orchards.

12.5   LACK OF POLLINIZERS AND POLLINATORS

Orchardists are not aware about the importance of pollinizers and pollinators and a very low proportion of these (about 8% pollinizers) exist in the orchards, thus affecting fruit yields and quality. Majority of the temperate fruit crops require cross pollination for fruitfulness. When the right pollinizers are missing, it affects the yield

TABLE 12.11
Per Cent Share in Area and Production of Major Temperate Fruits in J&K (2015–2016)

Fruit Crops	% of Total Area Under Fruits	% of Total Production
Apple	47.91	78.85
Pear	4.29	4.25
Apricot	1.81	0.57
Peach	0.77	0.24
Plum	1.27	0.47
Cherry	0.83	0.41
Walnut	26.33	10.67
Almond	2.11	0.28

Source: Anonymous (2016).

TABLE 12.12
Per Cent Share in Production of Major Temperate Fruits by Different Divisions of J&K (2015–2016)

Fruit Crops	Jammu	Kashmir	Ladakh
Apple	2.31	97.41	0.28
Pear	31.41	68.57	0.02
Apricot	31.27	33.87	34.86
Peach	39.85	59.97	0.18
Plum	36.58	63.40	0.02
Cherry	0.00	99.92	0.08
Walnut	34.26	65.69	0.05
Almond	0.08	99.88	0.04

Source: Anonymous (2016).

and quality of the fruits. In areas where the unfavorable climatic conditions coincide with the time of bloom, at least 25% pollinizers should be planted for adequate pollination. Honey bees are the main agents for pollination in the majority of the temperate fruit crops and other pollinators account for 10%–15% of the pollination. These bees should be in adequate numbers at the time of flowering. Placing 3–4 beehives per acre of an orchard at flowering will result in good pollination. Top working of a branch with pollinizers or placement of bouquets of blossoms of a pollinizer variety also facilitate pollination. In apple, for early varieties, the pollinizers are Tydeman's Early Worcester, McIntosh, Black Ben Davis, Allington Pippin, for mid season varieties Red Gold, Golden Delicious, King of Pippin, Spartan, Yellow Newton, Sunhari, Lord Lambourne and for late varieties Granny Smith, Wealthy, Golden Clone B. Crab apples (Snow Drift, Golden Hornet, Manchurian) are pollinizers. In pear; Beurre Hardy, Fertility, Flemish Beauty, Conference are good pollinizers.

TABLE 12.13

Districts With Highest Share in Production of Major Temperate Fruits by Different Divisions of J&K (2015–2016)

Fruit Crops	Districts	% of Total Production
Apple	Baramulla	26.92
Pear	Budgam	19.67
Apricot	Leh	22.55
Peach	Anantnag	9.96
Plum	Poonch	14.99
Cherry	Shopian	28.82
Walnut	Anantnag	17.17
Almond	Pulwama	50.25

Source: Anonymous (2016).

TABLE 12.14

Recommended Rootstocks and Planting Distance for Apple Under Conventional and High Density Plantings

Varieties	Rootstocks	Planting Distance (m)
Non-spur type	Crab or other commercially known rootstock	6×6 to 7×7
Non-spur type	MM-111	6.0
Non-spur type	M-7, MM-106	4.5
Non-spur type	M-9	1.5
Spur type	Seedling	5.0
Spur type	MM-111	3.5
Spur type	M-7, MM-106	3.0

Source: Anonymous (2011).

In almond, for Mukhdoom variety, Shalimar and Waris, for Parbat, Mukhdoom, for Waris, Makhdoom and Shalimar, for Shalimar, Makhdoom and Waris should be used as pollinizers. For other almond varieties, Ne Plus Ultra and Drake are good pollinizers. In cherry, Stella, Compact Stella, Rainier, Sunburst, in plum, Santa Rosa, Beauty, Mariposa and in kiwifruit, Tomuri and Matua should be used as pollinizers.

12.6 POOR CANOPY MANAGEMENT

Canopy management determines the yield and quality of temperate fruits. In stone fruits, trees are mostly trained in open vase system and apple and pears to modified central leader system. However in the state orchards, the systems are not properly developed, thus the full bearing potential of these trees is not fully exploited.

Pruning is faulty as large quantities of wood on which the fruits are formed are removed and water sprouts or criss-cross branches are retained. New training systems like tall spindle, spindle bush, slender spindle, Y-axis, cordon, espalier, HYTEC, dwarf pyramid, tatura trellis etc. have been developed and are found to improve the yield and quality of the fruits. When cutting or heading back a limb larger than 3 cm in diameter, these should be protected with some covering material to avoid rot-causing fungi.

12.7   POOR ORCHARD AND FERTILIZER MANAGEMENT

Scientific orchard management should be adopted so that the trees grow, flower, and set and mature fruits properly. Weeds in tree basins should be kept under check by hand weeding or with the use of herbicides and mulching. Green manuring crops like sunflower and beans can be used for improving the texture and nutritional status of the soil. Judicious nutrient management is necessary for regular production of quality fruits. Many orchards are never fertilized and in the absence of proper nutrition, their yields are very low. The organic matter content of the soils in the majority of the orchards has become very low as no effort is being made to improve it annually to build health status of the impoverished lands. Majority of the orchards are deficient in essential nutrients. The nutrition of other temperate as well as other fruits is more neglected than apple. Recommendations for the application of manures and fertilizers are made by the state agricultural universities and these should be adopted. On the apple and pear trees, deficiency of nitrogen, zinc, boron, manganese, and calcium elements may occur which can be corrected with the application of desired elements.

12.8   RAINFED ORCHARDING

In the temperate region of Jammu and Kashmir, more than 70% of the orchards are rainfed. As the trees do not receive water at the critical periods of their growth, there is considerable reduction in the yield and quality of the fruits. Rainwater harvesting and its application through drip irrigation holds promise in improving yield and quality of the produce. Management practices like mulching with plastic or dry grass, weed control in the tree basins, tree basin management by slightly digging the basins, and foliar sprays of anti-transpirants can improve the plant performance under rainfed conditions.

12.9   OLD AND SENILE ORCHARDS

As the majority of the orchards, particularly apple orchards, are quite old (>30–35 years old), their yields have started declining. They have passed their productive age and have turned senile, producing low grade fruits. These orchards require rejuvenation through top working with improved varieties, corrective pruning, and application of fertilizers. Replanting of these orchards should be carried out with improved varieties under high density followed by scientific orchard management which will ensure higher yields of better quality fruits.

12.10   INSECT-PESTS AND DISEASES

Insect-pests like shothole borer, stem borer, mites, hairy caterpillars, wooly aphids, Sanjose scale, root borer and aphids and diseases like scab, cankers, root rot, collar rot, powdery mildew and gummosis should be managed properly through the recommended practices for improving fruit yields and quality.

12.11   LACK OF GRADING, STORAGE & PROCESSING FACILITIES

In the Jammu and Kashmir, only about 30%–35% of apples and other fruits are of "A" grade, 35%–40% are of "B & C" grades, and 30% are culled. There is also a lack of grading, storage, and processing facilities. About 20%–30% post harvest losses in different temperate fruits also occur causing huge economic loss to the farmers.

12.12   DEPENDENCE ON UNSKILLED LABOR

All the operations in majority of the orchards are carried out by unskilled labor and are not supervised by skilled persons. The operations like training, pruning and preparation, and application of insecticides and fungicides, as well as nutrient sprays should be performed using trained skilled labor.

12.13   LIKELY IMPACT OF CLIMATE CHANGE

During the past 2 decades, the average rise in mean temperature in the Kashmir province was reported to be 1.45 °C, and in the Jammu region it was 2.32 °C. The likely impacts of climate change in Jammu and Kashmir are assumed to be early melting of glaciers, warmer and extended winters, erratic and reduced winter precipitation and snowfall, depletion of ground water and water scarcity, shift in temperate fruit cultivation toward higher altitude and cold arid areas, shift in ecological zones, poor fruit color development, sun burning, fruit cracking, incidence and resurgence of insect and diseases, insufficient fulfillments of chilling hours and heat units which disturbs the bud burst and fruit set, and crop failure of fruit crops which requires high chilling.

12.14   STRATEGIES FOR OPTIMIZING TEMPERATE FRUIT PRODUCTION UNDER CHANGING CLIMATE SCENARIO

The strategies to overcome the impact of climate changes will involve the introduction and evaluation of low chilling cultivars of different fruit crops, use of rootstocks with specific characters, breeding for development of climate resilient varieties, rainwater harvesting and moisture conservation strategies, adoption of insect and disease forecasting system, and technology transfer.

REFERENCES

Anonymous. 2011. Temperate Fruits, Package of Practices. SKUAST-Kashmir, Shalimar (J&K).
Anonymous. 2016. Area and production statement for the year 2015–16. Department of Horticulture. Jammu and Kashmir Government.

13 Impact of Climate Change on Quality Seed Production of Important Temperate Vegetable Crops

Pradeep Kumar Singh
Sher-e-Kashmir University of Agricultural Sciences
and Technology of Kashmir (SKUAST-K)

CONTENTS

13.1   INTRODUCTION

India has a wide spectrum of diverse agro-climatic conditions but vegetable cultivation practices in India have been generally restricted to regional and seasonal needs. Although the production of vegetables has increased to a level of 94 million tons, still the technology used and practices followed are predominantly traditional, resulting in low yields and inconsistent quality and quantity of produce supplies to the markets. In several parts of the country, especially in the northern plains, the soils are highly fertile but temperatures extremes ranging from 0°C to 48°C during a year do not allow year-round outdoor vegetable cultivation. Protected structures modify these extremes to a greater extent to grow vegetables almost year-round. In the upper reaches of the Himalayas, cold desert conditions prevail, where the temperature is extremely low (−5°C to −30°C) during winter season and the region remains cut off from the rest of the country from November to March due to heavy snowfall. It is, therefore, difficult to grow vegetables in such a climate, but some specific protected structures, called poly-trenches, have proved to be very useful for vegetable cultivation as the region receives an abundance of sunshine (Indian Society of Vegetable Science Souvenir, 1998; Phookan and Saikia, 2003). Similarly, in several parts of the country, biotic stresses during the rainy and post-rainy season do not allow successful vegetable cultivation. As a result, most of the vegetables are damaged by the severe incidence of viruses. Protected structures covered with insect proof nets (insect proof net houses) provide a big opportunity of virus-free vegetable cultivation even on a commercial scale. Protected conditions for vegetable cultivation are created by using different types of structures, which are season and location specific. These structures are designed as per the climate modification requirement of the area. Temperature, humidity, wind velocity, soil conditions, etc. also play a major role in the design of protected structures for growing vegetable crops, but there are several constraints and problems which restrict protected cultivation of vegetable crops (Rai et al., 2004; Singh, 1998; Singh et al., 1999; Anonymous, 2002). It is, therefore, necessary to work out the best-suited design of protected structures for different climatic conditions along with their techno-economic feasibility.

13.2 PRINCIPLE OF GREENHOUSE

A greenhouse is generally covered with a transparent material such as polythene or glass. Depending upon the cladding material and its transparency, a major fraction of the sunlight is absorbed by vegetable crops and other objects. These objects in greenhouse in turn emit long wave thermal radiations for which cladding material has lower transparency, with the result that solar energy is trapped (ENVIS Bulletin, 2004) and raises the temperature inside the greenhouse. This is popularly known as greenhouse effect. This rise in temperature in the greenhouse is responsible for growing vegetables in cold climates. During the summer months, the air temperature in the greenhouse is to be brought down by providing a cooling device. In commercial greenhouses, besides temperature-controlled humidity, carbon dioxide, photo-period, soil temperature, plant nutrients, etc. facilitate round the year production of the desired vegetable crops. Controlled climatic and soil conditions provide an opportunity to the vegetable crops to express their yield potentials.

13.2.1 Popularization of Off-Season Vegetable Production

There is a lot of potential in increasing the area under low-cost greenhouses in peri-urban areas of the valley for production of high-value vegetables during the off-season to take benefit of the produce. High altitude areas of the valley are the potential areas for the production of off-season vegetables for the plains. These areas need to be identified and involved for the production of off-season vegetables as off-season vegetables have a special significance in various metropolitan cities of the country because of their quality in terms of aroma, flavor, freshness, prolonged shelf-life, etc.

13.2.2 Post-Harvest Management

Because of the higher degree of perishability of vegetables, their poor shelf life calls for conducting vital research on efficient post-harvest management. It also reduces the gap between production and net availability. The 30% loss of vegetables in post-harvest handling needs to be minimized. Also, standardized proper packaging of different commodities of vegetables is of urgent need. Pesticide residue management and new product development will add value to the produce. There is also a need for a bulk handling system of vegetables including processing, controlled atmospheric storage, and modified atmospheric packaging.

13.2.3 Organic Farming

Organic farming systems differ fundamentally from commercial ones on management practices that promote and enhance ecological harmony. A survey report of organic vegetable growers of USA indicated that organic vegetable production is less experienced than conventional vegetable growing. There is a need to develop vegetable genotypes suited to organic production as well as standardizing the technology of organic production for different vegetables.

13.2.4 MICRO-IRRIGATION

Micro-irrigation is a system that provides high frequency application of water in and around the root zone of the plant with the help of emitters/drippers. In India, more than 70,000 ha of land area has already been covered in the states of Maharashtra, Tamil Nadu, Karnataka, and Andhra Pradesh. The system has the potential to produce significant energy and water savings along with improved crop quality and yield. Micro-irrigation vegetables are fast growing, and regions and most of the root systems are shallow. Thus, vegetables are very sensitive to water stress. Also, due to spaced planting, micro-irrigation may be an effective, efficient, and economically viable method for the irrigation of vegetable crops. The experiments on vegetables indicate that micro-irrigation can save considerable water with increase in the productivity and quality of the produce.

13.2.5 BIOTECHNOLOGY

Novel techniques or biotechnological tools form an important alternative to the traditional plant breeding in achieving the goals and targets of vegetable research in less time. Biotechnology has made a vital contribution in maintaining and characterizing biological and genetic diversity and their conservations. Direct isolation of genes and their incorporation in the elite cultivars of vegetables provide an additional and rapid approach to traditional methods of plant breeding. Nowadays, genetically engineered plants with genes of direct interest can be produced in a relatively shorter time and would be of direct value in the vegetable research. The long-term beneficial impact of biotechnology on vegetable production will be realized both directly and indirectly. It can also be applied for improving the sensory properties and shelf life of vegetables.

13.3 BENEFITS OF GREENHOUSE

13.3.1 VEGETABLE FORCING FOR DOMESTIC CONSUMPTION AND EXPORT

During winters in Northeast Himalayan (NEH) region, the temperature and solar radiations are sub-optimal for growing off-season vegetables, namely tomato, capsicum, brinjal, cucumber, okra, and chilli. In tomatoes, low temperature and low radiation cause puffiness and blotchy ripening. Hence during extreme conditions of the winter season (October–February) these vegetables will be cultivated under polyhouse. In a medium cost greenhouse, the yield of tomato and capsicum can be taken as 98.6–110.5 and 87.2 tons/ha, respectively. The protected environments would be well adapted in the field where winter is prolonged. A polyhouse can be made which will receive sunlight for growing chilli, tomato, brinjal, capsicum, and cucumber. The improved varieties and hybrids of these crops would be evaluated. The high-priced vegetables: asparagus, broccoli, leek, tomato, cucumber, and capsicum are the most important crops for production around metropolis and big cities during winter season or off-season. Thus, in the NEH region during winter it may be useful to grow tomato and capsicum in plastic tunnels as the plants which are protected from cold and frost will manifest faster and better growth resulting in earlier fruiting than the crops grown in the open.

13.3.2 RAISING OFF-SEASON NURSERIES

The cost of hybrid seeds is very high. So, it is necessary that every seed must be germinated. For 100% germination, it requires controlled conditions. The cucurbits are warm season crops. They are sown in the last week of March to April when the night temperature is around 18°C–20°C. But in polyhouses their seedlings can be raised during December and January in polythene bags. By planting these seedlings during the end of February and the 1st week of March in the field, their yield could be taken in one and a half months in advance than the normal method of direct sowing. This technology fetches a bonus price due to the marketing of produce in the off-season. Similarly, the seedlings of tomato, chilli, capsicum, brinjal, cucumber, cabbage, cauliflower, and broccoli can be grown under plastic cover protecting them against frost, severe cold, and heavy rains. The environmental conditions, particularly the increase in temperature, inside a polyhouse hastens the germination and early growth of warm season vegetable seedlings for raising early crops in spring-summer. Vegetable nursery raising under protected conditions is becoming popular throughout the country, especially in hilly regions. Management of vegetable nursery in protected structures is easier and early nursery can be raised. Needless to emphasize, this practice eliminates the danger of destruction of nurseries by hailstorms and heavy rains because the world's highest rains occur in this region and the period of the rainy season is also wide (April to October). Protection against biotic and abiotic stresses becomes easier.

13.3.3 PROTECTIVE STRUCTURES FOR SEEDLING PRODUCTION

Seedlings need care and nourishment and a protected enclosure is necessary to grow healthy and quality seedlings. Vegetable seedlings are being grown in low-cost polyhouses, net houses, cloches/low tunnels, cold frames, hot beds, lath houses, etc. which provide control of growing conditions, creating a micro environment congenial for the propagation and cultivation of vegetable crops.

13.3.4 HOT BEDS

The main objective of a hot bed is to raise seedlings earlier and protect them from weather hazards. A hot bed is one where heat is generated by the decomposition of fresh manure. The heat generated is utilized for seed germination, which results in early nursery raising, early supply of vegetable produce in the market, and more profits. First of all a trench 2 ft deep, 3 ft wide, and 6 ft long is prepared. The frame generally made of wood is filled in such a way that from the back side it extends up to 30–35 cm and from the front side 20–25 cm above the ground. The sides of the frame are covered with paddy straw to prevent the loss of heat. The trench is first filled with fresh manure up to 25–30cm in two layers, each separated with a layer of straw, followed by a 10–12cm thick layer of straw, followed by 10–12cm thick layer of light soil. The top of the frame is filled with polythene lined lids, used during night and rains.

13.3.5   CLOCHES/LOW TUNNELS

Cloches or low tunnels are also used for raising vegetable seedlings under unfavorable weather conditions. These cloches or tunnels are made curved and are covered with polythene. The end of theses cloches/tunnels can be closed with polythene sheets as per the climatic requirement. Cloches prevent both hardening and freezing of the land, thereby helping in sowing of the seeds earlier and when desired.

13.3.6   THATCHES

Thatches are traditional structures used to protect the vegetable nurseries from unfavorable weather conditions during both the winter and summer seasons. In winter, thatches are erected in a slanting manner at 45° angle from ground level and are oriented in south west directions. The slanting roof is covered with paddy straw or straw mats. The shade is removed when the seedlings have come up and have attained 1cm height with two or four leaves.

13.3.7   SEED PANES/BOXES

Seed panes/boxes are used to raise delicate kind of seeds. Seed panes are shallow, earthen pots about 4in. high and 14in. in diameter at the top, with a single hole at the bottom. Seed boxes are made of wood, 16in. wide, 24in. long and 3–4in. deep with 6–8 holes drilled at the bottom for effective drainage. Gravel stones or wood charcoal may also be put on the bottom of both panes and boxes to ensure proper and regular drainage. Then these panes/boxes are filled with fine soil up to the desired depth and the seeds are sown.

13.3.8   POLYBAG NURSERY RAISING

Nursery raising of cucurbits in polybags under protected structures is highly remunerative. Polybags of desired length (200 gauge, 20 × 10cm size) are taken and are perforated to ensure proper drainage. The polybag mixture consists of two parts garden soil, one part sand, and one part farm yard manure. The polybags are filled with this mixture upto the desired height leaving some space empty and then seeds are sown. Perforations are provided on all sides to ensure proper drainage and aeration.

13.3.9   POLYHOUSE/NET HOUSE

In a polyhouse, the main frame can be of steel pipes and wooden poles of 6 to 8ft height. It can also be erected with stone pillars to reduce the expenditure. Polyhouses are covered with 200µm ultra violet (UV) stabilized polyethylene film on the roof and the sides are covered with 40 mesh insect-proof nylon net. A refractable shade net is provided to bring down the temperature during the summer days. In a net house, stone pillars are erected as a main frame and the roof is generally covered with a shade net instead of polysheet. The sides are covered with the insect proof net. However, it is advisable to cover the roof also with the insect proof net above the retractable shade net to have better control over the entry of insect vectors like white flies. It is essential to harden the seedlings before transplanting. A retractable shade net will be useful to regulate the shade in the greenhouse depending on the light levels. Plastic pipes of ¾ inch are bent in an arch shape over the nursery beds and are covered with a plastic sheet to protect the seedlings from rain in a net house. The polyhouse and net house structure provides adequate light, shade, and humidity. It protects the seedlings from thrips and white flies which spread viral diseases. Farmers can also grow vegetable seedlings in plastic trays on a small scale in their farms in a low-cost net house measuring about 20 ft long, 10 ft wide, and 8 ft high.

Based on the type of cladding material used in covering the installed structures, the polyhouse can be broadly divided into the following groups:

1. Fiber-glass polyhouse.
2. Single or double polyethylene film polyhouse
3. Ordinary glass house
4. Poly-carbonate house
5. Ultraviolet stable polyethylene film house

13.3.10 SEEDLING TRAYS

Seedling trays are also called as pro-trays (propagation tray) or flats, plug trays or jiffy trays. The most commonly used are 98 celled trays for tomato, capsicum, cabbage, cauliflower, chilli, brinjal, and bitter gourd. The dimensions are 54 cm in length and 27 cm in breadth with a cavity depth of 4 cm. Trays are made of polypropylene and are reusable. The life of the tray depends on the handling. Before using every time it is necessary that these trays are thoroughly washed and disinfected with a fungicide. The holes at the bottom of the cells control the moisture properly. Equally spaced cells facilitate equal growth of the seedlings. Seedling trays have been designed in such a way that each seedling gets the appropriate quantity of growing media and the right amount of moisture. Trays have pre-pinched holes to each cavity for proper drainage of excess water and also have right spacing to facilitate equal growth of the seedlings.

13.4 INGREDIENTS USED AS A MEDIA FOR GROWING TRANSPLANTS

Well decomposed and sterilized medium is essential to grow disease-free seedlings. The traditional potting mix of soil manure sand has been replaced over the years by peat vermiculite sand or Perlite mix. The most commonly used growing medium is coco peat and it retains the optimum amount of moisture to support seed germination.

Coco peat is a byproduct of the coir industry and it has high water holding capacity. Neem cake (100 kg) and Trichoderma (1 kg) are added per ton of the coco peat to prevent seedling diseases. Vermicompost is also used as a growing medium in place of coco peat. These ingredients are mixed in a 3:1:1 ratio before filling the trays.

13.4.1 ORGANIC PRODUCTS

Sphagnum moss or peat moss contains 80%–90% organic matter, 4%–20% ash with a cation exchange capacity (CEC) of 60–120 meq/100g. Peat humus or brown peat contains 50% organic matter, 5.05% ash with a CEC of 250–350 meq/100g. Organic wastes like saw dust, pine bark, pine chips, paddy husk, and coir dust (coco peat) can be used to produce organic product.

13.4.2 INORGANIC PRODUCTS

Vermiculite: Holds and releases large quantities of water which reinforce similar properties of peat when mixed with it. It is neutral in reaction and has a relatively high CEC of 80 meq/100g, reducing the leaching of nutrients. The disadvantage is its high cost and early breakdown leading to compression of the substrates.

Perlite: Totally inert, has low CEC or buffering capacity and low water content. It provides air space to the medium, is neutral in pH, very light in weight, and is a good temperature stabilizer. The disadvantages are Aluminium (Al) toxicity in some seedlings at low pH, and limited capacity for water supply under conditions of high transpiration.

Sand: Porosity around 40% of the bulk volume, particle size 0.5 to 2 mm in diameter, contains no nutrient and has no buffering capacity. The CEC is 5.50 meq/100g. It is used together with organic materials.

Synthetic Products: Glass wool, polyurethane foam, etc. can be used to grow seedlings. The low pH of the substrate can be adjusted by adding lime (calcium carbonate) and dolomite (Ca – Mg) carbonate and with basic fertilizers like calcium nitrate, sodium or potassium nitrate. The high pH is adjusted by the adding of sulphur, gypsum, Epsom salt, and acidic fertilizers like urea, ammonium sulphate, ammonium nitrate, ammonium phosphate, and acids like phosphoric acid and sulphuric acid. Poor aeration of a highly decomposed black peat or of clay soil, which on the other hand has an appreciable water retention capacity, can be corrected by adding materials such as sand, polystyrene, perlite, or expanded clay in which the common characteristic is to increase aeration. Sand and peat or coir dust mixture, or peat and vermiculite or perlite are supplemented with balanced fertilizers from the best medium to grow seedlings.

Disinfection of the Medium: Seedlings are very vulnerable to soil-borne diseases and for seed and potting composts it is worthwhile sterilizing the soil before mixing the other ingredients of the compost. The soil of nursery beds (flat and raised beds) is disinfected by solarization or the beds and substrates of growing medium are sterilized with steam or formaldehyde to control soil/medium-borne diseases. Diseases such as Damping off can be arrested by chestnut compound. Pots and boxes should also be washed and dipped in a good disinfectant.

13.5 METHODS OF SEEDLING RAISING

Fill the seedling tray with an appropriate growing medium.

- Make a small depression (0.5 cm) with a fingertip, in the center of the cell sowing. Alternatively, a depression can be created by stacking 10 trays one over the other and pressing the trays together.
- Sow one seed per cell and cover with the medium.
- No irrigation is required before and after sowing if coco peat having 300%–400% moisture is used.
- Keep 10 trays one over the other for 3–6 days, depending on the crops. Cover the entire stack with polyethylene sheet. This ensures conservation of the moisture until germination. No irrigation is required till the emergence of the seedlings. Care must be taken while spreading the trays when the seedlings are just emerging, otherwise the seedlings will get etiolated.
- Seeds start emerging after about 3–6 days of sowing depending upon the crops. Shift the trays to a polyhouse or net house and spread them over a bed covered with polyethylene sheet.
- The trays should be irrigated lightly every day depending upon the prevailing weather conditions by using a fine sprinkling rose can or with a hose pipe fitted with rose. Never over-irrigate the trays, as it results in leaching of nutrients and building up of diseases.
- Drench the trays with a fungicide as a precautionary measure against seedling mortality.
- The media may need supplementation of nutrients if the seedlings show deficiency symptoms. Spray 0.3% (3 g/L) of 100% water soluble fertilizer (19 all with trace elements) twice (12 and 20 days after sowing).
- Protect the trays from rain by covering with polyethylene sheets in the form of a low tunnel.
- Hardening the seedlings by withholding irrigation and removing the shade for a week before transplanting.
- The seedlings will be ready in about 21–42 days for transplanting to the main field depending upon the crops.

13.5.1 ADVANTAGES OF SEEDLINGS PRODUCTION IN TRAY

- Seeds germinate properly and mortality of the seedlings is negligible.
- No loss of expensive seeds of hybrids.
- Adequate space for each seedling to grow properly.
- Damage due to pests and diseases is very rare.
- Promotes better root growth.
- Transplanting shock is negligible.
- Easy to handle and transport.
- Seedlings do not wither during transport.
- Uniform growth in nursery ensures better establishment and growth of plants in the main field.

13.5.2 Vegetable Seed Production

Seed production in vegetables is the limiting factor for cultivation of vegetables in the J&K region of India as well as in India. The vegetables require specific temperature and other climatic conditions for flowering and fruit setting. Seed production of brinjal, capsicum, cauliflower, and broccoli is very difficult in open conditions in this area due to high rainfall at the maturity stage. To reduce such micro-climatic conditions, a protected environment is essential. Therefore, the seed production of highly remunerative crops, namely tomato, capsicum, and cucumber is performed under protected environments. The maintenance and purity of different varieties/ lines can be achieved by growing them under greenhouse without giving isolation distance particularly in cross-pollinated vegetables, namely onion, cauliflower, and cabbage. Hence vegetable production for domestic consumption and export in low and medium cost greenhouse is a technical reality in India. Such a production system has not only extended the growing season of vegetables and their availability, but has also encouraged the conservation of different rare vegetables.

13.5.3 Hybrid Seed Production

In the 21st century, protected vegetable production is likely to be commercial practice not only because of its potential, but also out of sheer necessity. In vegetable production, hybrid seeds, transgenic, stress resistant varieties, micro propagated transplants, synthetic seeds are likely to replace the conventional varieties. Protected environments will be helpful in the production of hybrid seeds of cucumber and summer squash by using gynoecious lines. Gibberlic acid is used to maintain such lines followed by selfing. The desired pollen can be used for the production of hybrid seed of cucumber. Similarly, in summer squash, use of Ethaphone in inducing female flower at every node would help in the hybrid seed production by using desired pollen parent.

13.5.4 Maintenance and Multiplication of Self-Incompatible Line for Hybrid Seed Production

In the case of cauliflower, there is a problem of maintaining and multiplication of potential self-incompatible lines for the production of F_1 hybrid seed. Temporary elimination of the self-incompatibility with the use of CO_2 gas has solved this problem. For this purpose, the self-incompatible line is planted in a greenhouse and bees are allowed to pollinate the crop when it is in bloom. Then keeping the greenhouse closed tightly within 2–6 hours of pollination, it is treated with 2%–5% CO_2 gas which allows successful fertilization by temporarily eliminating the self-incompatibility.

13.5.5 Polyhouse for Plant Propagation

Asparagus, sweet potato, pointed gourd, and ivy gourd are sensitive to low temperature. The propagating materials of these vegetables can be well-maintained under polyhouse in winter season before planting their cuttings in early spring summer season for higher profit.

TABLE 13.1

Approximate Area (ha) Under Greenhouses

Country	Area
Japan	54,000
China	48,000
Spain	25,000
South Korea	21,000
Italy	18,500
Turkey	10,000
Holland	9,600
USA	4,000
Israel	1,500
India	525

Source: ENVIS Bulletin (2004).

13.5.6 STATUS

Commercial greenhouses with climate controlled devices are very few in the country. Solar greenhouses comprising glass and polyethylene houses are becoming increasingly popular both in temperate and tropical regions. In the early sixties, the Field Research Laboratory (FRL) of Defence Research and Development Organisation (DRDO) at Leh attempted solar greenhouse vegetable-production research and made an outstanding contribution to the extent that almost every rural family in the Leh valley possesses a polyhouse these days. Indian Petro Chemical Corporation Ltd (IPCL) boosted the greenhouse research and application for raising vegetables by providing UV stabilized cladding film and Aluminum polyhouse structures. Several private seed production agencies have promoted greenhouse production of vegetables. In comparison to other countries, India has very little area under greenhouses as is evident from Table 13.1.

The major share has been in the Leh & Ladakh region of Jammu and Kashmir where commercial cultivation of vegetables is being promoted. In NEH region, polyhouse cultivation is still a new emerging technology for raising nursery of vegetable crops. Assistance provided under the plasticulture scheme since the VIII & IX plan has helped in generating awareness about the importance of greenhouses in enhancing productivity and production, particularly of horticultural crops.

13.6 TYPES OF GREENHOUSE/POLYHOUSE

13.6.1 LOW-COST GREENHOUSE/POLYHOUSE

The low-cost polyhouse is a zero-energy chamber made of polythene sheet of 700 gauge supported on bamboos with sutli (ropes) and nails. It will be used for protecting the crop from high rainfall. Its size depends upon the purpose and availability of space. The structure depends on the sun for energy. The temperature within the polyhouse increases by 6°C–100°C more than outside. In a UV stabilized plastic-film covered pipe framed polyhouse, the day temperature is higher and the night temperature is lower than outside. The solar radiation entering the polyhouse is 30%–40% lower than that reaching the soil surface outside.

13.6.2 MEDIUM-COST GREENHOUSE/POLYHOUSE

With a slightly higher cost, a Quonset-shaped polyhouse (greenhouse) can be framed with a galvanized iron pipe (class B) of 15 mm bore. This polyhouse will have a single layer covering of UV stabilized polythene of 800 gauge. Thermostatically controlled exhaust fans are used for ventilation. A cooling pad is used for humidify-ing the air entering the polyhouse. The polyhouse frame and glazing material have a life span of about 20 years and 2 years, respectively.

13.6.3 HIGH-COST GREENHOUSE/POLYHOUSE

It is constructed on the structure (frame) made of iron/aluminum, designed as dome-shaped or cone-shaped (as per choice). Temperature, humidity, and light are auto-matically controlled as per the requirement of the users. The floor and a part of the walls are made of concrete. It is highly durable, about 5–6 times costlier, requires a qualified operator, and needs proper maintenance, care, and precautions while oper-ating. The low and medium-cost greenhouses have a wide scope in the production of domestic as well as export-oriented vegetables. NEH region records the highest rainfall in the world. The duration of the rainy season is also wide (April–October). During this period, the growing of vegetables such as cabbage, cauliflower, broccoli, tomato, brinjal, and French bean in open conditions is very difficult. Severe attacks of pests and diseases occur due to heavy rains. So, the growing of vegetable crops in low-cost polyhouses during this period is very profitable. Control of diseases and pests in polyhouses is also easy.

13.7 OTHER PLANT-PROTECTION STRUCTURES

13.7.1 PLASTIC LOW TUNNELS

Plastic low tunnels are miniature forms of greenhouses to protect the plants from rains, winds, low temperature, frost, and other vagaries of weather. These low tunnels are very simple structures requiring very limited skills to maintain, are easy to construct, and offer multiple advantages. For construction of low tunnels, a film of 100 μm would be sufficient. The cost of a 100 μm thick film would be about Rs. 10/m².

13.7.2 NET HOUSES

Net houses are used for raising vegetable crops in high rainfall regions. The roof of the structure is covered with a suitable cladding material. The sides are made of wire mesh of different gauges. Such structures are useful for NEH region.

13.7.3 CONSTRAINTS IN PROTECTED VEGETABLE PRODUCTION

In NEH region, the polyhouse culture is in infant stage and has not become popular as yet. High cost and non-availability of various components are the two major limiting factors in the adoption of polyhouse technology for commercial cultivation.

Many of the polyhouse components like fiber glass, cooling pads, fans, etc. have to be imported at high costs including freight and custom duty. The design of greenhouse and other structures for different agro-climatic of the region is not standardized. A lack of awareness among farmers pertaining to the potentials of protected vegetable production and lack of major research programs on protected vegetable farming are the other limiting factors.

13.7.4 PROSPECTS OF PROTECTED CULTIVATION OF VEGETABLES IN INDIA

There are a number of opportunities in the various agro-climatic zones in India for protected vegetable production. In temperate areas, vegetable growers can use low-cost protected structures for raising early crops to increase their income. Raising of vegetable nursery in protected structures has several benefits such as ease in management, off seasonality, and protection from biotic and abiotic stresses. Plastic low tunnel technology for off-season cultivation of cucurbits in the northern plains of the country has great potential for the future. Walk-in tunnels can be used on a large scale for growing off-season nursery and vegetables in the northern plains or hilly areas during winter months when the temperature is very low. For virus and pesticide-free tomato, chilli, sweet pepper, and okra cultivation during rainy season or late rainy season, insect proof net houses can be used on a large scale. Low-cost greenhouses can be used for protected vegetable cultivation in peri-urban areas of the country as consumer awareness for good nutrition and quality vegetables free from pesticides is increasing day by day. Protected vegetable cultivation is a boon to the cold desert areas of the country where this technology is suitable for the commercial cultivation of several vegetables, and for production during freezing winters when nothing can be produced outdoors.

The potential of protected vegetable cultivation to meet the demand should not be overlooked. Protected cultivation provides many advantages over open field vegetable cultivation. This technology is highly productive, amenable to automation, conserves water, fertilizer, and land. It is also eco-friendly and does not require much sophistication. In this century, protected vegetable cultivation is likely to be a common commercial practice, not because of its potential but out of its sheer necessity.

13.8 FUTURE THRUST

1. **Generation of a sound and accurate database**

   Vegetable production and development did not receive priority in agricultural planning and consequently the progress in improving vegetable productivity was slower than for cereals. There is an urgent need for the generation of a sound and accurate database on area, production, productivity, nutritional status, export, and marketing of vegetables and its utilization for proper planning and execution of the objectives.

2. **Strengthen research extension linkage**

   Efforts should be made by the development departments to transfer technology/research findings rapidly to the farmers through tight linkage between research and extension services in order to promote the adoption of new technology and varieties for higher productivity.

3. **Zonalization of the seed producing areas**

   To raise a huge quantity of quality seeds for home consumption and export to the neighboring countries, efforts should be made to improve the seed yield and quality by adopting scientific methods. The development departments should identify and perform:

   • Zonalization of the seed producing areas for a particular crop/variety to produce genetically pure seeds by providing natural isolation.
   • The seed act of 1966 should be strictly enforced for producing quality seeds.

4. **Identification of new areas for vegetable cultivation**

   Development department should identify new areas having the potential for vegetable cultivation and encourage development of vegetable belts/villages for organized production and marketing (Figure 13.1 and Table 13.2)

5. **Timely availability of quality seeds and inputs**

   Vegetable cultivation is basically a time bound program. The development department should ensure timely supply of inputs like seeds, fertilizers, and pesticides so that the program is not affected. The cropping program starts in the month of June, so the development department should devise a strategy to supply the inputs in time so that the program is not affected.

6. **Identify growers for voluntary multiplication of various seeds**

   The state department has limited resources and cannot cope with the increasing seed demand. A survey should be made to identify growers to take up the seed multiplication.

7. **Provision of financial support for various inputs**

   Vegetable cultivation is basically an input-intensive venture. Price hikes of fertilizers and improved seeds restrain the farmers from purchasing and applying the required fertilizers in proper ways and in appropriate time. The development department, besides making inputs available in time, should provide the necessary financial support to small and marginal farmers for increasing the infrastructural facilities like greenhouse, water pumps, etc. to boost the vegetable production.

8. **Regulation of pesticide/fertilizer trade**

   Sale of spurious pesticides/fertilizers has become one of the major problems. It becomes imperative on the development department to strengthen the quality control and the enforcement agency so that the effective and more toxic pesticides with long residual effects do not find a place in the market.

FIGURE 13.1 Different protected structures used for early nursery raising.

TABLE 13.2
Cropping Sequence Under Polyhouse

Pot cultivation of cabbage kale, lettuce, mint, coriander, methi

Jan	Feb After 15	Mar	Shelves	April	May	Jun	Jul ending	Aug	Sep	Oct	Nov after 3rd week sowing of leafy vegetable/peas	Dec
Spinach (+)	Nursery raising of tomato	+		Tomato	+	+	+	+	+	+	Spinach	+
Spinach (+)	Nursery raising of Chilli	+		Tomato	+	+	+	+	+	+	Spinach	+
Fenugreek (+)	Nursery raising of Capsicum	+		Tomato	+	+	+	+	+	+	Fenugreek	+
Fenugreek (+)	Nursery raising of Brinjal	+		Capsicum	+	+	+	+	+	+	Fenugreek	+
Orach (+)	Nursery raising of Cauliflower	+		Capsicum	+	+	+	+	+	+	Orach	+
Orach (+)	Nursery raising of Knol khol	+		Capsicum	+	+	+	+	+	+	Orach	+
Coriander (+)	Nursery raising of Cabbage	+		Cucumber	+	+	+	+	+	+	Coriander	+
Coriander (+)	Nursery raising of cherry tomato			Cucumber	+	+	+	+	+	+	Coriander	+
Mallow/Hand (+)	Nursery raising of Knol khol	+		Cucumber	+	+	+	+	+	+	Mallow/Hand	+

Raising of Cucurbits in poly bags/Nunar, Handh

Raising of cucurbits in poly bags

Pot cultivation of cabbage kale, lettuce, mint, coriander, methi

(Continued)

TABLE 13.2 (Continued)
Cropping Sequence Under Polyhouse

Broccoli (+)	Nursery raising of Cauliflower	+	Early Beans/Cherry tomato	+	+	Nursery raising of broccoli for late season	+	Transplant of broccoli	+	+	+
Lettuce (+)	Nursery raising of Cucurbits	Nunar	Early Beans/Cherry tomato	+	+	Nursery raising of lettuce for late season	+	Transplant of lettuce	+	+	+
Cauliflower/cabbage (+)	Nursery raising of Cucurbits	Handh	+		+	Nursery raising of cabbage/cauliflower for late season	+	Transplant of cauliflower/cabbage	+	+	+

Pot cultivation of cabbage kale, lettuce, mint, coriander, *methi*.

13.9 CONCLUSIONS

Though India is the largest producer of vegetables in the world, next only to China, its requirements of vegetables are rapidly increasing because of its continuously increasing population. Adverse climatic conditions, horticultural crop potentials, agro-inputs availability, small land holdings, and increasing demand of high-quality vegetables necessitate adoption of the protected cultivation of vegetable crops. Low-cost protected structures, viz., plastic low tunnels, walk-in tunnels, low-cost greenhouses, are suitable for off-season vegetable cultivation and nursery raising in the major vegetable growing areas and the peri-urban areas of the country. Insect proof net houses are also highly suitable for raising virus-free healthy seedlings and growing pesticide-free vegetables during rainy and post rainy season. Poly trenches are best suited to vegetable cultivation in the cold deserts of the country. Increased productivity and off-season ability under protected conditions favor its early adoption in the peri-urban areas of the northern plains and the hilly areas of India.

REFERENCES

Anonymous. 2002. Indian Council of Agricultural Research. Agricultural Research Data Book, ICAR, 2004.

ENVIS Bulletin. 2004. ENVIS Bulletin Himalayan Ecology 12(2), ISSN: 0971-7447 (Print), Centre on Himalayan Ecology with financial support from the Ministry of Environment, Forest & Climate Change (MoEF & CC, Government of India).

Indian Society of Vegetable Science Souvenir. 1998. National Symposium Dec 12–14, Varanasi, U.P., India, p. 90.

Phookan, D.B. and Saikia, S. 2003. Vegetable production under naturally ventilated plastic house cum rain shelter. In *Plasticulture Inervantion for Agriculture Development in North Eastern Region*, Edited by K.K. Satapathy and A. Kumar, pp. 127–141.

Rai, N., Nath, A., Yadav, D.S. and Patel, K.K. 2004. February. Effect of polyhouse on shelf-life of bell pepper grown in Meghalaya. In National Seminar on Diversification of Agriculture through Horticultural Crops, IARI Regional Station, Karnal, pp. 21–23.

Singh, B. 1998. December. Vegetable production under protected conditions: Problems and Prospects. In Indian Society of Vegetable Science Souvenir: Silver Jubilee, National Symposium Dec (Vol. 12, No. 14, p. 1998).

Singh, N., Diwedi, S.K. and Paljor, E. 1999. *Ladakh Mein Sabjion Kei Sanrakshit Kheti. Regional Research Laboratory of DRDO*, DRDO, Leh, p. 56.

14 Climate Change, its Impact and Mitigation Strategies with Reference to Vegetable Crops

Nayeema Jabeen and Ajaz A. Malik
Sher-e-Kashmir University of Agricultural Sciences
and Technology of Kashmir (SKUAST-K)

CONTENTS

14.1 INTRODUCTION

Worldwide, over 527.05 lakh ha of vegetables were harvested in 2007 for the total production of over 9,088.38 lakh tons (FAOSTAT, 2016). This represents a significant portion of the dietary needs of the world's human population. Vegetables are considered as "protective supplementary food" as they contain large quantities of minerals, vitamins, and essential amino acids which are required for the normal functioning of the human metabolic processes. They are important to neutralize the acids produced during digestion and are also useful as roughage. Peas and beans are enriched with proteins and tuber crops like potato are well-known as sources of carbohydrates. Calcium, phosphorus, and iron are the important minerals which are lacking in cereals and are available in abundant quantities in vegetables like peas, beans, spinach, and bhindi. Amaranthus, cabbage, and beans contain large quantities of cellulose which aids digestion. All the leafy and fruit vegetables possess the required quantities of vitamins. Above all, most of the vegetable crops possess high medicinal value in curing certain diseases which have now drawn the attention of plant biochemists and pharmacologists to manufacture medicines of biotic origin. In addition, many vegetables have been associated with protective effects against cancer and heart diseases (Table 14.1).

A significant variation, either in the state of climate or in its variability, persisting for an extended period is referred to as climate change (IPCC, 2001). Climate change, in Intergovernmental Panel on Climate Change (IPCC) usage, refers to a change in the state of the climate that can be identified (e.g., using statistical tests) by changes in the mean and/or the variability of its properties and that persists for an extended period, typically decades or longer. The United Nations Framework Convention on Climate Change (UNFCCC), in its Article 1, defines climate change as: "a change

TABLE 14.1

Vegetable Profile in South Asia (Year 2007)

Country	Area (lakh ha.)	Production (lakh tons)
India	59,040	77,243
China	23,717.2	4,516.33
Bangladesh	4.53	31.97
Nepal	2.08	23.75
Pakistan	4.19	55.09
Srilanka	0.81	7.34

Source: faostat.fao.org.

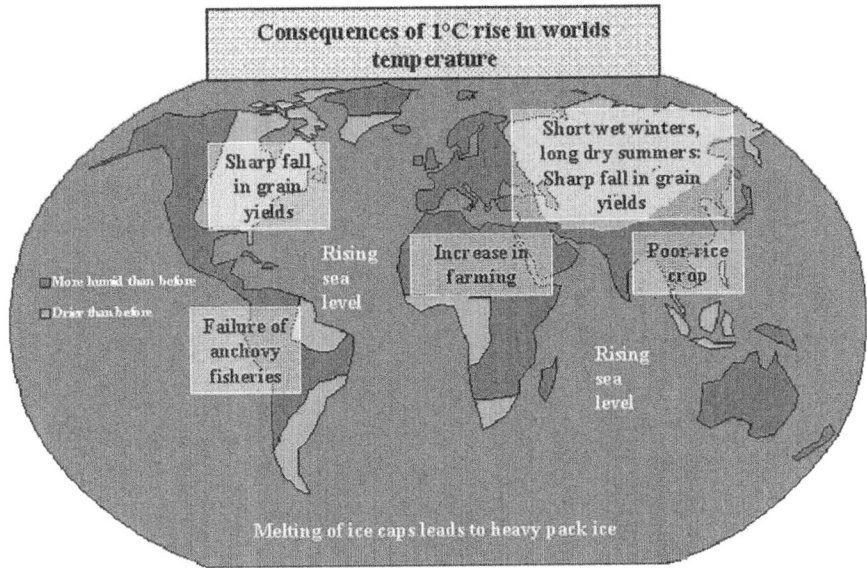

FIGURE 14.1   Consequences of 1°C rise in world temperature (Adapted from University of Southampton, 2000).

TABLE 14.2
Climate Change Likely on Key Parameters

		Temperature Change (°C)		Rainfall Change (%)	
Year	Season	Lowest	Highest	Lowest	Highest
2020s	Annual	1.00	1.41	2.16	5.97
	Rabi	1.08	1.54	−1.95	4.36
	Kharif	0.87	1.17	1.81	5.10
2050s	Annual	2.23	2.87	5.36	9.34
	Rabi	2.54	3.18	−9.22	3.82
	Kharif	1.81	2.37	7.18	10.52
2080s	Annual	3.53	5.55	7.48	9.90
	Rabi	4.14	6.31	−24.84	−4.50
	Kharif	2.91	4.62	10.10	15.18

Source: Lal M, 2001, *Current Science* 81, 1205.

of climate which is attributed directly or indirectly to human activity that alters the composition of the global atmosphere and which is in addition to natural climate variability observed over comparable time periods" (Figure 14.1 and Table 14.2). Over the 20th century there has been a consistent and large scale warming of the land and ocean surface. The global mean surface temperature has increased by 0.74°C (0.4°C–0.8°C) over the last 100 years (Figure 14.2).

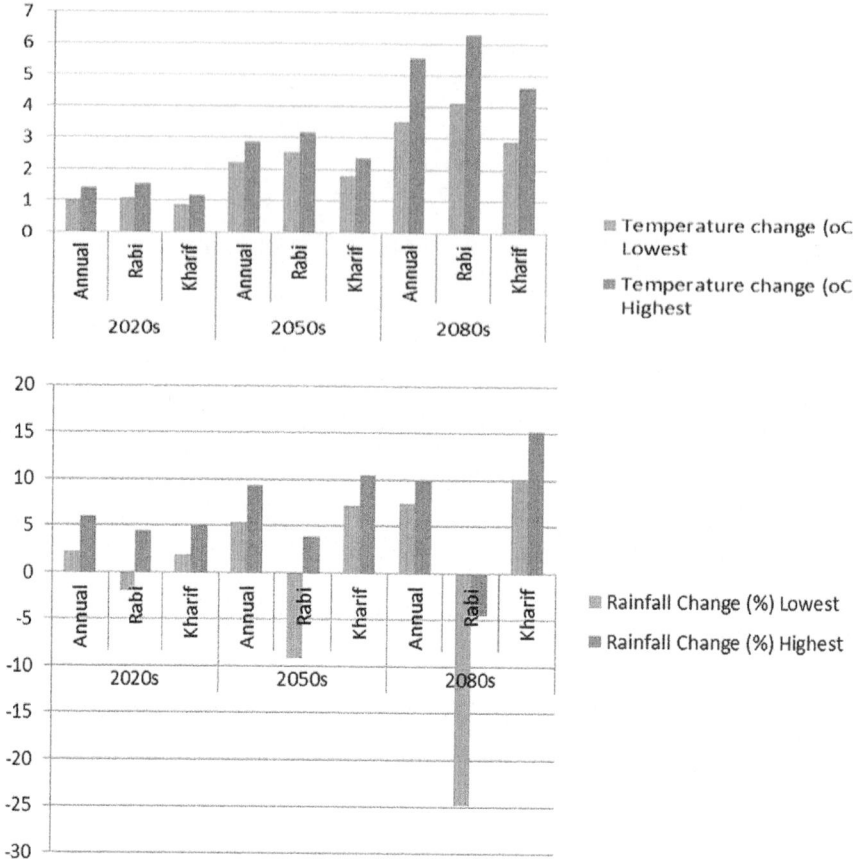

FIGURE 14.2 Components of climate change.

- Greenhouse gases (CO_2, CH_4, water vapor, O_3, nitrous oxide, halo carbons, industrial gases).
- Concentration of CO_2 has increased from about 280 ppm (pre-industrial era) to about 379 ppm (present), similarly methane and nitrous oxide levels have also increased markedly.
- Temperature (intensity and quality).

14.2   TEMPERATURE FLUCTUATION EFFECTS ON PLANT GROWTH AND DEVELOPMENT, AND SURVIVAL

Most biological activities become low to almost zero if the temperature falls below 5°C. Higher summer temperatures up to certain levels favor plant growth up to a threshold temperature referred to as base temperature, if other factors are not limiting. If the temperature exceeds the optimum value for any particular plant's growth rate or it falls to the point where it damages tissues, it leads to complete secession of growth or death of the plant. Due to climate change, temperatures may reach that level in the next 50–100 years at which they may cause direct damage to plants, however such a possibility cannot exist, especially in greenhouses. Spring is advancing by 2–6 days/decade and autumn is delayed by about 2 days/decade. Growing seasons experiencing high temperatures may affect the phenology of plants.

14.3   EFFECT OF TEMPERATURE ON FLOWERING

Kashmir being a temperate zone, seed production of most of the cole vegetables, root vegetables, and bulbous vegetables require subzero temperatures for vernalization. Mild winter temperature and warmer springs, higher average temperature leads to early blooming, and low temperature during spring affects cell division in the post bloom period, resulting in low seed production. High temperatures above normal during summer months result in a flower drop in most of the viable seed production vegetable crops like capsicum, chilli, and cucurbits.

14.4   EFFECT OF CLIMATE CHANGE ON DISEASE AND INSECTS

Wetter, warmer winters favor the incidence of mosses, algae, and diseases like (*Phytophthora, Fussarium*). Drier, warmer summers limit the growth of mosses and algae but cause diseases such as powdery mildew and early and late blight. In raised mean annual temperature of 2°C, butterflies will appear 2–3 weeks earlier and the range and distribution of butterflies will shift dramatically affecting pollination. Moreover, the aphid population will increase resulting in incidence of virus.

14.5   EFFECT OF CLIMATE CHANGE ON WEED INFESTATION

Higher carbon dioxide levels, higher winter temperatures, and perhaps greater water availability in early spring will favor early germination and the growth of highly competitive annual weeds, thus raising a need for increased maintenance. Drier summer conditions may reduce the weed growth, but will also reduce the effectiveness of glyphosate and hormone weed killers such as 2–4D and 2-methyl-4-chlorophenoxyacetic acid, which work best during active plant growth. Weeds become more competitive than desired plants due to their C_4 nature.

14.6   CLIMATE CHANGE AND INDIAN AGRICULTURE

The threat of climate change that led to the Framework Convention on Climate Change (FCCC) at Rio, is perceived differently by different countries. This fact has delayed any effective international agreement on how to deal with the problem. In the case of the Montreal Protocol covering ozone-depleting substances, there was a wide consensus and effective action was mobilized quickly. Thus, an understanding of the perceptions and positions of different countries makes it easier to explore the possibilities of effective action.

The FCCC objective states that greenhouse gas (GHG) concentrations should be stabilized at levels where food production is not threatened (UN, 1992). Thus, by examining the impact on agriculture of different climate change scenarios, one can get an idea of what is tolerable. In a detailed study of India, Kumar and Parikh (2001) examined the impact of climate change on agricultural crop yields, gross domestic product (GDP), and welfare. Considering a range of equilibrium climate change scenarios which project a temperature rise of 2.5°C–4.9°C for India, Kumar and Parikh (2001) estimated that: (i) without considering the carbon dioxide fertilization effects, the yield losses for rice and wheat vary between 32% and 40%, and 41% and 52%, respectively; (ii) GDP would drop by 1.8%–3.4%. Their study also showed that even with carbon fertilization effects, the losses would be in the same direction but somewhat smaller. Using an alternative methodology, Kumar and Parikh (2002) showed that even with farm-level adaptations, the impacts of climate change on Indian agriculture would remain significant. They estimated that with a temperature change of +2°C and an accompanying precipitation change of +7%, the farm level total net-revenue would fall by 9%, whereas with a temperature increase of +3.5°C and precipitation change of +15%, the fall in farm level total net-revenue would be nearly 25%. For a developing country, these are very large changes which can cause much human misery. From India's point of view, a 2°C increase would be clearly intolerable. Other developing countries may be even more vulnerable (possibly Bangladesh or Small Island States).

14.7   CLIMATE CHANGE AND KASHMIR

Kashmir is having extremely cool winters and mild summers. The growing period of vegetable ranges from <60 to >300 days with harsh terrain, undulating sloppy terraced land, high soil erosion, and low soil depth. The climate of Kashmir also provides the chilling requirement for temperate vegetables. Most potential crops for Kashmir region are cole crops, capsicum, asparagus, baby corn, carrot, Chinese cabbage, snow pea, radish, okra, saffron, onion, chilli, and cucumber. The state forms a transitional regime of diverse physical features between the weak monsoon zone of Punjab and the cold arid zone of Tibet. On the basis of altitude, climate state may be divided into four zones (Table 14.3).

The area and production of vegetables in Jammu and Kashmir in 2007–2008 was 58.6 thousand ha and 1,238.3 thousand tons, respectively. The factors related to climate change have resulted in low productivity, heavy incidence of pest and disease, and pre- and post-harvest losses.

The common theme in all these changes is water availability. Already, one third of the world's people live in river basins where they face water scarcity. Himalayan glaciers feed Asia's nine largest rivers. Their melting could jeopardize water supplies for the 1.3 billion people who live downstream. Global warming is causing the Himalayan glaciers to melt – small glaciers in many regions of Kashmir have completely disappeared, while others are a quarter of their original height. The reduction in glaciers is drying up hundreds of springs. According to a report, the last 40 years have seen a reduction in water levels of almost all the streams and rivers in Kashmir by two-thirds (Talib, 2007).

TABLE 14.3

Vegetable Growing Zones of Jammu and Kashmir

S. No.	Name of the Zone		Average Altitude in Meters	Precipitation
1.	Sub tropical zone	The outer plains and outer hills of Jammu province (Jammu, Kuthwa and parts of Udhampur Districts)	215–360	1,000 mm
2.	Intermediate zone	Parts of Doda, Udhampur, Poonch, and Rajouri districts of the state	750	The area has marked variation in temperature and precipitation
3.	Temperate zone	Whole of the Kashmir valley, higher reaches of Doda and Poonch	1,500–2,500	The area has marked variation in temperature and precipitation. The normal precipitation is 650 mm mostly received during March and April
4.	Arid zone	Ladakh	Habituated evaluation 2,400–4,000 mt.amsl. Peaks 7,200–8,400 mt.amsl.	Marked variation in temperature and precipitation

The combined effects of these changes are likely to lead to:

- Decreased soil moisture and increased evaporation
- Increased risk of soil erosion from more extreme rainfall events
- Changes in plant growth and productivity
- Changes in the distribution and abundance of pests, diseases, and weeds
- Changes in the suitability of certain crop varieties resulting from changed growing conditions in some regions
- Changes in growing seasons
- Changes in the optimum locations for vegetable industries.

14.7.1 TRANSPIRATION

In most crops, increased CO_2 improves the water use efficiency (WUE), because of declines in the stomatal conductance. However, the effect of decreased transpiration on vegetable crop yields is unlikely to be large since vegetables are irrigated in most production areas. Physiological disorders such as tipburn in lettuce (*Lactuca sativa* L. var. *capitata* L.) and cole crops, and blossom end rot in tomato [*Lycopersicon esculentum* Mill syn. *Lycopersicon lycopersicum* (L.) Karsten], pepper (*Capsicum annuum* L. Grossum group) and watermelon (*Citrulluslanatus* Thunb, Matsum, and Nakai) are sometimes associated with excessive transpiration, so the incidence of these disorders may be reduced.

14.7.2 Respiration

Respiration of leaves and roots in the dark slows within minutes of an increase in ambient CO_2, so night-time respiration would be lower at high CO_2. This direct, short-term, and readily reversible effect of CO_2 on respiration has been noted in tomatoes, lettuce, peppers, peas, and maize (Table 14.4). It is apparently caused by inhibited respiration per se, rather than stimulated carboxylase activity, but the specific mechanism(s) is not known.

14.7.3 Root Growth

In a recent review of 167 studies on root response to elevated CO_2, Rogers et al. (1994) found that root dry weight increased in about 87% of the studies and plant roots were longer or more numerous in 77% of the studies. In cassava (*Manibotesculenta* Grantz), a tropical root crop, there was not only a large increase in growth (150%), but partitioning to the root was also increased. Overall, however, effects of CO_2 enrichment on root/shoot ratio and partitioning have been highly variable and may differ between C_3 and C_4 crops (Wolfe et al., 1998).

TABLE 14.4

Physiological Disorders of Vegetables Caused or Exacerbated by High or Low Temperatures

Crops	Disorders	Aggravating factors
Asparagus	High fiber in stalks	High temperatures
Asparagus	Feathering and lateral branch growth	Temperatures >32°C, especially if picking frequency is not increased
Bean	High fiber in pods	High temperatures
Carrot	Low carotene content	Temperatures <10°C or >20°C
Cauliflower	Blindness, buttoning, ricy curds	Low temperatures
Cauliflower,	Hollow stem, leafy heads,	Low temperatures
Broccoli	No heads, bracting	High temperatures
Cole crops and lettuce	Tipburn	Drought, especially combined with high temperatures; high transpiration
Lettuce	Tipburn, bolting, loose, puffy heads	Temperatures > 17°C–28°C during day and 3°C–12°C during night
Maize	Poor kernel development, poor husk cover, tasselate ear	High temperatures, especially combined with draught
Onion	Bulb splitting	High temperatures
Pepper	Low seed production and off-shaped fruit	Low temperatures
Pepper	Sunscald	High temperatures
Potato	Secondary growth and heat sprouting	High temperatures
Tomato	Fruit cracking, sunscald	High temperatures
Tomato, pepper, watermelon	Blossom-end rot	High temperatures, especially combined with drought; high transpiration

---

14.7.4 Nutrition

With increasing plant size, elevated CO_2 increases the total nutrient uptake. Since nutrients are distributed over a larger plant, however, the concentration per unit weight is reduced. Nutrient use efficiency (unit of biomass produced per unit of nutrient) generally increases under elevated CO_2, while nutrient uptake efficiency (unit of nutrient per unit weight of root) generally decreases [e.g., Sritharan et al. (1992), in kohlrabi (*Brassica oleracea* L. var. *gongylodes* L.)].

14.7.5 Seedling Germination and Emergence

Rise or fall in temperature affects the germination and emergence of vegetables. Cool-Season vegetable crops on the basis of their temperature requirements for seed germination are shown in Table 14.5 below:

TABLE 14.5

Minimum Germination Temperature (T_{min}) and Heat Sum (S) in Degree-Days for Seedling Emergence, and the Applicable Temperature (T) Range for Germination of Various Vegetables

Group	Crops	Genus and species	(T_{min}) (°C)	S (degree days)	T (°C)
Leafy vegetables and brassica crops	Purslane	*Portulaca oleracea*	11.0	48	15–25
	Cress	*Lepidium sativum*	1.0	64	3–17
	Lettuce	*Lactuca sativa*	3.5	71	6–21
	Witloof, chicory	*Cichorium sativa*	5.3	85	9–25
	Endive	*Cichorium endive*	2.2	93	3–17
	Savoy cabbage	*B. oleriaciea var. sabauda*	1.9	95	3–17
	Turnip	*B compestris var. rapa*	1.4	97	3–17
	Borecole, kale	*B. oleracea var. acephala*	1.2	103	3–17
	Red cabbage	*B. oleraciea var. purpurea*	1.3	104	3–17
	White cabbage	*B. oleraciea var. capitat*	1.0	106	3–17
	Brussels sprouts	*B. oleraciea var. germmifera*	1.1	108	3–17
	Spinach	*Spinacea oleracea*	0.1	111	3–17
	Cauliflower	*B. oleracea var. botrytis*	1.3	112	3–17
	Corn salad	*Valerianella olitoria*	0.0	161	3–17
	Leek	*Allium porrum*	1.7	222	3–17
	Celery	*Apium graveolens*	4.6	237	9–17
	Parsley	*Petroselinum crispum*	0.0	268	3–17
Fruit vegetables	Tomato	*Lycopersicon esculentum*	8.7	88	13–25
	Aubergine	*Solanum melongena*	12.1	93	15–25
	Gherkin	*Cucumus sativus*	12.1	108	15–25
	Melon	*Cucumis melo*	12.2	108	15–25
	Sweet pepper	*Capsicum annuum*	10.9	182	15–25

(Continued)

TABLE 14.5 (*Continued*)

Minimum Germination Temperature (T_{min}) and Heat Sum (S) in Degree-Days for Seedling Emergence, and the Applicable Temperature (T) Range for Germination of Various Vegetables

Group	Crops	Genus and species	(T_{min}) (°C)	S (degree days)	T (°C)
Leguminous crops	Garden pea	*Pisum sativum*	3.2	86	3–17
	French sugar pea	*P. sativum var. sacharatum*	1.6	96	3–17
	Bean (French)	*Phaseolus vulgaris*	7.7	130	13–25
	Broad bean	*Viciafaba*	0.4	148	13–17
Root crops	Radish	*Raphanus sativus*	1.2	75	3–17
	Scorzonera	*Scorzonera hispanica*	2.0	90	3–17
	Beet	*Beta bulgaris*	2.1	119	3–17
	Carrot	*Daucus carota*	1.3	170	3–17
	Onion	*Allium cepa*	1.4	219	3–17

Source: From Taylor (1997).

Crops are ranked within groups by heat sum (*S*) in degree-days.

14.8 MITIGATION TECHNOLOGIES AND PRACTICES

Agriculture releases significant amounts of CO_2, CH_4, and N_2O to the atmosphere (Cole et al., 1997; IPCC, 2001; Paustian et al., 2004). CO_2 is released largely from microbial decay or the burning of plant litter and soil organic matter (Smith and Conen, 2004; Janzen, 2004). CH_4 is produced when organic materials decompose in oxygen-deprived conditions, notably from fermentative digestion by ruminant livestock, from stored manures, and from rice grown under flooded conditions (Mosier et al., 1998). N_2O is generated by the microbial transformation of nitrogen in soils and manures, and is often enhanced where the available nitrogen (N) exceeds plant requirements, especially under wet conditions (Oenema et al., 2005; Smith and Conen, 2004). Agricultural GHG fluxes are complex and heterogeneous, but the active management of agricultural systems offers possibilities for mitigation. Many of these mitigation opportunities use current technologies and can be implemented immediately.

Opportunities for mitigation of climate change fall into three broad categories, based on the underlying mechanism:

a. **Reducing Emissions:** The fluxes of these gases can be reduced by more efficient management of carbon and nitrogen flows in agricultural ecosystems. For example, practices that deliver added N more efficiently to crops often reduce N_2O emissions (Bouwman, 2001), and managing livestock to make most the efficient use of feeds often reduces the amount of CH_4 produced (Clemens and Ahlgrimm, 2001). The approaches that best reduce emissions depend on the local conditions, and therefore, vary from region to region.

b. **Enhancing Removals:** Agricultural ecosystems hold large carbon reserves (IPCC, 2001), mostly in soil organic matter. Historically, these systems have lost more than 50 Pg C (Paustian et al., 1998; Lal and Bruce, 1999), but some of this carbon lost can be recovered through improved management, thereby withdrawing atmospheric CO_2. Any practice that increases the photosynthetic input of carbon and/or slows the return of stored carbon to CO_2 via respiration, fire, or erosion will increase carbon reserves, thereby "sequestering" carbon or building carbon "sinks". Many studies, worldwide, have now shown that significant amounts of soil carbon can be stored in this way, through a range of practices, suited to local conditions (Lal, 2004). Significant amounts of vegetative carbon can also be stored in agro-forestry systems or other perennial plantings on agricultural lands (Albrecht and Kandji, 2003). Agricultural lands also remove CH_4 from the atmosphere by oxidation (but less than forests; Tate et al., 2006), but this effect is small compared to other GHG fluxes (Smith and Conen, 2004).

c. **Avoiding (or Displacing) Emissions:** Crops and residues from agricultural lands can be used as a source of fuel, either directly or after conversion to fuels such as ethanol or diesel (Schneider and McCarl, 2003; Cannell, 2003). These bio-energy feedstocks still release CO_2 upon combustion, but now the carbon is of recent atmospheric origin (via photosynthesis), rather than from fossil carbon. The net benefit of these bio-energy sources to the atmosphere is equal to the fossil-derived emissions displaced, less any emissions from producing, transporting, and processing. GHG emissions, notably CO_2, can also be avoided by agricultural management practices that forestall the cultivation of new lands now under forest, grassland, or other non-agricultural vegetation (Foley et al., 2005).

There are many agricultural practices that may mitigate the emission of more than one GHG. These practices include: cropland management; grazing land management/pasture improvement; management of agricultural organic soils; restoration of degraded lands; and manure/bio-solid management.

14.8.1 CROPLAND MANAGEMENT

Because they are often intensively managed, croplands offer many opportunities to impose practices that reduce net GHG emissions. Mitigation practices in cropland management include the following partly overlapping categories:

a. **Agronomy:** Improved agronomic practices that increase yields and generate higher inputs of carbon residue can lead to increased soil carbon storage (Follett, 2001).

b. **Nutrient Management:** Practices that improve nutrient use efficiency include: adjusting application rates based on precise estimation of crop needs (e.g., precision farming); using slow- or controlled-release fertilizer forms or nitrification inhibitors (which slow the microbial processes leading to N_2O formation); applying N when least susceptible to loss, often just

prior to plant uptake (improved timing); placing the N more precisely into
the soil to make it more accessible to crops' roots; or avoiding N applications in excess of immediate plant requirements (Robertson et al., 2004;
Dalal et al., 2003; Paustian et al., 2004; Cole et al., 1997; Monteny et al.,
2006).

c. **Tillage/Residue Management:** Advances in weed control methods and
farm machinery now allow many crops to be grown with minimal tillage (reduced tillage) or without tillage (no-till). These practices are now
increasingly used throughout the world.

d. **Water Management:** About 18% of the world's croplands now receive supplementary water through irrigation (Millennium Ecosystem Assessment,
2005). Expanding this area (where water reserves allow) or using more
effective irrigation measures can enhance carbon storage in soils through
enhanced yields and residue returns (Follett, 2001; Lal, 2004).

14.8.2 GRAZING LAND MANAGEMENT AND PASTURE IMPROVEMENT

Grazing lands occupy much larger areas than croplands and are usually managed
less intensively. The following are examples of practices to reduce GHG emissions
and to enhance removals:

a. **Grazing Intensity:** The intensity and timing of grazing can influence the
removal, growth, carbon allocation, and flora of grasslands, thereby affecting the amount of carbon accrual in soils (Conant et al., 2001).

b. **Species Introduction:** Introducing grass species with higher productivity,
or carbon allocation to deeper roots, has been shown to increase soil carbon. For example, establishing deep-rooted grasses in savannahs has been
reported to yield very high rates of carbon accrual (Fisher et al., 1994).

14.8.3 MANAGEMENT OF ORGANIC/PEATY SOILS

Organic or peaty soils contain high densities of carbon accumulated over many centuries because decomposition is suppressed by the absence of oxygen under flooded
conditions. To be used for agriculture, these soils are drained, which aerates the soil,
favoring decomposition and therefore, high CO_2 and N_2O fluxes. Methane emissions
are usually suppressed after draining, but this effect is far outweighed by pronounced
increases in N_2O and CO_2 (Kasimir-Klemedtsson et al., 1997).

14.8.4 RESTORATION OF DEGRADED LANDS

A large proportion of agricultural lands have been degraded by excessive disturbance, erosion, organic matter loss, salinization, acidification, or other processes
that curtail productivity (Foley et al., 2005; Lal, 2004). Often, carbon storage in
these soils can be partly restored by practices that reclaim productivity including:
re-vegetation (e.g., planting grasses); improving fertility by nutrient amendments;
applying organic substrates such as manures, biosolids, and composts; reducing

tillage and retaining crop residues; and conserving water (Olsson and Ardö, 2002; Paustian et al., 2004). Where these practices involve higher nitrogen amendments, the benefits of carbon sequestration may be partly offset by higher N_2O emissions.

14.8.5 MANURE MANAGEMENT

Animal manures can release significant amounts of N_2O and CH_4 during storage, but the magnitude of these emissions varies. Methane emissions from manure stored in lagoons or tanks can be reduced by cooling, use of solid covers, mechanically separating solids from slurry, or by capturing the CH4 emitted (Amon et al., 2006; Clemens and Ahlgrimm, 2001; Paustian et al., 2004).

14.9 APPROACHES FOR ADOPTION TO THE CLIMATE CHANGES

Potential impacts of climate change on the agricultural production will depend not only on the climate itself per se, but also on the internal dynamics of the agricultural systems, including their ability to adapt to the changes. Success in mitigating climate change depends on how well agricultural crops and systems adapt to the changes and concomitant environmental stresses of those changes on the current systems. The farmers in developing countries of the tropics need tools to adapt and mitigate the adverse effects of climate change on agricultural productivity, and particularly on vegetable production, quality, and yield. Current, and new, technologies being developed through plant stress physiology research can potentially contribute to mitigate threats from climate change on vegetable production. However, farmers in developing countries are usually small-holders, have fewer options and must rely heavily on resources available in their farms or within their communities. Thus, technologies that are simple, affordable, and accessible must be used to increase the resilience of farms in less developed countries. Asian Vegetable Research and Development Center (AVRDC) – The World Vegetable Center has been working to address the effect of environmental stress on vegetable production. Germplasm of the major vegetable crops which are tolerant of high temperatures, flooding, and drought has been identified and advanced breeding lines are being developed. Efforts are also underway to identify nitrogen-use efficient germplasm. In addition, development of production systems geared towards improved WUE and expected to mitigate the effects of hot and dry conditions in vegetable-production systems are the top research and development priorities.

14.10 ENHANCING VEGETABLE-PRODUCTION SYSTEMS

Various management practices have the potential to raise the yield of vegetables grown under hot and wet conditions of the lowland tropics. AVRDC – The World Vegetable Center has developed technologies to alleviate production challenges such as limited irrigation water and flooding, to mitigate the effects of salinity, and also to ensure appropriate availability of nutrients to the plants. Strategies include modifying fertilizer application to enhance nutrient availability to plants, direct delivery of water to roots (drip irrigation), grafting to increase flood and disease tolerance, and use of soil amendments to improve soil fertility and enhance nutrient uptake by plants.

14.11   WATER-SAVING IRRIGATION MANAGEMENT

If water is scarce and supplies are erratic or variable, then timely irrigation and conservation of soil moisture reserves are the most important agronomic interventions to maintain yields during drought stress. There are several methods of applying irrigation water and the choice depends on the crop, water supply, soil characteristics, and topography. Application of irrigation water could be through overhead, surface, drip, or sub-irrigation systems.

14.12   CULTURAL PRACTICES THAT CONSERVE WATER AND PROTECT CROPS

Various crop management practices such as mulching and the use of shelters and raised beds help to conserve soil moisture, prevent soil degradation, and protect vegetables from heavy rains, high temperatures, and flooding. The use of organic and inorganic mulches is common in high-value vegetable-production systems. These protective coverings help reduce evaporation, moderate soil temperature, reduce soil runoff and erosion, protect fruits from direct contact with soil, and minimize weed growth.

14.13   DEVELOPING CLIMATE-RESILIENT VEGETABLES

Improved, adapted vegetable germplasm is the most cost-effective option for farmers to meet the challenges of a changing climate. However, most modern cultivars represent a limited sampling of available genetic variability including tolerance to environmental stresses (Pereira and Chavez, 2007). Breeding new varieties, particularly for intensive, high input production systems in developed countries, under optimal growth conditions may have counter-selected for traits which would contribute to adaptation or tolerance to low input and less favorable environments. Superior varieties adapted to a wider range of climatic conditions could result from the discovery of novel genetic variation for tolerance to different biotic and abiotic stresses. Genotypes with improved attributes conditioned by superior combinations of alleles at multiple loci could be identified and advanced.

14.14   TOLERANCE TO HIGH TEMPERATURES

AVRDC – The World Vegetable Center has developed tomatoes and Chinese cabbage with general adaptation to hot and humid tropical environments and low-input cropping systems since the early 1970s. This has been achieved by developing heat-tolerant and disease-resistant breeding lines. The Center has made significant contributions to the development of heat-tolerant tomato and Chinese cabbage lines and the subsequent release of adapted, tropical varieties worldwide. Indeed, AVRDC – The World Vegetable Center's heat-tolerant hybrids have resulted in the successful cultivation of Chinese cabbage in the lowland tropics.

14.15 DROUGHT TOLERANCE AND WUE

Plants resist water or drought stress in many ways. In slowly developing water deficit, plants may escape drought stress by shortening their life cycle. Transfer and utilization of genes from these drought resistant species will enhance the tolerance of tomato cultivars to dry conditions, although wide crosses with *Solanum pennellii* produce fertile progenies. *Solanum chilense* is cross-incompatible with *Solanum lycopersicum* and embryo rescue through tissue culture is required to produce progeny plants. Research at AVRDC – The World Vegetable Center and other institutions is in progress to identify the genetic factors underlying drought tolerance in *S. chilense* and *S. pennellii*, and to transfer these factors into cultivated tomatoes (La Peña and Hughes, 2007).

14.16 CLIMATE-PROOFING THROUGH GENOMICS AND BIOTECHNOLOGY

Increasing crop productivity in unfavorable environments will require advanced technologies to complement traditional methods which are often unable to prevent yield losses due to environmental stresses. In the past decade, genomics has developed from whole genome sequencing to the discovery of novel and high throughput genetic and molecular technologies. Genes have been discovered and gene functions understood. This has opened the way to genetic manipulation of genes associated with tolerance to environmental stresses. These tools promise more rapid, and potentially spectacular, returns but require high levels of investment. Many activities using these genetic and molecular tools are in place, with some successes. National and international institutes are re-tooling for plant molecular genetic research to enhance traditional plant breeding and benefit from the potential of genetic engineering to increase and sustain crop productivity.

REFERENCES

Albrecht, A. and Kandji, S.T. 2003. Carbon sequestration in tropical agroforestry systems. *Agriculture, Ecosystems & Environment*, 99(1–3), pp. 15–27.

Amon, B., Kryvoruchko, V., Amon, T. and Zechmeister-Boltenstern, S. 2006. Methane, nitrous oxide and ammonia emissions during storage and after application of dairy cattle slurry and influence of slurry treatment. *Agriculture, Ecosystems & Environment*, 112(2–3), pp. 153–162.

Anonymous. 2007. Vegetable profile in South Asia. Cited from faostat.fao.org.

Bouwman, A. 2001. *Global Estimates of Gaseous Emissions from Agricultural Land*. FAO, Rome, p. 106.

Cannell, M.G., 2003. Carbon sequestration and biomass energy offset: theoretical, potential and achievable capacities globally, in Europe and the UK. *Biomass and Bioenergy*, 24(2), pp. 97–116.

Clemens, J. and Ahlgrimm, H.J., 2001. Greenhouse gases from animal husbandry: mitigation options. *Nutrient Cycling in Agroecosystems*, 60(1–3), pp. 287–300.

Cole, C.V., Duxbury, J., Freney, J., Heinemeyer, O., Minami, K., Mosier, A., Paustian, K., Rosenberg, N., Sampson, N., Sauerbeck, D. and Zhao, Q. 1997. Global estimates of potential mitigation of greenhouse gas emissions by agriculture. *Nutrient Cycling in Agroecosystems*, 49(1–3), pp. 221–228.

Conant, R.T., Paustian, K. and Elliott, E.T. 2001. Grassland management and conversion into grassland: effects on soil carbon. *Ecological Applications*, 11(2), pp. 343–355.

Dalal, R.C., Wang, W., Robertson, G.P. and Parton, W.J. 2003. Nitrous oxide emission from Australian agricultural lands and mitigation options: a review. *Soil Research*, 41(2), pp. 165–195.

Field, C.B. and Raupach, M.R. (eds.) 2004. *The Global Carbon Cycle: Integrating Humans, Climate, and the Natural World* (Vol. 62). Island Press.

Fisher, M.J., Rao, I.M., Ayarza, M.A., Lascano, C.E., Sanz, J.I., Thomas, R.J. and Vera, R.R. 1994. Carbon storage by introduced deep-rooted grasses in the South American savannas. *Nature*, 371(6494), pp. 236–238.

FAOSTAT. 2016. Statistiscal Year Book of Food and Agricultural Organization of United Nations 2016.

Foley, J.A., DeFries, R., Asner, G.P., Barford, C., Bonan, G., Carpenter, S.R., Chapin, F.S., Coe, M.T., Daily, G.C., Gibbs, H.K. and Helkowski, J.H. 2005. Global consequences of land use. *Science*, 309(5734), pp. 570–574.

Follett, R.F. 2001. Organic carbon pools in grazing land soils. In The Potential of U.S. Grazing Lands to Sequester Carbon and Mitigate the Greenhouse Effect. Follett, R.F., Kimble, J.M. and Lal, R., (eds.), Lewis Publishers, Boca Raton, FL, pp. 65–86.

IPCC. 2001. Contribution of working group I to the third assessment report of the intergovernmental panel on climate change. In Climate Change 2001: The Scientific Basis. Houghton, J.T., Y. Ding, D.J. Griggs, M. Noguer, P.J. van der Linden, X. Dai, K. Maskell, and C.A. Johnson, (eds.), Cambridge University Press, Cambridge, UK, p. 881.

Janzen, H.H. 2004. Carbon cycling in earth systems – a soil science perspective. *Agriculture, Ecosystems & Environment*, 104(3), pp. 399–417.

Kasimir-Klemedtsson, A., Klemedtsson, L., Berglund, K., Martikainen, P., Silvola, J. and Oenema, O. 1997. Greenhouse gas emissions from farmed organic soils: a review. *Soil Use and Management*, 13, pp. 245–250.

Kumar, K.K. and Parikh, J.K. 2001. Indian agriculture and climate sensitivity. *Global Environmental Change*, 11(2), pp. 147–154.

Kumar, K.K. and Parikh, J.K. 2002. Socio-economic impacts of climate change on Indian agriculture. Indira Gandhi Institute of Development Research.

La Pena, R.D. and Hughes, J. 2007. Improving vegetable productivity in a variable and changing climate. *An Open Access Journal published by ICRISAT*, 4(1), pp. 1–22.

Lal, R. 2004. Soil carbon sequestration impacts on global climate change and food security. *Science*, 304(5677), pp. 1623–1627.

Lal, R. and Bruce, J.P. 1999. The potential of world cropland soils to sequester C and mitigate the greenhouse effect. *Environmental Science & Policy*, 2(2), pp. 177–185.

Millennium Ecosystem Assessment. 2005. *Ecosystems and Human Well- Being: Current State and Trends. Findings of the Condition and Trends Working Group. Millennium Ecosystem Assessment Series*. Island press, Washington D.C., pp. 815.

Monteny, G.J., Bannink, A. and Chadwick, D. 2006. Greenhouse gas abatement strategies for animal husbandry. *Agriculture, Ecosystems & Environment*, 112(2–3), pp. 163–170.

Mosier, A.R., Duxbury, J.M., Freney, J.R., Heinemeyer, O., Minami, K. and Johnson, D.E. 1998. Mitigating agricultural emissions of methane. *Climatic Change*, 40(1), pp. 39–80.

Oenema, O., Wrage, N., Velthof, G.L., van Groenigen, J.W., Dolfing, J. and Kuikman, P.J. 2005. Trends in global nitrous oxide emissions from animal production systems. *Nutrient Cycling in Agroecosystems*, 72(1), pp. 51–65.

Olsson, L. and Ardö, J. 2002. Soil carbon sequestration in degraded semiarid agroecosystems – perils and potentials. *AMBIO: A Journal of the Human Environment*, 31(6), pp. 471–477.

Paustian, K., Cole, C.V., Sauerbeck, D. and Sampson, N. 1998. CO2 mitigation by agriculture: an overview. *Climatic Change*, 40(1), pp. 135–162.

Paustian, K., Babcock, B.A., Hatfield, J., Lal, R., McCarl, B.A., McLaughlin, S., Mosier, A., Rice, C., Robertson, G.P., Rosenberg, N.J., Rosenzweig, C., Schlesinger, W.H., and Zilberman, D. 2004. Agricultural mitigation of greenhouse gases: science and policy options. CAST (Council on Agricultural Science and Technology) Report, R141 2004, ISBN 1-887383-26-3, p. 120.

Pereira, J.S. and Chaves, M.M. 2007. Plant responses to drought under climate change in Mediterranean-type ecosystems. In *Global Change and Mediterranean-Type Ecosystems*. Springer, New York, NY, pp. 140–160.

Robertson, G.P., Field, C. and Raupach, M. 2004. Abatement of nitrous oxide, methane, and the other non-CO_2 greenhouse gases: the need for a systems approach. *Scope-Scientific Committee on Problems of the Environment International Council of Scientific Unions*, 62, pp. 493–506.

Rogers, J., Dowsett, A.B., Dennis, P.J., Lee, J.V., and Keevil, C.W. 1994. Influence of Plumbing Materials on Biofilm Formation and Growth of Legionella pneumophila in Potable Water Systems. *Applied and Environmental Microbiology*, 60(6), pp. 1842–1851.

Schneider, U.A. and McCarl, B.A. 2003. Economic potential of biomass based fuels for greenhouse gas emission mitigation. *Environmental and Resource Economics*, 24(4), pp. 291–312.

Smith, K.A. and Conen, F. 2004. Impacts of land management on fluxes of trace greenhouse gases. *Soil Use and Management*, 20(2), pp. 255–263.

Sritharan, R., Caspari, H. and Lenz, F. 1992. Influence of Co2 enrichment and phosphorus supply on growth carbohydrates and nitrate utilization of kohlrabi plants. *Gartenbauwissenschaft*, 57, pp. 246–251.

Talib. 2007. *On the Brink a Report on Climate Change and Its Impact in Kashmir*. Action Aid India, Srinagar, www.actionaidindia.org

Tate, K.R., Ross, D.J., Scott, N.A., Rodda, N.J., Townsend, J.A. and Arnold, G.C. 2006. Post-harvest patterns of carbon dioxide production, methane uptake and nitrous oxide production in a Pinus radiata D. Don plantation. *Forest Ecology and Management*, 228(1–3), pp. 40–50.

United Nations. 1992. *Framework Convention on Climate Change*. United Nations, New York.

Wolfe, N.D., Escalante, A.A., Karesh, W.B., Kilbourn, A., Spielman, A., and Lal, A.A. Apr–Jun 1998.Wild primate populations in emerging infectious disease research: the missing link? *Emerging Infectious Diseases*, 4(2), pp. 149–158.

15 Remote Sensing and GIS
A Tool of Precision Agriculture

Mushtaq A. Wani
Sher-e-Kashmir University of Agricultural Sciences
and Technology of Kashmir (SKUAST-K)

CONTENTS

15.1   PRECISION FARMING (PF)

PF, also called Precision Agriculture (PA) or site-specific crop management (SSCM), is an integrated information- and production-based farming system that is designed to increase long term, site specific, and whole farm production efficiency, productivity, and profitability while minimizing unintended impacts on wildlife and the environment (Earl et al., 1996). The basic principle of PF is to maximize the efficiency of the inputs as measured by the outputs, which is to optimize the inputs according to field variability in order to maximize the yields, diminishing production costs and environmental impacts of agricultural practices, by giving the right amount of input at the right place and the right time. In this sense, PF can relate to any agricultural production system and can be considered as the application of information technologies, together with production experience, to:

1. Optimize production efficiency
2. Optimize quality
3. Minimize environmental impact
4. Minimize risk

All, at the site-specific level (Earl et al., 1996), these are not particularly new concepts in agriculture. There are essays on this topic dating from the early 18th century. What is new is the scale at which we are able to implement these goals. The development and application of PF has its roots on the spatial and temporal variation normally observed in fields and yields under the same treatments (seed density, fertilizer, herbicide, etc.).

If spatial or temporal variability does not exist then PF does not make sense and a uniform management system is both the cheapest and the most effective management strategy. Based on these considerations, PF is at present operating on a zonal rather than a completely site-specific basis (Figure 15.1).

PA management practices can significantly reduce the amount of nutrients and other crop inputs used while boosting yields. Farmers thus obtain a return on their investment by saving on phytosanitary and fertilizer costs. The second, larger-scale benefit of targeting inputs—in spatial, temporal, and quantitative terms—concerns environmental impacts. Applying the right amount of inputs in the right place and at the right time benefits crops, soils, and groundwater, and thus the entire crop cycle. Consequently, PA has become a cornerstone of sustainable agriculture, since it respects crops, soils, and farmers. Sustainable agriculture seeks to assure a continued supply of food within the ecological, economic, and social limits required to sustain production. PA therefore seeks to use high-tech systems in pursuit of this goal. The adoption of PF is based on the following basic premises:

- Significant within-field variability in factors that influence crop yield;
- Causes of this variability can be identified and measured;
- Information obtained can be used to effectively modify crop management practices;
- Improvement in economic yield justifies the cost.

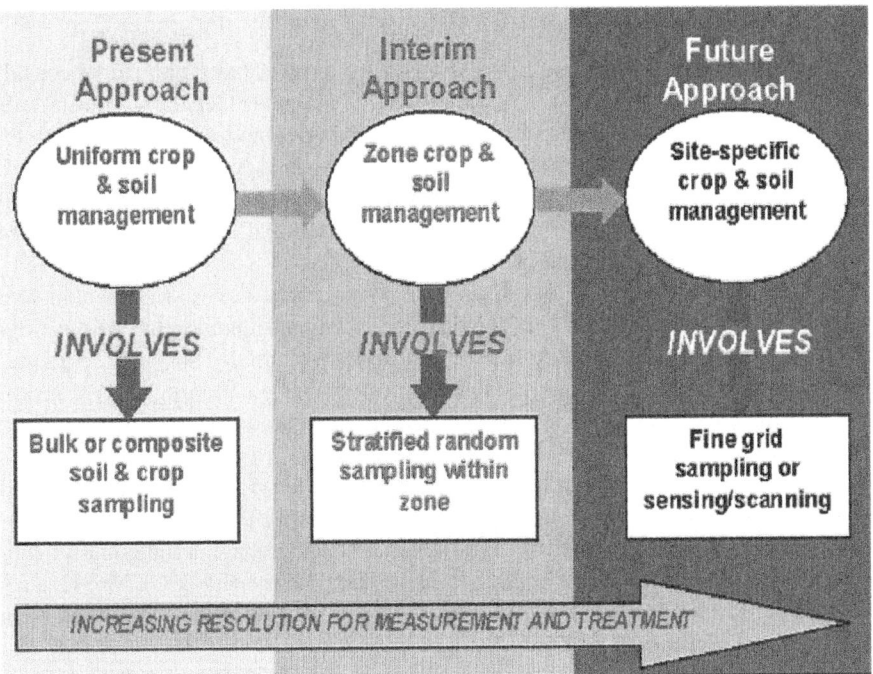

FIGURE 15.1   The evolving timeline of precision farming.

15.2   OBJECTIVES OF PF

15.2.1   Optimizing Production Efficiency

In general, the main aim of PA is to optimize the returns across a field. Unless a field has a uniform yield potential, the identification of variability in yield potential may offer possibilities to optimize the production quantity at each site or within each "zone" using differential management.

15.2.2   Optimizing Quality

Production efficiency is measured in terms of yield (quantity), mainly because yield and biomass sensors are the most reliable and common place sensors. The ability to site-specifically collect grain quality data will allow growers to consider production efficiency from the perspective of either yield, quality, or their interaction.

15.2.3   Minimizing Environmental Impact

If better management decisions are made to adapt inputs in order to meet production needs then by default there must be a decrease in the net loss of any applied input to the environment.

15.2.4 MINIMIZING RISK

Risk management is a common practice today for most farmers and can be considered from two points of view: income and environmental. Generally, minimizing income risk is seen as more important than minimizing environmental risk but PF attempts to offer a solution that may allow both positions to be considered in risk management. This improved management strategy will come about through a better understanding of the environment-crop interaction and a more detailed use of emerging and existing information.

Though the concept of PF has been around since very early in the development of agriculture, it can be said that the practice of PF was enabled by the development of technology that allowed to quantify and differentially manage the natural variability of fields. Particularly, the advent of global positioning system (GPS) and global navigation satellite systems (GNSS) triggered this process (Goddard et al., 1995).

The farmer's and/or researcher's ability to locate their precise position in a field allows for the creation of maps of the spatial variability of as many variables as can be measured [e.g., crop yield, terrain features/topography, organic matter content, moisture levels, nitrogen levels, pH, electrical conductivity (EC), Mg, K, etc.]. Furthermore, these maps can be interpolated onto a common grid for comparison (Whelan and McBratney, 2003).

Spatial and temporal variability of crop variables are at the heart of PF, while the spatial and temporal behaviors of that variability are key to defining the amendment strategies.

PA has also been enabled by technologies like crop yield monitors mounted on GPS equipped harvesters, the development of variable rate technology (VRT) like seeders, sprayers, etc., the development of an array of real-time vehicle mountable sensors that measure everything from chlorophyll levels to plant water status, multi- and hyper-spectral aerial and satellite imagery, information technology, and geospatial tools. Altogether, they enable farmers to use electronic guidance aids to direct equipment movements more accurately, provide precise positioning for all equipment actions and chemical applications and, analyze all the data in association with other sources of data (agronomic, climatic, etc.).

These advances add up to a new and powerful toolbox of management tools for the progressive farm manager. PF technologies affect the entire production and management functioning of a farm.

The technological tools often include the GPS, geographical information system (GIS), yield monitor, VRT, and remote sensing (RS).

1. The **GPS** is a network of satellites developed for and managed by the U.S. Defense Department. The GPS constellation orbiting the earth transmits precise satellite time and location information to ground receivers. The ground receiving units are able to receive this location information from several satellites at a time and thus, determine its exact location. This information is provided in real-time, meaning that continuous position information is provided while in motion. Having precise location information at

any time allows crop, soil, and water measurements to be mapped. GPS receivers, either carried to the field or mounted on implements, allow users to return to specific locations to sample or treat those areas. GPS-based applications in PF are being used for farm planning, field mapping, soil sampling, tractor guidance, crop scouting, variable rate applications, and yield mapping. GPS allows farmers to work during low visibility field conditions such as rain, dust, fog, and darkness. GPS equipment manufacturers have developed several tools to help farmers and agribusinesses become more productive and efficient in their PF activities.

2. Today, many farmers use GPS-derived products to enhance operations in their farming businesses.

3. Location information is collected by GPS receivers for mapping field boundaries, roads, irrigation systems, and problem areas in crops such as weeds or diseases. The accuracy of GPS allows farmers to create farm maps with precise acreage for field areas, road locations, and distances between points of interest. GPS allows farmers to accurately navigate to specific locations in the field, year after year, to collect soil samples, or monitor crop conditions. The ability to georeference activities gives producers the option to map and visually display farm operations. This provides insights into both production variability as well as inefficiencies in crop production and farm operations.

4. **GIS** consists of a hardware-software database system used to capture, store, retrieve, manipulate, analyze, and display, in map like form, spatially referenced geographical information (Figure 15.2). In the simplest terms, GIS is the merging of cartography, statistical analysis, and database technology. GIS maps are interactive. On the computer screen, map users can scan a GIS map in any direction, zoom in or out, and change the nature of the information contained in the map. Balancing the inputs and outputs on a farm is fundamental to its success and profitability. The ability of GIS to analyze and visualize agricultural environments and workflows has proved to be very beneficial to those involved in the farming industry (Orellana et al., 2006). While natural inputs in farming cannot be controlled, they can be better understood and managed with GIS applications such as crop yield estimates, soil amendment analysis, and erosion identification and remediation. Enhancing a GIS with land-cover data layers has proved helpful to crop growers' associations, crop insurance companies, seed and fertilizer companies, farm chemical companies, libraries, universities, federal and state governments, and value-added remote-sensing/GIS companies.

5. *Yield monitors* are crop yield measuring devices installed on harvesting equipment. The yield data from the monitor is recorded and stored at regular intervals along with positional data received from the GPS unit. They also track other data such as distance and bushels per load, number of loads, and fields. Using GIS software it is possible to produce yield maps (Figure 15.3).

6. **VRT** consists of farm field equipment with the ability to precisely control the rate of application of crop inputs and tillage operations. Variable rate controllers are available for granular, liquid, and gaseous fertilizer materials

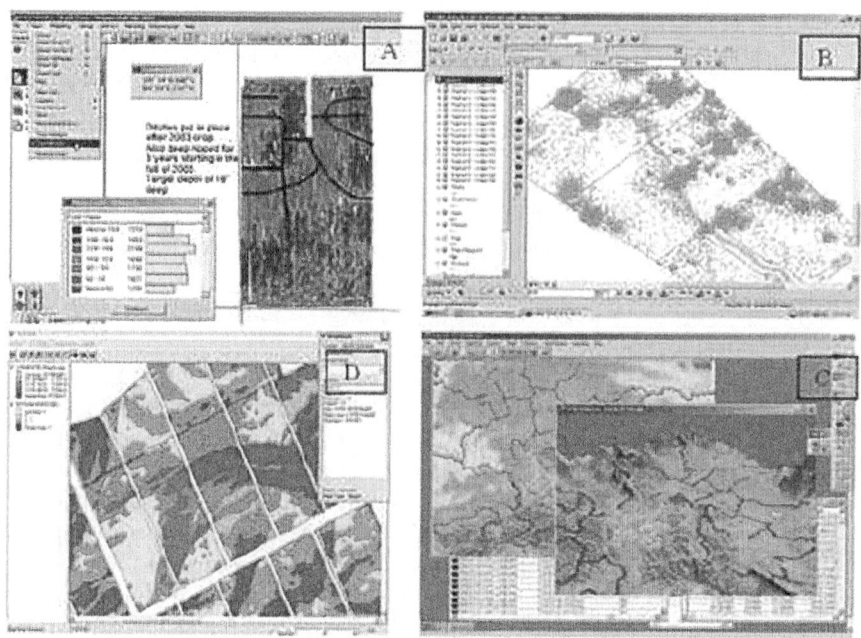

FIGURE 15.2 Screenshots of different proprietary GIS available on the market. (a) Farm Works; (b) ESRI ArcView; (c) MapInfo; (d) CorView.

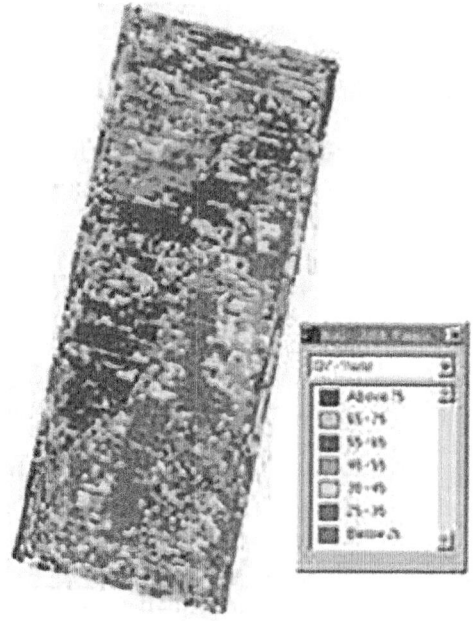

FIGURE 15.3 Example of a yield map for a maize crop.

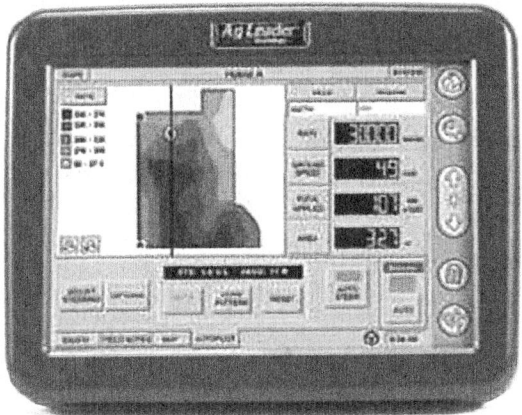

FIGURE 15.4   Variable rate application monitor for chemicals.

(Figure 15.4). Variable rates can either be manually controlled by the driver or automatically controlled by an on-board computer with an electronic prescription map. This technology package allows the grower to apply the quantity of crop inputs needed at a precise location in the field based on the individual characteristics of that location. Crop inputs that can be varied in their application commonly include tillage, fertilizer, weed control, insect control, plant variety, plant population, and irrigation. Typical VRT system components include a computer controller, GPS receiver, and GIS map database. The computer controller adjusts the equipment application rate of the crop input applied. The computer controller is integrated with the GIS database, which contains the flow rate instructions for the application equipment. A GPS receiver is linked to the computer. The computer controller uses the location coordinates from the GPS unit to find the equipment location on the map provided by the GIS unit. The computer controller reads the instructions from the GIS system and varies the rate of the crop input being applied as the equipment crosses the field. The computer controller will record the actual rates applied at each location in the field and store the information in the GIS system, thus maintaining precise field maps of materials applied (Bragachini et al., 2006; Bongiovani, 2006).

7. **RS** image data from the soil and crops is processed and then added to the GIS database. There are 3 sources of RS data commonly used in agriculture: proximal sensors, airborne sensors, and satellite sensors. Proximal hand-held sensors are mainly used for basic research, i.e., data obtained by this kind of sensors is used to establish relationships between spectral behavior and biophysical parameters of crops under certain stress (nutritional, thermal, and water). Field research using spectral sensors addresses two approaches: establish direct relationships between spectral reflectance and green biomass, or estimate the proportion of incident solar energy that is absorbed by the crop canopy and then relate it to the production of biomass

or grain (Rudorff et al., 1997; Oviedo and Rudorff, 2000). In general, these field data are transformed into vegetation indexes that are then related to agronomical parameters such as foliar area index, productivity, biomass, etc. Among airborne sensors, the most used are cameras, scanners, video cameras, and radars [light detection and ranging (LiDAR) sensors are also a possibility in some cases but are quite expensive]. The most commonly used is the photographic camera that obtains high-quality data in a region going from 350 to 900 nm. Satellite remote-sensing images are used to monitor and map agricultural areas. These images have some advantages over aerial images, i.e., periodicity, lower cost, wider vision, and spectral analysis.

The developing and availability of hyper-spectral and high resolution images is promising for one of the most important issues in PF, the determination of homogeneous management zones. However, the drawbacks of these images are their high cost and the small coverage area.

According to various studies, physical features, such as organic matter content, texture, and permeability of soil may be correlated with the spectral response recorded by remote-sensing images (King et al., 1995; Leone et al., 1995). Land surface temperature (obtained from thermal images) has also been used to study water content and soil compaction. In any case, the most widely used RS-derived tool is Normalized Difference Vegetation Index (NDVI), which has been related to different crop variables though mainly yield, and has been used to identify variability in field conditions. All these RS information is useful to guide soil and crop samplings and assist decision-making process regarding property management.

15.3   STAGES AND IMPLEMENTATION OF PF

The PF cycle is illustrated in Figure 15.5 and detailed in Table 15.1. It is important to remember, however, that we are talking of a continuous management strategy. Initially, some form of monitoring and data analysis is needed to form a decision.

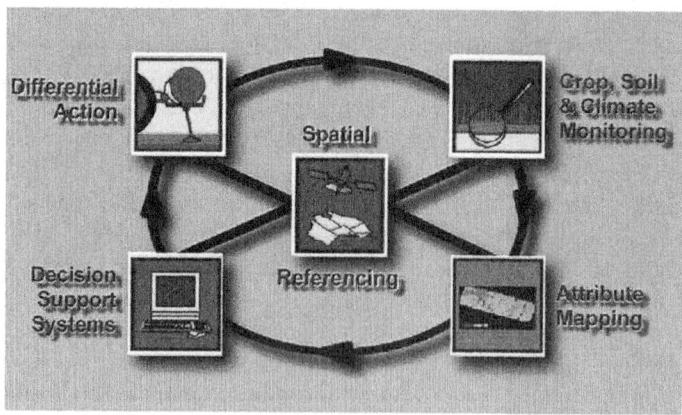

FIGURE 15.5 Precision farming cycle indicating spatial referencing as the enabling technology that drives the other parts of the cycle.

TABLE 15.1
Summary of PF Stages

Stage	Technology Involved	Activities
Collecting and importing data	GPS – DGPS GIS Topographic instruments Remote sensors Direct sensors	Topographic measurement Soil sampling by zones Weeds and diseases detection Direct measurement of soil and crop properties Satellite yield monitoring Remote-sensing images from crop and soil
Analysis, processing, and interpreting information	GIS Statistical software Technician and operator experience	Map digitizing Spatial autocorrelation analysis Maps of evaluation and prescription
Differential application	VRT GPS-assisted sprayer Specialized software	Variable application of nutrients Variable application of chemicals Differential sowing

Nevertheless, it is just as important to continue to monitor and analyze the effect of decisions and use this information to feed subsequent management decisions. The concept is the same as for adaptive management. With some variations according to different authors, PF is a five-stage process (Whelan and Taylor, 2013):

15.3.1 GEO-LOCATION OF DATA

Geo-locating a field enables the farmer to overlay information gathered from analysis of soils and residual nitrogen, and information on previous crops and soil resistivity. Geo-location is done in two ways:

a. The field is delineated using an in-vehicle GPS receiver as the farmer drives a tractor around the field.
b. The field is delineated on a base map derived from aerial or satellite imagery. The base images must have the right level of resolution and geometric quality to ensure that geo-location is sufficiently accurate.

15.3.2 CROP, SOIL, AND CLIMATE MONITORING

Intra- and inter-field variability may result from a number of factors. These include climatic conditions (hail, drought, rain, etc.), soils (texture, depth, and nitrogen levels), cropping practices (no-till farming), weeds, and disease. Permanent indicators-chiefly soil indicators-provide farmers with information about the main environmental constants. Point indicators allow them to track a crop's status, i.e., to see whether diseases are developing, if the crop is suffering from water stress, nitrogen stress, or lodging, whether it has been damaged by ice, and so on. This information may come from weather stations and other sensors (soil electrical resistivity, detection with the naked eye, satellite imagery, etc.). Soil resistivity measurements combined with soil analysis make it possible to precisely map agro-pedological conditions.

Many sensors and monitors already exist for in-situ and on-the-go measurement for a variety of crop, soil, and climatic variables. These include yield sensors, bio-mass, and crop response sensors (aerial and space borne multi- and hyper-spectral cameras), radio or mobile phone networked weather stations, soil apparent EC sensors, and gamma-radiometric soil sensors.

A huge amount of research is currently being carried out to identify how to make use of the outputs from these sensors to improve production. Agricultural scientists also need to continue to assess how and which multiple crop and production indicators can be measured.

Two very useful inputs are: topographic maps made by means of differential global positioning system (DGPS), that are useful when interpreting yield maps and weed maps, as well as planning for grassed waterways and field divisions; and salinity maps (GPS can be coupled to a salinity meter which is towed behind an ATV or pickup) which are valuable in interpreting yield maps and weed maps as well as tracking the change in salinity over time.

15.3.3 ATTRIBUTE MAPPING

Crop, soil, and climate sensors often produce large, intensive data sets. The observations are usually irregularly spaced and need to be "cleaned" and interpolated onto a surface to allow for statistical analysis. The software for mapping and displaying data from different sources on a common platform is improving annually. The development of GIS specifically for agriculture is allowing this to occur, however the adaptation and adoption of this technology for use in PF on individual farms is still in its infancy. The main issues still to be resolved are the development of a user friendly advanced data filtering system and the determination of initial and future sampling schemes to ensure that the variability of the system is properly characterized.

15.3.4 DECISION SUPPORT SYSTEMS (DSS)

PF may produce an explosion in the amount of records available for farm management. Electronic sensors can collect a lot of data in a short period of time. Lots of disk space is needed to store all the data as well as the maps resulting from the data. Electronic controllers on-board of seeders, fertilizers, and sprayers can also be designed to provide signals that are recorded electronically. Therefore, a lot of new data are generated every year (yields, weeds, etc.). Farmers will want to keep track of the yearly data to study trends in fertility, yields, salinity, and numerous other parameters. Techniques for data presentation, storage and analysis, such as GIS, are already available and should be relatively easily applied, with none or a few modifications, to agriculture.

However, DSS are not so flexible and it is in this area that much research needs to be done. DSS use agronomic and environmental data, combined with information on possible management techniques, to determine the optimum management strategy for production. Most commercial DSS are based on "average" crop response across a field. The majority of engineering companies currently supplying SSCM technology are not producing DSS to support the differential use of their equipment in a production system. Therefore the burden is falling on individual industry bodies, and to a lesser extent on government agencies, to fill the gap. Initially it may be sufficient to adapt existing agricultural DSS such as WHEATMAN, COTTONLOGIC, or APSIM to site-specific situations. In the long run, however, a DSS that is able to site-specifically model plant-environment interactions in terms of yield and quality will be needed.

15.3.5 DIFFERENTIAL ACTION

The differential application of inputs using variable rate application (VRA) technology is essentially an engineering problem. Due to the commercial potential of VRA technology, much of this engineering development is being driven by the private sector. The main input required for VRA implements is accurate information on required application rates and associated locations or times for the applications. VRA equipment should also record the actual application procedure for quality control. The biggest barrier to adoption is the lack of information from a DSS on where, and by how much, inputs should be varied. Controllers are available to electronically turn booms on and off, and alter the amount of herbicide applied. Moreover, several manufacturers are currently producing guidance systems using high precision DGPS that can accurately position a moving vehicle within a foot or less. These guidance systems may replace conventional equipment markers for spraying or seeding and may be a valuable field scouting tool. The PF stages are summarized in Table 15.1.

15.4 RS AND PF

RS offers the opportunity of mapping and monitoring crop and soil variability and an efficient way of mapping and monitoring the effects of any condition that affects plant health, yield, or quality of a crop. Mainly based on reflectance differences among red and near infrared bands, different analysis are done regarding biomass, stress condition, growth rate, among others. But above all, RS imagery is used as an input in GIS-based analysis to guide field samplings that aid in the understanding of observed patterns and assist in decisions about present or future management of fields. Therefore, RS is a very useful tool for general analysis to characterize the field, and used in combination with other management tools as crop simulation models, producers can answer questions about past events and make predictions for a certain combination of circumstances in the future. Besides, temporal analysis of RS data may help to get an image of the behavior of a field in yield terms. This can be done with yield maps if available, but if they are not, yield has proven to be highly correlated to near infrared (NIR) reflectance, so these images may be used as a control tool to make management decisions even before the harvest.

RS applications in PA began with sensors for soil organic matter, and have quickly diversified to include satellite, aerial, and hand-held or tractor-mounted sensors. Wavelengths of electromagnetic radiation initially focused on a few key visible or near infrared bands. Today, electromagnetic wavelengths in use range from the ultraviolet to microwave portions of the spectrum, enabling advanced applications such as LiDAR, fluorescence spectroscopy, and thermal spectroscopy, along with more traditional applications in the visible and near infrared portions of the spectrum. Spectral bandwidth has decreased dramatically with the advent of hyperspectral RS, allowing improved analysis of specific compounds, molecular interactions, crop stress, and crop biophysical or biochemical characteristics. A variety of spectral indexes now exist for various PA applications, rather than a focus on only NDVI. Spatial resolution of aerial and satellite RS imagery has improved from 100s of meter to sub-meter accuracy, allowing evaluation of soil and crop properties at fine spatial resolution at the expense of increased data storage and processing requirements. Temporal frequency of RS imagery has also improved dramatically. At present there is considerable interest in collecting RS data at multiple times in order to conduct near real-time soil, crop, and pest management (Mulla, 2013).

RS applications in agriculture are based on the interaction of electromagnetic radiation with soil or plant material. Typically, RS involves the measurement of reflected radiation, rather than transmitted or absorbed radiation. RS refers to non-contact measurements of radiation reflected or emitted from agricultural fields. The platforms for making these measurements include satellites, aircraft, tractors, and hand-held sensors. Measurements made with tractors and hand-held sensors are also known as proximal sensing, especially if they do not involve measurements of reflected radiation. In addition to reflectance, transmittance and absorption, plant leaves can emit energy by fluorescence (Apostol et al., 2003) or thermal emission (Cohen et al., 2005).

Thermal RS for water stress in crops is based on emission of radiation in response to temperature of the leaf and canopy, which varies with air temperature and the rate of evapotranspiration. RS applications in agriculture are typically classified according to the type of platform for the sensor, including satellite, aerial, and ground based platforms. These platforms and their associated imaging systems can be differentiated based on the altitude of the platform, the spatial resolution of the image, and the minimum return frequency for sequential imaging. Spatial resolution affects the area that can be identified. As spatial resolution improves, the area of the smallest pixel decreases, and the homogeneity of soil or crop characteristics within that pixel increases. Poor spatial resolution implies large pixels with increased heterogeneity in soil or plant characteristics. Return frequency is important for assessment of temporal patterns in soil or plant characteristics. The availability of RS images from satellite and aerial platforms is often severely limited by cloud cover (Moran et al., 1997), whereas ground based RS is less affected by this limitation.

15.5 SATELLITE RS

Satellites have been used for RS imagery in agriculture since the early 1970s. They were mainly used to perform large scale crop classifications (Panigrahy and Sharma, 1997). These applications of RS in conventional agriculture soon led to applications in PA. The first application of RS in PA occurred when it used Landsat imagery of bare soil to estimate spatial patterns in soil organic matter content, which were then used as auxiliary data along with ground based measurements to estimate the spatial patterns in soil phosphorus and wheat grain yield (Mulla and Bhatti, 1997). The spatial resolution of Landsat, SPOT, and IRS satellites is fairly coarse (20–30 m) for current applications in PA. As new and improved spatial and spectral sensors appeared, better correlations were found. For example, Seelan et al. (2003) used IKONOS images to identify N deficiencies in sugarbeet, fungicide performance efficiency in wheat and field sites that had inadequate artificial drainage in wheat. Bausch and Khosla (2010) showed that QuickBird estimates of Normalized Green Normalized Difference Vegetation Index (NGNDVI) were strongly correlated with spatial patterns in nitrogen sufficiency in irrigated maize. Research showed that QuickBird images of olive orchards in Spain could be used to estimate areas of olive plantations, numbers of trees, and spatial patterns in a projected area of tree canopies, and olive yields.

Satellite and/or aerial imagery is frequently used to estimate spatial patterns in crop biomass (Yang et al., 2000) and potential crop yield (Doraiswamy et al., 2003) using the NDVI. The advent of yield monitors provided finer scale resolution yield measurements across large spatial areas that could be used to improve the capacity of RS to predict crop structural characteristics such as leaf area index (LAI), biomass, and yield.

Moran et al. (1997) and Yao et al. (2011) summarized the major challenges for using satellite RS for PA. Satellite imagery in the visible and NIR bands are limited to cloud-free days, and are most usable when irradiance is relatively consistent across time. Only radar imagery collected using satellites or airplanes is unaffected by cloud cover. Other challenges include calibrating raw digital numbers to true surface reflectance, correcting imagery for atmospheric interferences and/or off-nadir view angles, and georectifying pixels using GPS-based ground control locations.

15.6   PROXIMAL RS OF CROPS FOR PF

There has been significant interest in proximal RS techniques to assess crop growth and crop stress. Proximal RS involves sensors mounted on tractors, spreaders, sprayers, or irrigation booms, which allow real-time site specific management of fertilizer, pesticides, or irrigation. The foundation for a transition from RS to proximal sensing-based assessment of crop status was established by using a Minolta soil plant analysis development (SPAD) meter to measure leaf greenness (chlorophyll) in maize crops at the silking stage under a range of applied N fertilizer rates. After that, a whole set of sensors and spectral indexes were developed to record different crop properties that were then related to N stress in plants and set the basis for VRT.

15.7   HYPER-SPECTRAL RS IN PF

Hyper-spectral RS collects reflectance data over a wide spectral range at small spectral increments (typically 10 nm). It provides the ability to investigate spectral response of soils and vegetated surfaces in narrow spectral bands (10 nm wide) across a wide spectral range. When collected across large spatial extents at fine spatial resolution, hyper-spectral imaging provides powerful insight into the spatial and spectral variability in reflectance for a bare or vegetated surface. Hyper-spectral imaging differs from multi-spectral imaging in the continuity, range, and spectral resolution of bands.

In theory, it offers the capability of sensing a wide variety of soil and crop characteristics simultaneously, including moisture status, organic matter, nutrients, chlorophyll, carotenoids, cellulose, leaf area index, and crop biomass (Goel et al., 2003; Haboudane et al., 2002; Zarco-Tejada et al., 2005). Advanced statistical methods for chemometric analysis of reflectance spectra have been used to interpret hyperspectral RS data, including partial least squares (PLS) (Viscarra Rossel et al., 2006), principal components analysis (PCA) (Geladi, 2003), and pattern classification and recognition techniques including object oriented (Frohn et al., 2009) and decision tree (Wright and Gallant, 2007) classification techniques. PLS regression is perhaps more powerful than PCA in that PLS (like PCA) not only identifies the factors that describe spectral variance, but also eliminates the spectral bands that contain redundant information (Alchanatis and Cohen, 2016).

Potential applications of hyper-spectral RS in PA have recently been reviewed by Yao et al. (2010). These applications include: (i) bare soil imaging for management zone delineation, (ii) weed mapping, (iii) crop N stress detection, (iv) crop yield mapping, and (v) pest and disease detection. Perhaps of greatest interest for PF is using hyper-spectral RS for variable rate, in-season management of nitrogen fertilizer based on spatial patterns in chlorophyll content.

15.7.1 KNOWLEDGE GAPS FOR RS IN PF

Rapid advances in RS for PA have occurred over the last 25 years. Satellite imagery has improved in spatial resolution, return visit frequency, and spectral resolution. Aerial hyperspectral imagery has revolutionized the ability to distinguish multiple crop characteristics, including nutrients, water, pests, diseases, weeds, and biomass and canopy structure. Ground-based sensors have been developed for on-the-go monitoring of crop and soil characteristics such as N stress, water stress, soil organic matter, and moisture content. There is a significant potential in PA for combining archived RS data with real-time data for improved agricultural management (Thenkabail, 2003). Historical archives of satellite RS data are available at many locations for Landsat, SPOT, IRS, IKONOS, and Quick Bird. These data typically include reflectance in the B, G, R, and NIR bands, at spatial resolutions of from 0.6 to 30 m spatial resolution. Images at a fixed location could be analyzed across multiple crop growth stages, seasons and years in order to identify relatively homogeneous sub-regions of fields that differ from one another in leaf area index, NDVI, and potential yield. Auxiliary data at these same sites, including crop yield maps, digital elevation models and soil series maps could be combined, via GIS-related techniques, with historical RS data to identify potential management zones where PA input operations can be implemented. With this in mind, there are several needs for future research in PF.

These include the following:

More emphasis is needed on chemo-metric or spectral decomposition/derivative methods of analysis since spatial and spectral resolution of hyper-spectral sensing systems are now adequate for most PA applications

Sensors are needed for direct estimation of nutrient deficiencies without use of reference strips

Spectral indexes should be developed that simultaneously allow assessment of multiple crop characteristics (e.g., LAI, biomass) and stresses (e.g., water and N; weeds and insects, etc.)

Historical archives of satellite RS data should be integrated with real-time RS data at high spatial and spectral resolution for improved decision-making in PA.

15.7.2   The Future in PF: Unmanned Aircraft System (UAS)

High-resolution satellite imagery is now quite commonly used to study variations of crop and soil conditions. However, the availability and the often prohibitive costs of such imagery would suggest an alternative product for this particular application in PF. Specifically, images taken by low altitude RS platforms, or small UAS, are shown to be a potential alternative given their low cost of operation in environmental monitoring, high spatial and temporal resolution, and their high flexibility in image acquisition programming (Hunt et al., 2013). Results of recent studies indicate that, to provide a reliable end product to farmers, advances in platform design, production, standardization of image georeferencing and mosaicing, and information extraction workflow are required. Moreover, it is suggested that such endeavors should involve the farmer, particularly in the process of field design, image acquisition, image interpretation, and analysis (Zhang and Kovacs, 2012). This cutting edge technology is powerful and cost efficient in time-critical, repetitive, and locally operated RS applications. The satellite imagery taken by different mode is summarized in Figure 15.6.

15.8   GIS IN PF

GIS are an important element in the management of data generated by PF. However, information stored in the GIS should be also treated with other software applications that make data interpretation and appropriate management-taking decisions possible. These are programs which are relatively easy to use and which allow yield monitor data manipulation and appropriate file specification for VRT equipment and machinery. However, their routine use is too simple and difficult to guarantee suitable interpretation and management decision-taking. The development of DSS in PF remains a pending assignment.

GIS allow for the association of a group of graphic information (plans and maps) with digital databases. This means that GIS allows for an integrated management of graphic and alphanumeric information to address high complexity issues. In PF, the idea of having the proper information in the proper time and place is crucial to make management decisions when facing an emergency as well as to increase the productivity and yield of parcels. Therefore, GIS is a basic tool to administrate and manage field information, since it enables the building, storing, updating, integration, and visualization of geographical information from different sources (Orellana et al., 2006).

The importance of GIS is rooted in that solutions for several problems require the access to several types of information that may only be related by geographic

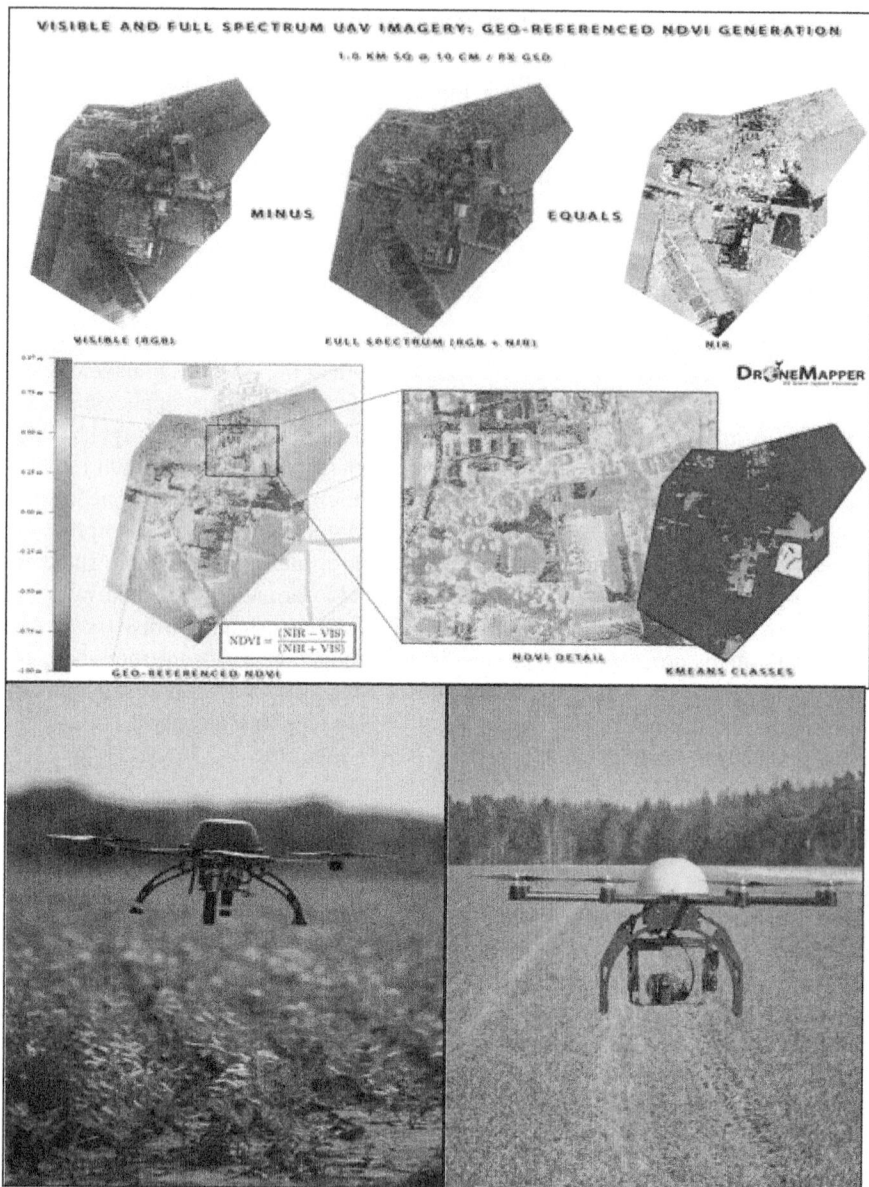

FIGURE 15.6 Satellite imagery.

position (spatial association). Then, only GIS technology allows to store and manipulate information based on the geography of elements to analyze patterns, relationships, and tendencies that aid and assist the decision-making process.

Several tools to quantitatively assess the spatial relationships within and between layers of environmental information have been developed in the last years. These tools allow to quantitatively establish if a certain variable has some sort of spatial pattern or structure, or if it can be related to other(s), and in that way explain and/or predict the productive and quality behavior of a crop. Undoubtedly, the mere graphic representation of data has important implications in our capacity to understand or visualize probable associations between, for example, environmental variables and yield. However, we are neither able to see if those associations are meaningful, nor if associations or patterns are obscured by different sources of error or stochasticity. Therefore, here is where the statistical analysis plays the most important role, allowing to quantify and numerically characterize the spatial associations present in the field.

15.8.1 Geo-Statistics

There is another technique, however, that assumes a continuous and normal distribution of values of a variable in the geographic space of a field. This technique is called *Geo-statistics*. Mapping of the variables sampled using geo-statistical methods and a reference grid, is a recommendable measure (Plant, 2001). Geo-statistics is basically a probabilistic method of spatial interpolation. The final construction of a map corresponding to a parcel is performed based on the estimation of the values of a variable at non sampled points, using the spatial variability structure of the sampled data (variogram) and an interpolation method (kriging). It is based on the study and modeling of the correlation or semivariance of a variable according to distance among sample points. The tool most widely used for the spatial description of data sets is the variogram or semi-variogram, which is a mathematical description of the relationship between variance among pairs of observations and the distance among them. Several models may then be fit to the experimental variogram which are then used as a basis for kriging interpolation methods (Best and Leon, 2006).

Kriging is an optimal interpolation based on regression against observed values of surrounding data points, weighted according to spatial covariance values. All interpolation algorithms (inverse distance squared, splines, radial basis functions, triangulation, etc.) estimate the value at a given location as a weighted sum of data values at surrounding locations. Almost all of them assign weights according to functions that give a decreasing weight with increasing separation distance. Kriging, instead, assigns weights according to a (moderately) data-driven weighting function, rather than an arbitrary function. In any case, it is still just an interpolation algorithm and will give very similar results to others in many cases.

If the data locations are fairly dense and uniformly distributed throughout the study area, you will get fairly good estimates regardless of the interpolation algorithm. If the data locations fall in a few clusters with large gaps in between, you will get unreliable estimates regardless of the interpolation algorithm. However, some advantages of kriging are that:

It helps to compensate for the effects of data clustering, assigning individual points within a cluster rather than isolated data points. It gives an estimate of the estimation error (kriging variance or residual map). The availability of estimation error provides a basis for stochastic simulation of possible realizations of the variable.

15.8.2 Spatial Econometry

The generalized spread of GIS has generated the need for a methodology that allows to manage the spatial models and autocorrelation. Anselin (2001) has defined Spatial Econometry as the collection of techniques that addresses the peculiarities caused by space in statistical modeling. This technique differs from geo-statistics in its basic principles. Geo-statistics assumes that the spatial variation is continuous while spatial econometry assumes that the covariance is a result of the interaction among discrete objects. This premise requires the specification of a spatial stochastic process represented by a matrix of spatial weights (Anselin, 2001).

15.8.2.1 Spatial Regression

In order to be able to generate proper recommendations for site-specific management, it is important to consider site-specific variation and spatial structure of data in models used to understand and explain the relationship between yield and crop conditions. The spatial regression provides the tools to diagnose and manage spatial autocorrelation.

15.8.2.2 Management Zones

One of the objectives of PF is to identify and delineate the areas or zones of different productive potential within a parcel in order to apply different management strategies (Bramley et al., 2005). For zone delineation of a parcel, information about the yield variation pattern is a very interesting starting point. These areas, called management zones, normally differ in terms of soil properties, slope, and microclimate. The final goal is to zone the parcel taking into consideration the classes provided by cluster analysis.

15.9 CONCLUSION

It is clear that PF has several advantages, but it can make farm planning and management both easier and more complex. It does not happen as soon as a farmer purchases a GPS unit or a yield monitor. It occurs over time as the farmer increases his level of knowledge regarding PF technologies and realizes that PF is an integrative approach to manage the whole farm and not only increase yields. What is perhaps most important for the success of PF is the increased knowledge that a farmer needs of his natural resources in the field. This includes a better understanding of the soil types, hydrology, microclimates, and aerial photography. A farmer should identify the variance of factors within the fields that affect crop yield before a yield map is acquired. A yield map should only serve as verification data to quantify the consequences of the variation that exists in a field. Management strategies and prescriptions will likely rely on sources other than yield maps. Yield maps, however, have been of fundamental importance for the development of PF. A high degree of variation will mean higher VRA of inputs and, therefore, greater economic and environmental benefit in comparison to uniform management. However, decision-making remains the cornerstone of PF. The parameters and methods used for zoning, the variables to be sampled on site, or the actions to be adopted in each zone, are some examples of the decisions that have to be taken and on which the success or failure of the proposed site-specific management will depend.

REFERENCES

Alchanatis, V. and Cohen, Y. 2016. Spectral and spatial methods of hyperspectral image analysis for estimation of biophysical and biochemical properties of agricultural crops. In *Hyperspectral remote sensing of vegetation*. CRC Press, pp. 324–343.

Anselin, L. 2001. Spatial econometrics. In Baltagi, B. (ed.), *A Companion to Theoretical Econometrics*. Basil Blackwell, Oxford, pp. 310–330.

Apostol, S., Viau, A.A., Tremblay, N., Briantais, J.M., Prasher, S., Parent, L.E. and Moya, I. 2003. Laser-induced fluorescence signatures as a tool for remote monitoring of water and nitrogen stresses in plants. *Canadian Journal of Remote Sensing*, 29(1), pp. 57–65.

Bausch, W.C. and Khosla, R. 2010. QuickBird satellite versus ground-based multi-spectral data for estimating nitrogen status of irrigated maize. *Precision Agriculture*, 11(3), pp. 274–290.

Best, S., and Leon, L. 2006. Geoestadistica. In Bongiovanni, R., Montovani, E.C., Best, S. and Roel, A. (eds.), *Agricultura de Precision: Integrando conocimientos para una agricultura moderna y sustentable*. PROCISUR/IICA, Montevideo, pp. 147–161.

Bongiovanni, R. 2006. Econometria Espacial. In Bongiovanni, R., Montovani, E.C., Best, S. and Roel, A. (eds.), *Agricultura de Precision: Integrando conocimientos para una agricultura moderna y sustentable*. PROCISUR/IICA, Montevideo, pp. 162–167.

Bragachini, M., Mendez, A., Scaramuzza, F. and Proietti, F. 2006. Cultivos Tradicionales. In Bongiovanni, R., Montovani, E.C., Best, S. and Roel, A. (eds.), *Agricultura de Precision: Integrando conocimientos para una agricultura moderna y sustentable*. PROCISUR/IICA, Montevideo, pp. 45–54.

Bramley, R.G.V., Lanyon, D.M. and Panten, K. 2005. Whole-of-vineyard experimentation – An improved basis for knowledge generation and decision making. Proc VECPA-Eur Conf on Precision Agriculture. Uppsala, Sweden, June 8–11, pp. 883–890.

Cohen, Y., Alchanatis, V., Meron, M., Saranga, Y. and Tsipris, J. 2005. Estimation of leaf water potential by thermal imagery and spatial analysis. *Journal of Experimental Botany*, 56(417), pp. 1843–1852.

Doraiswamy, P.C., Moulin, S., Cook, P.W. and Stern, A. 2003. Crop yield assessment from remote sensing. *Photogrammetric Engineering & Remote Sensing*, 69(6), pp. 665–674.

Earl, R., Wheeler, P.N., Blackmore, B.S., and Godwin, R.J. 1996. Precision farming – the management of variability. *The Journal of the Institution of Agricultural Engineers*, 51, pp. 18–23.

Frohn, R.C., Reif, M., Lane, C. and Autrey, B. 2009. Satellite remote sensing of isolated wetlands using object-oriented classification of Landsat-7 data. *Wetlands*, 29(3), pp. 931–941.

Geladi, P. 2003. Chemometrics in spectroscopy. Part 1. Classical chemometrics. *Spectrochimica Acta Part B: Atomic Spectroscopy*, 58(5), pp. 767–782.

Goddard, T.W., Lachapelle, G., Cannon, M.E., Penney, D.C. and McKenzie, R.C. 1995. The potential of GPS and GIS in precision agriculture. Proc. Geomatique V: "La Route De L'Innovation". November, pp. 9–10.

Goel, P.K., Prasher, S.O., Landry, J.A., Patel, R.M., Bonnell, R.B., Viau, A.A. and Miller, J.R. 2003. Potential of airborne hyperspectral remote sensing to detect nitrogen deficiency and weed infestation in corn. *Computers and Electronics in Agriculture*, 38(2), pp. 99–124.

Haboudane, D., Miller, J.R., Tremblay, N., Zarco-Tejada, P.J. and Dextraze, L. 2002. Integrated narrow-band vegetation indices for prediction of crop chlorophyll content for application to precision agriculture. *Remote Sensing of Environment*, 81(2–3), pp. 416–426.

Hunt, E.R., Daughtry, C.S., Mirsky, S.B. and Hively, W.D. 2013, August. Remote sensing with unmanned aircraft systems for precision agriculture applications. In Agro-Geoinformatics (Agro-Geoinformatics), 2013 Second International Conference, IEEE, pp. 131–134.

King, B.A., Brady, R.A., McCann, I.R. and Stark, J.C. 1995, January. Variable rate water application through sprinkler irrigation. In Site-specific management for agricultural systems (No. sitespecificman). American Society of Agronomy, Crop Science Society of America, Soil Science Society of America. pp. 485–493.

Leone, A.P., Wright, G.G. and Corves, C. 1995. The application of satellite remote sensing for soil studies in upland areas of Southern Italy. *Remote Sensing*, 16(6), pp. 1087–1105.

Moran, M.S., Inoue, Y. and Barnes, E.M. 1997. Opportunities and limitations for image-based remote sensing in precision crop management. *Remote Sensing of Environment*, 61(3), pp. 319–346.

Mulla, D.J. and Bhatti, A.U. 1997. An evaluation of indicator properties affecting spatial patterns in N and P requirements for winter wheat yield. Spatial Variability in Soil and Crop. In *Precision Agriculture '97: Spatial Variability in Soil and Crop*. Vol. 1. BIOS Science Publishers, Oxford, UK, pp. 145–154.

Mulla, D.J. 2013. Twenty five years of remote sensing in precision agriculture: Key advances and remaining knowledge gaps. *Biosystems Engineering*, 114(4), pp. 358–371.

Orellana, J., Best, S., and Claret, M. 2006. Capitulo 7. Sistemas de Informacion Geografica (SIG). In Bongiovanni, R., Montovani, E.C., Best, S. and Roel, A. (eds.), *Agricultura de Precision: Integrando conocimientos para una agricultura moderna y sustentable*. PROCISUR/IICA, Montevideo, pp. 131–144.

Oviedo, A.F.P., and Rudorff, B.F.T. 2000. Indice de Area Foliar e resposta espectral da cultura do trigo (Triticum aestivum L.) submetida ao estresse hidrico. *Revista Biociencia, Taubate*, 6, pp. 39–47.

Panigrahy, S., and Sharma, S.A. 1997. Mapping of crop rotation using multidate Indian remote sensing satellite digital data. *ISPRS Journal of Photogrammetry and Remote Sensing*, 52, pp. 85–91.

Plant, R.E. 2001. Site-specific management: the application of information technology to crop production. *Computers and Electronics in Agriculture*, 30, pp. 9–29.

Rossel, R.V., Walvoort, D.J.J., McBratney, A.B., Janik, L.J. and Skjemstad, J.O. 2006. Visible, near infrared, mid infrared or combined diffuse reflectance spectroscopy for simultaneous assessment of various soil properties. *Geoderma*, 131(1–2), pp. 59–75.

Rudorff, B.F.T., Moreira, M.A., Oviedo, A., and DeFreitas, J.G. 1997. Efeito do nitrogenio e do deficit hidrico na resposta espectral de cultivares de trigo. CD-ROM. Simp. Latino-Americano de Percepcao Remota, SELPER, Merida, Venezuela, 1, pp. 1–9.

Seelan, S.K., Laguette, S., Casady, G.M. and Seielstad, G.A. 2003. Remote sensing applications for precision agriculture: A learning community approach. *Remote Sensing of Environment*, 88(1–2), pp. 157–169.

Thenkabail, P.S. 2003. Biophysical and yield information for precision farming from near-real-time and historical Landsat TM images. *International Journal of Remote Sensing*, 24(14), pp. 2879–2904.

Whelan, B.M. and McBratney, A.B. 2003. Definition and interpretation of potential management zones in Australia. *Proceedings of the 11th Australian Agronomy Conference*, 107, pp. 3823–3830.

Whelan, B., and Taylor, J. 2013. *Precision Agriculture for Grain Production Systems*. Collingwood: CSIRO Publishing. p. 208.

Wright, C. and Gallant, A. 2007. Improved wetland remote sensing in Yellowstone National Park using classification trees to combine TM imagery and ancillary environmental data. *Remote Sensing of Environment*, 107(4), pp. 582–605.

Yang, C., Everitt, J.H., Bradford, J.M. and Escobar, D.E. 2000. Mapping grain sorghum growth and yield variations using airborne multispectral digital imagery. *Transactions of the ASAE*, 43(6), p. 1927.

Yao, H., Tang, L., Tian, L., Brown, R.L., Bhatnagar, D. and Cleveland, T.E. 2011. Using hyperspectral data in precision farming applications. *Hyperspectral Remote Sensing of Vegetation*, pp. 591–607.

Zarco-Tejada, P.J., Berjón, A., López-Lozano, R., Miller, J.R., Martín, P., Cachorro, V., González, M.R. and De Frutos, A. 2005. Assessing vineyard condition with hyperspectral indices: Leaf and canopy reflectance simulation in a row-structured discontinuous canopy. *Remote Sensing of Environment*, 99(3), pp. 271–287.

Zhang, C. and Kovacs, J.M. 2012. The application of small unmanned aerial systems for precision agriculture: A review. *Precision Agriculture*, 13(6), pp. 693–712.

16 Farm Machinery for Conservation Agriculture

Shiv Kumar Lohan and Mahesh Kumar Narang
Punjab Agricultural University

CONTENTS

16.1 FARM MECHANIZATION IN PUNJAB STATE

Farm mechanization is dependent mainly upon the size of land holding, sources, and availability of farm power. Farm mechanization is the application of engineering & technology in agricultural operation to do a job in a better way to improve the productivity. The proper use of mechanized inputs in agriculture has a direct

285

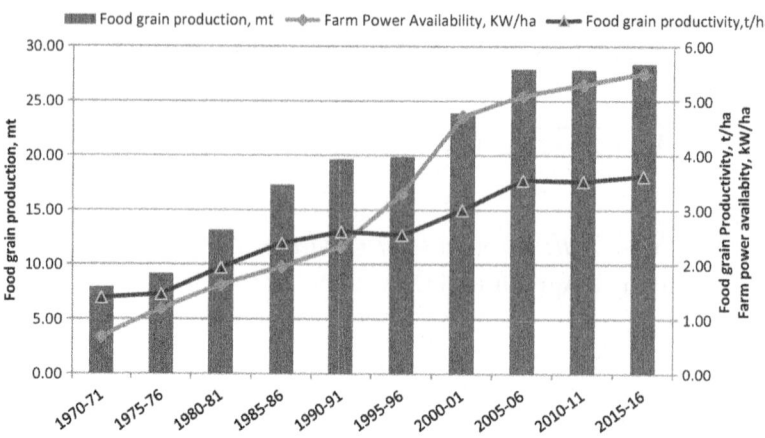

FIGURE 16.1   Trends of farm power availability, food grain production, and productivity.

and significant effect on the production, productivity, and profitability on agricultural farms, along with labor productivity and the quality of life of people engaged in agriculture (Singh et al., 2003; Singh et al., 2006; Dixit et al., 2014). Since the cultivated area cannot be increased, the increased production will be possible only by increased productivity and increased crop intensity. Out of the total 5,036,000 ha geographical area of Punjab state of India, the net sown area has been increased from 3.75 m ha to 4.25 m ha, whereas the gross cropped area has increased from 4.73 m ha to 7.88 m ha resulting in an increase in the cropping intensity from 112% to 196% from 1960–1961 to 2012–2013 (Lohan et al., 2015). The productivity of farms depends greatly on the availability and judicious use of farm power by the farmers (Yadav and Lohan, 2006). Farm mechanization has made a significant contribution in enhancing the agricultural productivity and production.

The increase in unit power per hectare from 0.37 to 5.68 kW/ha led to a total food grain productivity increase from 668 to 3,638 kg/ha and a production increase from 3.16 to 28.58 million tons (mt) during the same period (Figure 16.1). The timeliness of operations has assumed greater significance in obtaining optimal yields from different crops, which has been possible through the use of mechanization. This will call for precision farming, resource conservation technologies, crop diversification, design of gender friendly machines and equipments, and safety and comfort of the farm workers to reduce the hazards and drudgery.

16.2   CONSERVATION AGRICULTURE (CA)

CA is a concept designed for optimizing crop yields, and reaping the economic and environmental benefits. "Conservation agriculture is recognized as agriculture of the future, the future of agriculture" (Jat et al., 2009; Jat et al., 2011). The key elements of CA are minimum disturbance of soil, rational organic soil cover using crop residues or cover crops, and the adoption of innovative and economically viable cropping systems and measures undertaken to reduce soil compaction through controlled traffic (Bisen and Rahangdale, 2017) (Table 16.1).

TABLE 16.1

Strengths, Weaknesses, Opportunities and Threats (SWOT) Analysis of Conservation Agriculture in Punjab

Strengths	• Assured irrigation facilities
	• High farm power availability, cropping intensity, crop production, and productivity
	• Machinery manufacturing hub in Punjab state
	• Good liaison with farm machinery manufacturers
	• Eagerness of farmers to adopt new technologies
Weaknesses	• Degradation of natural resources (soil, water, & environment)
	• Stagnation in the crop yield
	• Lack of standardization and quality control of equipments
Opportunities	• Global market for need/farm size–based machinery/technology
	• Liberal policy of government regarding subsidy and employment generation
	• Capacity building of scientists/engineers/manufactures/farmers
	• Eco friendly environment
Threats	• Adequate exposure to farm machines
	• Scarcity of farm labor
	• Climate change and sustainability of agriculture

CA shows a lot of promise in using these residues for improving soil health, increasing productivity, reducing pollution, and enhancing the sustainability and resilience of agriculture. CA-based RCTs including laser-assisted precision land leveling, zero/reduced tillage, direct drilling into the residues, direct seeded rice, unpuddled mechanical transplanted rice, raised bed planting, and diversification/intensification are practised all over the state (Sidhu et al., 2015, Lohan et al., 2018). The options for crop residue management may include developing systems to plant into residue, mulching, baling, and removal for use as animal feed or for industry.

16.3 MACHINERY FOR FIELD PREPARATION/LEVELING

16.3.1 LASER LAND LEVELER

Leveled land helps in the mechanization of various field operations. Unevenness in land level results in uneven water coverage, uneven crop stands, increased weeds, and uneven maturing crops. All of these factors result in reduced yields and reduced grain quality. Proper and precise land leveling saves irrigation water, facilitates field operations, and increases yield and quality of the produce.

Laser leveling is a laser-guided precision leveling technique used for achieving fine leveling with the desired grade on the agricultural field (Figure 16.2). Laser leveling uses a laser transmitter unit that constantly emits a 360° rotating beam, parallel to the required field plane. The signal received is converted into cut and fill level adjustments, and the corresponding changes in scraper level are carried out automatically by a two-way hydraulic control valve. Laser leveling maintains

FIGURE 16.2   Laser land leveler in operation.

the grade by automatically performing the cutting and filling operations. The field is cultivated and planked before using the laser land leveler. It improves the water coverage from better land leveling and reduces weeds by upto 40%. This reduction in weeds results in less time required for weeding. A reduction from 21 to 5 labor days per hectare is achieved. Leveling reduces the time taken for planting, transplanting, and direct seeding. Land leveling provides greater opportunity to use direct seeding. The possible reduction in labor by changing from transplanting to direct seeding is approximately 30 person-days per hectare (Lohan et al., 2014).

16.3.2   Tractor-Operated Rotary Weeder

Tractor-operated rotary weeder has been developed by the Punjab Agricultural University, Ludhiana and has been used for weeding in wider row crops, especially sugarcane (Figure 16.3). This can be operated with any tractor above 35 hp. The row-to-row spacing between rotary weeding assemblies is adjustable. Generally, it has been used for weeding three rows of crops sown at a row-to-row spacing of 67.5 cm. The field capacity of this machine varied from 0.3–0.5 ha/h and damage to the plants was 1%–4%. The machine cuts and buries the weeds in the cutting width of rotary assemblies. However, the weeds near and around the plants have to be removed manually. This machine saves about 55% cost of weeding and 75% labour cost.

16.4   MACHINERY FOR SOWING/PLANTING/TRANSPLANTER

16.4.1   Lucky Seed Drill [Direct Seeding of Rice (DSR) with Spraying Attachment]

DSR avoids puddling and does not need continuous submergence and thus reduces the overall water demand for rice cultivation. Traditional fluted roller type seed drills do not facilitate maintaining spacing between the plants and also damage the rice

FIGURE 16.3 Tractor-operated rotary weeder.

seed. Therefore, to maintain spacing between plants and reduce seed rate, the use of planters having inclined plate metering device or cupped metering system is suitable. The Punjab Agricultural University, Ludhiana,developed a direct seeded rice drill with spraying attachment (Figure 16.4). The drill consisted of an inclined plate seed metering mechanism with notched cells and having 9 furrow openers, a tank, hydraulic pump, and nozzles mounted on boom. The drill plants the rice seeds and simultaneous sprays the weedicide. Thus there is a saving of labour and timely spraying helps in better control of weeds.

16.4.2 TRACTOR MOUNTED MULTI CROP PLANTER

It is used for sowing bold grains like maize, groundnut, peas, cotton, sunflower, etc. (Figure 16.5). The planting discs for different crops can be changed without dismantling the shaft. The capacity of the machine is about 0.4 ha/h and the desired seed rate can be achieved for different crops.

FIGURE 16.4 Lucky seed drill.

FIGURE 16.5   Tractor-mounted multi crop planter.

16.5   SPRAYING MACHINERY

16.5.1   Tractor-Operated Boom Sprayer

It is suitable for spraying on cotton crop or any other wider row crops. It consists of a centrifugal pump, a tank, pressure regulator valve, and a boom with nozzles and spray gun fitted on a frame (Figure 16.6). The sprayer is mounted on the 3-point linkage of the tractor and drive is given through from tractor power take off (PTO) through a set of gears. The boom height can be adjusted from 10-225 cm from the ground to suit different crop heights. It can cover up to 1,200 cm width and has a capacity of about 2.0 ha/h at a field speed of 3.0 km/h.

FIGURE 16.6   Tractor-operated boom sprayer.

16.5.2 PAU Multipurpose High Clearance Sprayer

It is most suitable for spraying on tall crops like cotton or wider row crops. The machine has three attachments for spraying, like auto rotating guns, boom fitted with nozzles, and dropdown fitted with nozzles which are used for spraying from the top as well as under cotton leaves (Figure 16.7). There is a provision to start or close the spray among these attachments. The boom's height can be adjusted between 66–225 cm to suit different crops and at different stages of crop growth. The width of coverage with boom spraying is 9.5 m, and when both the boom and spray gun are working, then the working width is 20 m. It has a capacity of about 2.0–2.5 ha/h at a field speed of 2.5–3.5 km/h. The mechanical damage caused by the movement of the high clearance sprayer is less in cotton crop.

16.5.3 Electrostatic Sprayer

The electrostatic sprayer has a better spray deposition and minimum losses due to the charging of spray particles (Figure 16.8). The spray range is 5–6 m and the droplet size varies from 40–50 micron. The electrostatic power sprayer has a flow rate of 9.5 L/h. The field capacity of the machine was 0.12 ha/h.

FIGURE 16.7 PAU multipurpose high clearance sprayer.

FIGURE 16.8 Electrostatic sprayer.

16.6 MACHINERY FOR HARVESTING AND THRESHING

16.6.1 Multi Crop Thresher

Commercially available spike tooth-type thresher has been used for threshing moong and mash after incorporating few modifications (Figure 16.9). The threshing cylinder has 36 spikes placed, six in each row. For threshing pulses, six spikes are retained on the cylinder in six rows, i.e., one in each row. The arrangement of spikes on the cylinder periphery is axial. The threshing efficiency of the thresher is around 99% and its cleaning efficiency is around 98%–99%. The output capacity is around 2.5 q/h.

16.6.2 Harambha (High Capacity) Thresher

Harambha thresher is suitable for threshing the wheat crop and is highly popular (Figure 16.10). The threshing material passes through the concave and light materials like chopped straw (*Bhusa*) are blown away with the aspirator blower while the heavier materials like grains, nodes, etc., fall on a set of reciprocating sieves. The sieves clean the grain and there is an optional attachment of an auger to elevate the grains and convey them directly on to a trolley. Feeding of the crop is manual by standing on a platform provided with the thresher. Generally 3–4 persons are required for continuously feeding the crop. The capacity of the thresher varies from 15–20 q/h. The labour required for threshing with the *harambha* thresher, including the transportation of the crop, varies from 30 to 35 man-h/ha.

FIGURE 16.9 Multi crop thresher.

FIGURE 16.10 Harambha (high capacity) thresher.

16.6.3 Grain Combine Harvester

The Combine Harvester is a machine designed for harvesting, threshing, separating, cleaning, and collecting the grain in a single operation. The present day combine harvesters are being mostly used for harvesting two major crops, namely wheat and paddy. Other crops can also be harvested with combines, like sunflower, maize, soybean, pulses, etc., with slight changes in the combine. Based on the sources of power the combine harvesters may be classified as (i) self-propelled combine harvester (Figure 16.11) and (ii) tractor-operated combine harvester (Figure 16.12). Tractor-operated combines can harvest 0.375–0.5 ha/h, whereas self-propelled combines

FIGURE 16.11 Self-propelled combine harvester.

FIGURE 16.12   Tractor-mounted combine harvester.

cover 5–6 ha/day. These machines are highly popular and large numbers of manufacturers are manufacturing these machines.

16.7   MACHINERY FOR CROP RESIDUE MANAGEMENT

New variants of zero-till seed-cum-fertilizer drill/planters such as Happy Seeder, spatial zero till drill has been developed for the direct drilling of seeds in surface residue retention (loose and anchored up to 10 t/ha) plots. These machines are very useful for managing crop residues for conserving moisture and nutrients and controlling weeds, as well as moderating the soil temperature.

16.7.1   HAPPY SEEDER

The Happy Seeder represented a breakthrough for farmers across north-west India in the rice-wheat cropping zone, both in terms of CA benefits and other benefits directly to farmers (Singh et al, 2009). The Happy Seeder machine enables the direct drilling of wheat seed in the combine harvested paddy residue mulched field (Figure 16.13). This machine is very useful for managing crop residues for conserving soil moisture, soil temperature, micro-nutrients, and controlling weeds. Happy Seeder consists of a straw managing unit and a sowing unit in one composite machine. The hinged flails mounted on the rotating shaft cuts the standing stubbles and loose straw coming in front of the furrow opener with simultaneous tine cleaning (for proper seed placement) and places the residue in between the sowing tines. This PTO operated machine can be operated with 45 hp tractors (Double clutch) and can cover 0.3–0.4 ha/h (Singh et al., 2015).

16.7.2   STRAW BALER

For the collection of straw after combine harvesting and using the residues for off farm works, straw baler machines are a very promising technology and are commercially available (Figure 16.14). These balers, however, recover only about 25%–30% of the potential straw yield after combining, depending upon the height of the plant cut by the combines. The Straw Baler makes rectangular or round bales by collecting the

FIGURE 16.13　Happy Seeder in operation.

FIGURE 16.14　Straw bailer for bailing of straw in field.

loose straw from the ground. This machine can recover about 200–250 bales, weighing between 15–30 kg (depending upon the moisture and field condition) with a size of 460 × 360 mm bale from a combine harvested field. The speed of the operation can vary between 2–3 km/h in combine harvested fields depending upon the field condition. The fuel consumption varies between 8.75–11.25 l/ha. The energy requirements vary widely from 0.6 to 1 kWh/tonne and the cost of the operation is 5,288 Rs./ha.

16.7.3　Wheat Straw Combine/Straw Reaper

Wheat straw combine is a machine which cuts and bruises the leftover wheat stubbles after wheat combining operations by a grain combine and also collects the bruised wheat straw into a moving trolley attached behind the straw combine (Figure 16.15). This trolley is covered with a wire mesh net.

FIGURE 16.15   Wheat straw combine in operation.

16.7.4   PADDY STRAW CHOPPER

This machine in a single operation harvests the paddy stubbles left after combining, chops them into small pieces, and spreads them on the ground (Figure 16.16). The chopped straw stubbles are easily buried in the soil by the use of a single operation of rotavator or rotary tiller with light irrigation. Subsequently, wheat sowing is done as usual by the use of a zero till drill.

16.7.5   SUPER STRAW MANAGING SYSTEM FOR COMBINE HARVESTER (SUPER SMS)

Combine harvested rice fields are left with windrows of loose chaff and straw coming from straw walkers and sieves of the combine harvester which creates uneven distribution of straw in the field. Evenly spread loose straw is a precondition for the smooth operations of all second generation drills and it takes around 8–13 man-h/ha for spreading of the loose straw (Sidhu and Singh, 2016).

An Super SMS was developed as an attachment to the existing combine for cutting, chopping, and evenly spreading the loose straw coming out of the harvester during harvesting of paddy (Figure 16.17). The machine will facilitate the operations of other in-situ straw management machinery like the Happy Seeder, spatial no till drill mould board plough, etc., which will ultimately stop the burning of paddy straw and help farmers enhance soil productivity with in-situ paddy straw management.

FIGURE 16.16   Paddy straw chopper.

FIGURE 16.17 Super Straw Management System (Super SMS) attached to the rear of combine harvester.

TABLE 16.2

Machine-wise Cost of Operation of Various Straw Management Machines

S. No.	Name of Machine	Approximate Initial Cost (Rs.)	Effective Field Capacity (ha/h)	Operational Cost (Rs./h)	Operational Cost (Rs./ha)	Energy Required (MJ/ha)
1.	Happy Seeder	150,000	0.30	805	2,682	1,554.3
2.	Zero till drill	65,000	0.62	750	1,209	749.5
3.	Stubble shaver	30,000	0.59	705	1,195	757.7
4.	Bailer	1,200,000	0.40	1,475	3,688	1,223.7
5.	Chopper/ mulcher	160,000	0.30	785	2,620	1,553.3
6.	Super SMS	112,000	1.16	117	101	758.7

The cost of operation and energy requirement of particular machines includes the cost and energy of prime mover except Super SMS.

The approximate cost, effective field capacity, cost of operation per unit area, and energy requirement of individual machines used for managing paddy residues are presented in Table 16.2.

16.8 CONCLUSIONS

The principle mechanization for CA is a means to increase labour and land productivity, and is one element in an array of inputs needed for a successful farming enterprise. This will help in increased agriculture production, timely sowing and harvesting of crops which avoids loss of grains. There are a number of common drivers positively influencing the expansion of mechanization and CA. Shortage of agricultural labor is the major constraint that forces farmers to opt for mechanization or any agricultural practice that could save labor. Depending upon the farm size and the stake holders in the farming community, the need of mechanization for CA is to be identified and adopted.

REFERENCES

Bisen, N. and Rahangdale, C.P. 2017. Crop residues management option for sustainable soil health in rice-wheat system: A review. *International Journal of Chemical Studies*, 5(4), pp. 1038–1042.

Dixit, J., Dixit, A.K., Lohan, S.K. and Kumar, D. 2014. Importance, concept and approaches for precision farming in India. In Ram T, Lohan SK, Singh R, Singh P. (Eds.) *Precision Farming; A New Approach*. Astral International Pvt Ltd., New Delhi, pp. 12–35.

Jat, M.L., Gathala, M.K., Ladha, J.K., Saharawat, Y.S., Jat, A.S., Kumar, V., Sharma, S.K., Kumar, V. and Gupta, R. 2009. Evaluation of precision land leveling and double zero-till systems in the rice–wheat rotation: Water use, productivity, profitability and soil physical properties. *Soil and Tillage Research*, 105(1), pp. 112–121.

Jat, M.L., Saharawat, Y.S. and Gupta, R. 2011. Conservation agriculture in cereal systems of South Asia: Nutrient management perspectives. *Karnataka Journal of Agricultural Sciences*, 24, pp. 100–105.

Lohan, S.K., Jat, H.S., Yadav, A.K., Sidhu, H.S., Jat, M.L., Choudhary, M., Peter, J.K. and Sharma, P.C. 2018. Burning issues of paddy residue management in north-west states of India. *Renewable and Sustainable Energy Reviews*, 81, pp. 693–706.

Lohan, S.K., Narang, M.K., Manes, G.S. and Grover, N. 2015. Farm power availability for sustainable agriculture development in Punjab state of India. *Agricultural Engineering International: CIGR Journal*, 17(3), pp. 196–207.

Lohan, S.K., Singh, M. and Sidhu, H.S. 2014. Laser guided land leveling and grading for precision farming. In Ram T, Lohan SK, Singh R, Singh P. (Eds.) *Precision Farming; A New Approach*. Astral International Pvt Ltd., New Delhi, pp. 12–35. ISBN; 978-93-5130-258-2: pp. 148–158.

Sidhu, H.S. and Singh, M. 2016. Mechanization; A need for sustainable intensification. Lead Papers Vol. 4: 4th International Agronomy Congress, Nov. 22–26, New Delhi, India. pp. 158–162.

Sidhu, H.S., Singh, M., Singh, Y., Blackwell, J., Lohan, S.K., Humphreys, E., Jat, M.L., Singh, V. and Singh, S. 2015. Development and evaluation of the Turbo Happy Seeder for sowing wheat into heavy rice residues in NW India. *Field Crops Research*, 184, pp. 201–212.

Singh, K.K., Lohan, S.K., Jat, A.S. and Rani, T. 2003. Influence of different planting methods on wheat production after harvest of rice. *Ama, Agricultural Mechanization in Asia, Africa & Latin America*, 34(4), pp. 18–19.

Singh, K.K., Lohan, S.K., Jat, A.S. and Rani, T. 2006. New Technologies of growing rice for higher production. *Research on Crops*, 7(2), pp. 369–371.

Singh, Y., Sidhu, H.S., Singh, M., Dhaliwal, H.S., Blackwell, J., Singh, R., Humphreys, L., Singla, N., Thind, H.S., Lohan, S.K. and Sran, D.S. 2009. Happy Seeder – A conservation agriculture technology for managing rice residues. Technical Bulletin No. 01, Department of Soils, Punjab Agricultural University Ludhiana.

Yadav, S. and Lohan, S.K. 2006. Tractor and Implement Owenership and Utilization of Haryana. *Ama, Agricultural Mechanization in Asia, Africa & Latin America*, 37(3), p. 15.

Index